AF718525

Pluralism in Mathematics: A New Position in Philosophy of Mathematics

LOGIC, EPISTEMOLOGY, AND THE UNITY OF SCIENCE

VOLUME 32

Editors
Shahid Rahman, *University of Lille III, France*
John Symons, *University of Texas at El Paso, USA*

Managing Editor
Ali Abasnezhad, *University of Lille III, France*

Editorial Board
Jean Paul van Bendegem, *Free University of Brussels, Belgium*
Johan van Benthem, *University of Amsterdam, the Netherlands*
Jacques Dubucs, *University of Paris I-Sorbonne, France*
Anne Fagot-Largeault, *Collège de France, France*
Göran Sundholm, *Universiteit Leiden, The Netherlands*
Bas van Fraassen, *Princeton University, U.S.A.*
Dov Gabbay, *King's College London, U.K.*
Jaakko Hintikka, *Boston University, U.S.A.*
Karel Lambert, *University of California, Irvine, U.S.A.*
Graham Priest, *University of Melbourne, Australia*
Gabriel Sandu, *University of Helsinki, Finland*
Heinrich Wansing, *Technical University Dresden, Germany*
Timothy Williamson, *Oxford University, U.K.*

Logic, Epistemology, and the Unity of Science aims to reconsider the question of the unity of science in light of recent developments in logic. At present, no single logical, semantical or methodological framework dominates the philosophy of science. However, the editors of this series believe that formal techniques like, for example, independence friendly logic, dialogical logics, multimodal logics, game theoretic semantics and linear logics, have the potential to cast new light on basic issues in the discussion of the unity of science.

This series provides a venue where philosophers and logicians can apply specific technical insights to fundamental philosophical problems. While the series is open to a wide variety of perspectives, including the study and analysis of argumentation and the critical discussion of the relationship between logic and the philosophy of science, the aim is to provide an integrated picture of the scientific enterprise in all its diversity.

For further volumes:
http://www.springer.com/series/6936

Michèle Friend

Pluralism in Mathematics: A New Position in Philosophy of Mathematics

Springer

Michèle Friend
The George Washington University
Washington, DC, USA

ISBN 978-94-007-7057-7 ISBN 978-94-007-7058-4 (eBook)
DOI 10.1007/978-94-007-7058-4
Springer Dordrecht Heidelberg New York London

Library of Congress Control Number: 2013953900

© Springer Science+Business Media Dordrecht 2014
This work is subject to copyright. All rights are reserved by the Publisher, whether the whole or part of the material is concerned, specifically the rights of translation, reprinting, reuse of illustrations, recitation, broadcasting, reproduction on microfilms or in any other physical way, and transmission or information storage and retrieval, electronic adaptation, computer software, or by similar or dissimilar methodology now known or hereafter developed. Exempted from this legal reservation are brief excerpts in connection with reviews or scholarly analysis or material supplied specifically for the purpose of being entered and executed on a computer system, for exclusive use by the purchaser of the work. Duplication of this publication or parts thereof is permitted only under the provisions of the Copyright Law of the Publisher's location, in its current version, and permission for use must always be obtained from Springer. Permissions for use may be obtained through RightsLink at the Copyright Clearance Center. Violations are liable to prosecution under the respective Copyright Law.
The use of general descriptive names, registered names, trademarks, service marks, etc. in this publication does not imply, even in the absence of a specific statement, that such names are exempt from the relevant protective laws and regulations and therefore free for general use.
While the advice and information in this book are believed to be true and accurate at the date of publication, neither the authors nor the editors nor the publisher can accept any legal responsibility for any errors or omissions that may be made. The publisher makes no warranty, express or implied, with respect to the material contained herein.

Printed on acid-free paper

Springer is part of Springer Science+Business Media (www.springer.com)

*I dedicate this book to my teachers:
human and equine.*

Preface

I came to write about this position as a result of my professional experience. I attended and participated in a lot of conferences, mostly in Europe. They were philosophy, mathematics and logic conferences. I observed that, for the most part, mathematicians and logicians did not behave as though they adhered to a philosophy of mathematics. In particular, with some exceptions, they did not seem to show adherence to one foundation of mathematics in a philosophical way. Some were working on some issues in a foundation, and were wedded to it as a result of their invested time and energy, not so much (again with some exceptions) for philosophical reasons. Yet, my studies in the philosophy of mathematics would have me believe that it is imperative that one have a philosophical outlook or position, and one should work within it. I was puzzled.

Oversimplifying: the philosophers seemed to be convinced that mathematics is *one thing* and that to show this one just pointed to the foundation of mathematics, and this was a particular theory in mathematics. The philosophers seemed to be completely ignoring the fact that there are several rival foundations, and none has a completely privileged position, except maybe Zermelo-Fraenkel set theory – but even that could not support the philosophical claims, since there were all sorts of equi-consistency proofs around. There would be no point in making such proofs if the other 'rival' foundations were for nought. Mathematicians and logicians in their presentations and in casual speech were quite willing to take seriously other theories that conflicted with the ones they were working in. In fact that is one of the reasons they go to conferences: to find out what is going on in other fields, to see how results in one area of mathematics share features with their own. They would quite happily talk of rival foundations in the same breath, and not be casting one away. Instead, they embraced the lot.

I was convinced that if one wanted to give a philosophy of all of today's working mathematics, one had to give a philosophy that was not foundational. I was going to call the position Meinongian structuralism, but Bill Griffiths convinced me that the name was too baroque. It later occurred to me that 'pluralism' would work as a name. Once I fastened on 'pluralism', I noticed the word used by a few philosophers of mathematics such as Shapiro and Maddy. In contrast to the philosophers, mathematicians for the most part behave in a pluralist way. I conclude that pluralism is 'in the air'. But if we look at how the word is used, we find it is used in so many different ways as to be almost useless! It occurred to me that it would be a useful service to develop a philosophical account of pluralism as a philosophy, as opposed to 'pluralism' being used to gesture towards a vague and ambiguous attitude of tolerance.

I confess to feeling I am a bit of a philosophical charlatan, since I hardly think I am doing anything original, again, since the idea is already very much in the air. At other times I think I am a charlatan on the grounds that the position is so obvious, as to be platitudinous. It seems to hardly qualify as a position at all, since it is just an articulation of the prevailing attitude of practicing mathematicians. But, then I quickly realise that this is not at all the case. Once developed in its entirety, I discovered how radical the position is. It is deeply radical. As such, if my arguments are persuasive, then the book will either convert readers, or act as a strong warning to treat the word 'pluralism' with care, use it sparingly, or only in the negative. One person's *modus ponens* is another's *modus tollens*.

Washington, DC, USA Michèle Friend

Acknowledgments

First, I should like to thank my co-conspirators. Andrea Pedeferri is co-author of two of the chapters of this book. It is through discussions with him, and collaboration in writing, that the chapters really took shape. Joe Mourad acted as my main 'mathematical consultant'. Second, I should like to thank the editorial board of Springer, and the managing team who realised the book. Third, I should like to thank my two academic hosts during my sabbatical year in Europe. The Arché group working on the foundations of logical consequence at the University of St. Andrews received me as a visiting fellow. István Németi and Hajnal Andréka of the Alfréd Rényi Institute of Mathematics, Hungarian Academy of Sciences, received me as a guest.

Close colleagues who gave gentle nudges, corrections and encouragement include: Andrew Aberdein, Norma B. Goethe, Tatyana Romanovskaya, Göran Sundholm and Tadeusz Zawidzki. In the next circle of influence, we find: Nimrod Bar-Am, Diderik Batens, Roy Dyckhoff, Valentina Harizanov, Benedikt Löwe, Colin McLarty, Graeme Priest, Stephen Read, Stewart Shapiro and Neil Tennant. Not all of these will want to be associated with the contents of the book, by the way. I leave the reader to guess who would. The following might not even think of themselves as having had an influence, partly because of their charming sense of modesty. Nevertheless, our discussions and their kindness have had an effect on at least a few sentences in the book: Hajnal Andéka, Paul Churchill, José Felix Costa, Bill Griffith, Nicholas Joll, Brendan Larvor, Bob Lockie, João Marcos, Michael Morreau, István Németi, Olga Opombo, Mark Powell, Guiseppe Primeiro, Georges Rey, Dirk Schlimm, Stewart Shapiro, Michael Sigrist, Gergeli Székely, Sara Uckleman, Petr Vopěnka, Alan Weir and Crispin Wright.

Pluralism in mathematics was neither invented by me, nor was it invented yesterday. It has been in the air for a while, although it has never been articulated as a fully developed position in the philosophy of mathematics. I would have been unable to formulate the position without long deliberation, so less directly, I have been influenced by: Sherry Ackerman, Otávio Bueno, Luiz-Carlos Peirreira, Peter Caws, Peter Clark, David Davies, Bill Demopoulos, Mic Detlefsen, Albert Dragalin, Bob Hale, Michael Hallett, Joachim Lambeck, Michael Makkai, Mathieu Marion,

Marcin Mostowski, Alex Oliver, Jeff Paris, Rebecca Pates, Michael Potter, Jehuda Rav, Peter Simons, Jairo daSilva, Hartry Slater, Peter Sullivan, Vítěslav Švejdar and Michael Wright. Again, some of these might deeply regret being mentioned here! They might regret that they did not have *more* influence in guiding me *away* from the pluralist path.

Academics who are not specialists in the philosophy of logic or mathematics but who also were helpful include all of my colleagues from George Washington University. It is a remarkably collegial department, and they have all been morally supportive, and more concretely they helped to reduce my administrative duties. They include: Jeff Brand, Peter Caws, Paul Churchill, David DeGrazia, Lloyd Eby, Bill Griffith, Josh Miller, Tom Morris, Andrea Pedeferri, Tatyana Romanovskaya, Eric Saidel, Athenassis Samaras, Geert Van Cleemput, Michael Sigrist, (Sigrist, Miller and I shared many a lovely beery evening), Chris Venner, Gail Weiss and Tadeusz Zawidzki.

On a less academic level, I must acknowledge the unbridled support of my parents: Tony and Henriette Friend. My friends tolerated my unsociable behaviour while at home writing and correcting the manuscript. I wrote most of the first draft of the book while on a sabbatical year in Europe. Separately I should like to thank my hosts during that year, in chronological order: my host in England, Graeme Hall; my hostess in France, Odile Geoffroy-Gibson; my host in Scotland, Roy Dyckhoff; and my host in Hungary, György Pándi.

Contents

List of Figures

Chapter 1
Introduction

Abstract The introduction is meant as a guide to reading the book. I briefly describe the parts and individual chapters of the book. I also outline some conventions adopted in the book.

1.1 Introduction

There are four parts to this book. The first is motivational. I give motivations for adopting pluralism from four separate starting points: realism, Maddy's naturalism, Shapiro's structuralism and formalism. For reading this part of the book, I suggest reading the first chapter on realism in order to gain some orientation concerning pluralism, and as an introduction to some vocabulary which is used idiosyncratically. There is a glossary for further reference, or to use as a reminder.

The three other chapters of the first part are self-contained, and are directed towards philosophers with certain inclinations. That is, if one has naturalist inclinations, one should read the naturalism chapter. If one has structuralist inclinations, one should read the structuralism chapter and if one has formalist inclinations, one should read the formalism chapter. If the reader is none of the above, then she can read these chapters only to become acquainted with some motivations for adopting pluralism. This part of the book is not exhaustive in discussing all possible motivations for pluralism. Not only are there only a few non-pluralist positions discussed, but even within the motivational chapters on naturalism and structuralism I target one philosopher's philosophy in this area, not all of the well received versions. The philosophers in question are Maddy and Shapiro, respectively. Motivating pluralism from other starting points is part of the greater pluralist programme. Similarly, comparing and contrasting other positions with pluralism is part of the greater programme. I return to it in one section of Chap. 14.

Pluralism is not just one position in the philosophy of mathematics, it is a family of positions. This is one of the reasons I call it a 'programme'. This book gives a starting push to the programme. Different members of the family are distinguished

M. Friend, *Pluralism in Mathematics: A New Position in Philosophy of Mathematics*, Logic, Epistemology, and the Unity of Science 32, DOI 10.1007/978-94-007-7058-4_1,
© Springer Science+Business Media Dordrecht 2014

along the dimensions of: degree (of pluralism), underlying logic and sort of pluralism. Examples of sorts are: foundational, methodological, epistemological and alethic. The pluralist not only distinguishes himself from other positions in the philosophy of mathematics, he is inspired by other positions. In particular, the pluralist retains lessons from the realist, the naturalist, the structuralist, the formalist and the constructivist. The last source of inspiration will be put to work in the fourth part of the book.

The second part of the book concerns the details of how to cope with the inevitable conflicts and contradictions which surface when entertaining very different philosophical positions and mathematical theories under one theory. This part concerns reasoning in the light of contradiction and conflict. I first present pluralism as a philosophical position in its own right. I make reference to a paraconsistent formal system as a guide to reasoning about conflicting ideas *without* necessarily having to decide that one idea is correct and the other is not, or that one position 'wins' over another. Sometimes one does win, but in more sophisticated arguments, there will not be a clearly correct position. Since I am presenting a philosophical position, I can only make *reference to* a formal system of logic, as opposed to *using* a formal system. This is because the pluralist philosopher is not comparing propositions or well-formed formulas and reasoning from these to theorems or conclusions. This is why I write about using a formal logic metaphorically in Chap. 7. The logic is used to set parameters and to sanction and guide the reasoning about whole theories. For this reason, the notions of rigour of argument and the idea of communication become very important to the pluralist. These are discussed in Chaps. 8, 9, and 14. We shall see in these chapters a tension and a struggle with meaning, ontology and truth. These are traded for the more practice reflecting: communication, rigour and protocol. The struggle is the struggle of the pluralist. It is the cost of taking on board the task of explaining what mathematics is about without compromising on the real subtleties in operation in mathematics.

In the third part of the book I work with the paradoxes of tolerance and the idea of transcending one's own position. The paradoxes of tolerance surface when we ask the questions: 'does it makes sense to be tolerant towards those who are intolerant of our own tolerant position?' and 'are there not some things the pluralist is intolerant towards?' In Chap. 10, I discuss the paradoxes, and explain that the pluralist is not a *global* pluralist, but a *maximal* pluralist. He would be a global pluralist if he were tolerant of everything. He is a maximal pluralist if he is as tolerant as possible without his position becoming self-defeating. The maximal pluralist is intolerant towards dogmatism, and particular moves made by, say, realists, naturalist and structuralists against other positions, and pluralism.

In Chap. 11 we visit the more subtle question of whether the pluralist is pluralist towards himself. Another way to ask this is to ask if the pluralist is dogmatic in the ways identified in the previous chapter. To answer this question, we first explore Meyer's collapsing lemma. I use this to a very modest end, to indicate that the paraconsistent logician wedded to LP (a particular paraconsistent logic) will have to be pluralist about interpretations of his logic. The result generalises to anyone who is both pluralist and fixes on a particular logic to underpin his pluralism.

In contrast, if we are logical pluralists, then we have another reason to be pluralist about pluralism. Using other logics will give a different flavour to pluralism. The pluralist is pluralist towards himself just in virtue of admitting alternative logical formal systems to underpin pluralism. Again, *qua* programme, here we see that we can make different versions of pluralism by adopting different underlying formal logical systems.

The fourth part of the book puts the pluralist to work. I indicate some sample pluralist exercises. The first concerns the notion of proof in mathematics. The pluralist analyses the notion of proof as it is used by the working mathematician and draws conclusions about the role of proof in mathematics. In Chap. 13, I undertake a different sort of exercise. This concerns pluralism about conflicting *philosophies* of logic. In Chap. 14, I launch three sorts of project, one is to take a feather from Maddy's hat, and identify an aspiration of some mathematicians, articulate and define the aspiration and put it to work to partly resolve a technical problem, and to deepen our understanding concerning the problem. The second project is to explore the notion of rational reconstruction, to see what they can teach us. The third discusses the issue of working in a trivial setting. This part of the book demonstrates how pluralism is programmatic. There is a lot of work for the pluralist philosopher of mathematics. For the reader who is interested in reading as little as possible, while still forming a view of pluralism, I suggest reading Chaps. 2, 6, and 11.

1.2 A Note on Conventions

Definitions for technical terms are usually given at their first mention, but not invariably, for example in this introduction I have used many such words without giving a definition. Technical terms are given a definition in the glossary. The index should provide further guidance.

'The pluralist' is used to name a character who takes on some sort of pluralist philosophy of mathematics. The definite article is used in the same way as when we say 'the logician' and are referring, not to an individual (person) but to a species, or type of person. More technically, 'the pluralist' is not a first-order singular term, but a second-order singular term. Pluralism is a family of positions. As such, the different pluralisms have many features in common, and can all avail themselves of most of the same arguments against other positions.

I use 'he' throughout for the pluralist. This is because I am a 'she' and I do not want to show prejudice. Other philosophical or mathematical characters might be given the preposition 'he', 'she' or 'it'. I use 'it' for the more obscure, remote or extreme positions, which are just philosophical constructs. In these cases it is possible that no one ever did or ever will hold the position. It is supposed that if someone were to hold the position, such a person would not hold it for long. It is more a position to be temporarily entertained than seriously defended. An example is the trivialist. An example is the trivialist position. A trivialist is an 'it'.

Foreign words, and phrases which I wish to emphasise, are italicised. There should not be very much confusion resulting from italics playing two roles. 'Or' is taken as inclusive throughout the text.

There are two chapters which were co-written with Pedeferri. Therefore, in both these chapters I use the first person plural. In other chapters, I use the first person singular. After the acknowledgments, preface and introduction, names of philosophers or mathematicians are only ever written using the family name.

My punctuation might also raise eyebrows. I part company with Fowler and Gowers, and put a comma after 'for' when it is used in the sense close to that of 'since'. I part company with the conventions of the grammar check on my computer, and do not always precede 'which' with a comma. Single quotation marks, or inverted commas, are use to show that a term is a technical term. They are also sometimes used to show mild irony. For the most part, they could be replaced by the words: 'as it were' or 'so called', but this would be more tedious than using the elegant single inverted comma.

Finally, I should add a word about the index. The index is lengthy, and has some odd entries. The purpose of the index is twofold. One purpose is for a reader interested in, say, finding out what I have to say about realism, and nothing else. But the other use is when a reader wants to re-read, say, an example, and remembers that it concerned an unusual phrase such as: 'the inconsistency of UN declarations'. For this second reason, odd entries, such as, 'UN declaration' are in the index.

Part I
Motivations for the Pluralist Position; Considerations from Familiar Positions

Chapter 2
The Journey from Realism to Pluralism

Abstract In this chapter I take the reader on a journey from a naïve realist position through to the beginnings of pluralism. Some simplifying assumptions are made, but this is done in order to *introduce* some of the concepts we find in pluralism, not to defeat all realist positions. In particular, in order to set the stage, the naïve realist will take Zermelo Fraenkel set theory to be the foundation for mathematics in a philosophically robust sense of capturing the essence, ontology and absolute truth of mathematics. The reader is given several reasons to abandon the naïve realist conception and to consider a more pluralist conception. The main aspect of pluralism discussed here is pluralism in foundations. 'Pluralism in foundations' is an oxymoron, and therefore, is unstable. Some other aspects of pluralism are then introduced: pluralism in perspective, pluralism in methodology and pluralism in measure of success.

2.1 Introduction: ZF Monism

Since this is the beginning of the of book, I should issue a warning. Especially in this chapter, I tell some lies. Or, rather, I begin with oversimplifications. This will be alarming for the more sophisticated readers. However, rest assured that as we proceed through the book, most of these oversimplifications will be re-expressed, refined, honed and made more explicit. The reason for the oversimplifications is that since this is a new position in the philosophy of mathematics, I prefer to start with some very naïve ideas.

In explaining a philosophical position, it is sometimes useful to start from a quite different, but easily recognised position, even if we think no one occupies it.[1] Realism is a familiar position in the philosophy of mathematics. However, since

[1]As we shall see in the subsequent chapters, I shall take the reader through journeys with other starting points: naturalism, structuralism and formalism. Balaguer (1998) works through different versions of realism, and teaches us to use the word carefully. It is quite possible that no one

M. Friend, *Pluralism in Mathematics: A New Position in Philosophy of Mathematics*, Logic, Epistemology, and the Unity of Science 32, DOI 10.1007/978-94-007-7058-4_2,
© Springer Science+Business Media Dordrecht 2014

'realism' is such a broad term with so many connotations and aspects, I shall fix the term and restrict 'realism' to a 'monist foundationalist' position, where Zermelo-Fraenkel set theory (henceforth: 'ZF') is The Foundation.[2]

Explaining and defining the terms just used: ZF is an axiomatic theory. Zermelo developed most of the axioms and Fraenkel added the axiom of replacement (Potter 2004, 296). The theory is very general. In ZF we study sets of objects, combinations of sets, the comparison of sets with each other, and the creation of one set from another, or of a new set from several others: for example, by taking their intersection or union.

Definition[3] *The Foundation* is an axiomatically presented mathematical theory to which all or most of successful existing mathematics can be reduced. It can be used normatively to exclude from *bona fide* mathematics any *purported* mathematics which cannot be reduced to the axiomatic theory.

Definition *Successful existing mathematics* is the body of mathematical theories and results about those theories that are currently judged by the mathematical community to be 'good mathematics' (as indicated by publication, reference in discussion, use in classrooms and study groups, airing at conferences and so on). This will include past mathematics not presently under mathematical investigation, but, for all that, not dismissed as bad mathematics.

What counts as successful existing mathematics is revisable. We might find out that what we thought was a good mathematical theory turns out not to be. Thus, 'successful existing mathematics' is a vague term, but the imprecision of the boundaries of application of the term need not concern us in the present context.

Definition The *monist foundationalist* believes that there is a unique correct, or true, foundation for mathematics, and uses The Foundation normatively to determine what is to count as successful existing mathematics.

What might motivate someone to adopt monist foundationalism? In the late nineteenth century and the early twentieth century, we developed various set theories. They were not all fully axiomatised at first. Cantor's set theory was not presented as a fully axiomatised theory at all. Despite they're not being presented as fully axiomatised, we discovered that set theories were very powerful. By 'powerful' we mean that a great deal of mathematics can be reduced to set theory. That is, we can translate, say, arithmetic, into the language of, say, ZF, avail ourselves of the axioms and inference rules of the proof theory of ZF, contribute some definitions in the language of ZF, and obtain, through proof, a number of theorems or 'results'.

---

presently holds the position I give here. It is, admittedly, a caricature. That does not matter for present purposes, since (1) the point is to start from a familiar position, not an occupied and carefully defended position, and (2) this chapter is not meant as a knock-down argument against rea[...] in all its forms. Rather, we begin with a naïve and familiar view in order to introduce [...]n.

[...]uite different from a 'full-blooded realism' (Balaguer 1998, 5).

[...]nitions in this chapter are to be read as working definitions. As such, in a more exacting [...]ey might require further refinement. Definitions are repeated in the glossary.

We can then translate back into the language used in the original arithmetic. If we compare the results we obtained in the original arithmetic to those we obtain in the ZF version of arithmetic, then we can prove that we can in principle reproduce all of the results of the original arithmetic in ZF. That is, there is no theorem of arithmetic, which does not have an analogue in ZF. Therefore, in principle, there is a complete reduction of arithmetic into ZF. The *power* of set theory consists in the fact that not only arithmetic, but also analysis and geometry, and therefore most of working mathematics can all be reduced to set theory. Because of the power of ZF, it can be presented as a candidate for founding mathematics.

ZF was not the only set theory developed. There were (and still are) rival theories of sets, and there arose problems with some of the theories with the discovery of paradoxes. Even apart from the paradoxes, other conceptual puzzles surfaced such as how to conceive of very large totalities, which many of us now think of as proper classes. These were both philosophical and technical problems. The paradoxes and puzzles produced a crisis in mathematics (Giaquinto 2002) and hailed the foundational and axiomatic movements. It was thought that mathematics needed a 'secure' foundation, since it was clear that some mathematical activity was deeply flawed. Philosophers, or professional mathematicians assuming a philosophical role, (henceforth: philosophers)[4] contributed in culling some set theories, such as the so-called naïve theories. The culling did not eliminate all but one set theory. So we had no clear unique founding set theory, we had several. Nevertheless, we can say that presently we have honed in on one. Under the received view today, we can say that ZF is the 'orthodoxy' of mathematics (Maddy 1997, 22). 'Orthodoxy' can be taken to mean 'the most accepted theory' or reference point for mathematics. Or, it can mean that ZFC (ZF with the axiom of choice) 'codifies current mathematical practice' (Hrbacek et al. 2009, 2). How might a philosopher interpret such phrases? There are conceptually distinct roles that ZF can play as 'orthodoxy'. Let us start at one extreme. The position: monist foundationalism reads 'orthodoxy' to mean that ZF sets the parameters for what is to count as mathematics. The reason for starting with this is not plausibility, but familiarity and conceptual simplicity. Philosophers are all familiar with some (less extreme) version of realism. At this extreme end of realism, The Foundation plays the following four roles.

1. All of what is counted as 'mathematics' has to be reducible to, or can be faithfully[5] translated into, The Foundation. The Foundation gives the scope of

---

[4]Obviously, being a professional mathematician does not preclude one from having philosophical thoughts or from writing quite philosophically about mathematics. The distinction here is not professional but conceptual, in the sense of philosophical and mathematical problems or puzzles requiring different sorts of solution.

[5]By 'faithfully' I mean that the language being translated into, here the language of ZF, has the expressive power to capture the nuances of the original concepts as expressed in the original language. A test for loyalty of a definition, say, would be that analogues of all of the same theorems can be derived when the definition is expressed in ZF as can be derived using the definition of the original language, all other definitions, theorems, lemmas and proof techniques remaining equal. In contrast, a reduction would be unfaithful if fewer or more (non-equivalent) theorems could be derived. Ancient Latin can only make an unfaithful translation of a modern computer manual.

the correct use of the word 'mathematics', or, we might say, the foundational theory determines the extension of the term 'mathematics'. We might call this 'the semantic determining role of The Foundation'.

Another, but related, role is that

2. the foundational theory tells us what the basic ontology of mathematics is: what it is that mathematicians ultimately study. In the case of ZF, it is sets, and not, for example, lines, planes, numbers or cuts. We might call this 'the ontological role of The Foundation'.

As a corollary,

3. what counts as correct, or legitimate methodology is also determined by ZF. This is 'the methodological role' played by The Foundation. We give some axioms, elaborate definitions and then prove theorems within ZF.

We might even

4. confer an epistemic role to ZF, by saying that to really understand and know mathematics, we have to study set theory. The rest of mathematics, written in other languages, is a pale imitation, and studying mathematics, not presented as a part of set theory, might even mislead us into thinking that we know an area of mathematics when we do not. Call this 'the epistemological role of The Foundation'. All these roles meant to have normative force over the practice of mathematics.

To hold that ZF plays all four roles is quite extreme, but this is where we shall begin our journey. The monist foundationalist who confers all four roles on the foundational theory is also the most extreme opponent to pluralism. So the journey from monist foundationalism to pluralism is long. In the course of the journey, we shall meet considerations that trouble the extreme position. Since considerations are not full arguments, each elicits different legitimate reactions. There will be better arguments in the chapter on structuralism. Thus, before we see the considerations we should add a note about how to think of them.

Upon thinking about the considerations, a reader might be prompted to muster arguments *against* the pluralist, or she might modify, or even change, her position. Thus, the considerations, can be thought of as: (a) points of re-entrenchment for the convinced monist foundationalist, (b) calls for conservative modification of the monist foundationalist view or (c) points of rejection or doubt towards monist foundationalism. The last leads us closer to the pluralist position. Since not every reader will be willing to follow me for the whole journey, we can think of the journey as an exercise in mapping out the philosophical territory and discovering where one stands *ab initio* (Fig. 2.1).

The monist foundationalist holds an extreme position because it has a vision of reformation.

Definition *The reformation* is a movement to constrain successful existing mathematics by The Foundation.

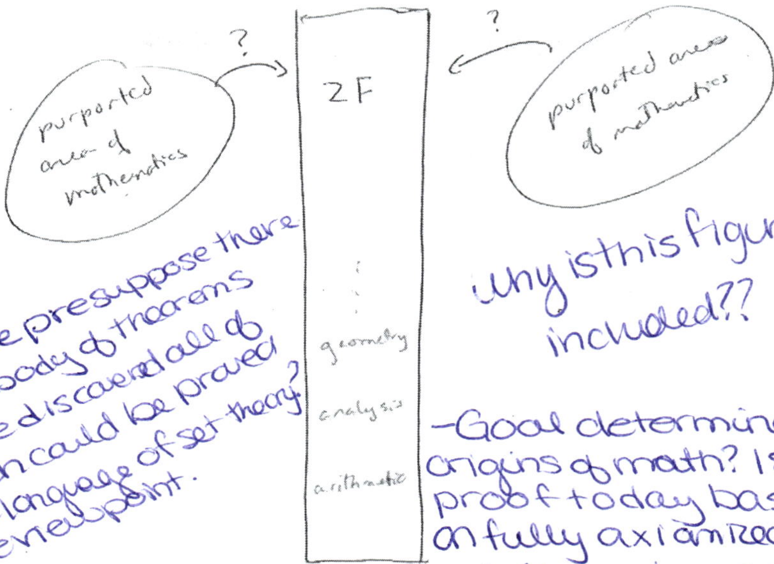

Fig. 2.1 ZF set theory as a foundation

If ZF is the orthodoxy of mathematics in a strong sense and is a good candidate for The Foundation, then we might think that we have an equal trade-off between good mathematics, as practiced, and set theory. If we have an equal trade-off, then we can do mathematics in the language of set theory or we can do mathematics in the original language developed for that theory, and the two are equivalent. In other words, we could set up two communities of mathematicians. One would continue to work in the languages of our plethora of mathematical theories: Euclidean geometry, topology, calculus, algebra and so on, but not set theory directly (in the sense of using only the language of set theory). The other would simply work in set theory. According to our reducing results concerning The Foundation, in the long run, the plethora community would produce a number of results; and the set theory community would produce analogues of all of the results of the plethora community. The order of producing the results would be different – because notation is differently suggestive, but 'at the end of the day' equivalent results would be generated. If this is an accurate prediction of what would happen, then this confirms (in a somewhat circular manner) that we have an equal trade-off.

However, the prediction would probably not be met. This is because the set theoretic community will produce some results not produced by the plethora community. The latter would concern results unique to set theory. It would seem that the set theory community is, therefore, *better off*, at least in the long run.

We now have good reason to initiate the reformation. It would be a wonderful feat to reform mathematical practice by stipulating that we *only* do mathematical work in the language of set theory, and we confine ourselves to the axioms of set theory and are allowed to introduce definitions only in the language of set theory. This would clear up misunderstandings, cut down on the time spent learning new symbols and vocabulary, and cut out all of the work which we do showing that two theorems in different areas of mathematics are equivalent in some respects since this would be clear and explicit if our work was all done within set theory in the first place. Under the reformation, mathematics would become an explicitly unified discipline.

Unfortunately, the suggested reformation would incur considerable loss. It is not at all clear that the set theory community simply reproduces all of the results of the plethora community, plus some more. Sometimes we use one area of mathematics to inform us about another area. We translate from the first area into the second, make a proof in the second and re-translate back to the original area. The reason for taking this circuitous route is that the execution of the proof, and what to look for, are much more obvious in the second area than in the first. For example, Arrighi and Dowek (2010) turn to quantum computing to make some sense of the notion of computable function in a space of infinite dimensions. Note that, strictly speaking, they could have generated the same results in the original classical theory.[6] However, it would not have been obvious, and it would not have been at all evident in the classical logic framework. Lobachevsky, whose work we shall discuss later in the book, turns to hyperbolic geometry to make sense of the notion of an indefinite integral in Euclidean geometry. So there is a heuristic advantage, and maybe even an epistemic advantage to working in different frameworks or theories. Moreover, it would be a mistake to think that these are isolated cases. Therefore, this is one reason to be cautious about the reformation.

However, the reformer would be quite right to retort that this is not a serious objection. The difference between the plethora community and the set theory community influences the *order* in which results are discovered, not in the body of results themselves. Furthermore, strictly speaking, and as we noted earlier, the set theory community will produce some results not produced in the other community. Moreover, their work will be more efficient, since they are not doubling up on results, and then have to prove the equivalence of theorems.

Notice that this retort is strongly underpinned by the sort of realism that emphasises monism in the sense of pre-supposing that there is a unique body of truths, or theorems, to be discovered. If we do not hold this pre-supposition, then the retort carries no weight. But it is a difficult presupposition to give up altogether, as witnessed by the very fact that philosophers are at pains to give arguments as to why non-uniqueness should not worry us. For example, see Balaguer (1998). Let us introduce some more vocabulary, and re-express the two antagonistic positions in that new vocabulary.

[6]This point was made in the oral presentation of the material, (Computability in Europe 2010) but is not obvious in the written version.

Definition *Realism* in mathematics has two conceptually distinct versions: realism in ontology and realism in truth-value (Wright 1986, 9).

Definition *Realism in ontology* is the position that the ontology of the subject we are realists about is independent of our investigations or knowledge.[7]

Some pluralists are anti-realist in ontology.

Definition *Anti-realists* are all those who are not realists. Following Wright (1986, 2), there are two sorts. Anti-realists can assume just a negative view *vis-à-vis* the realist and be sceptics. On the positive side they can be idealists, who believe that it is our ideas that shape the world around us, and determine our ontology. The idealist anti-realist is someone who epistemically constrains truth (rejects verification-transcendent truths).[8]

Definition A *realist in truth-value* of the sentences of a theory holds that the truth-value of sentences of a theory is independent of our ability to judge or establish or discover them.

Some pluralists are anti-realist in truth-value.

Note that one person can be a realist about one area of discourse and an anti-realist about another; for example, it is common to be a realist about physical objects, but an anti-realist about humour. Such split positions are fairly common. It is less common to be a global realist or anti-realist. Returning to mathematics, the anti-realist thinks of the mathematician as a sort of creator. In contrast, the realist thinks of the mathematician as a discoverer, who then enters the discoveries in a well-organised form in, what is suggestively called 'The Book of Proofs'.

Definition *The Book of Proofs* is a unique book that records all of the proofs of mathematics made in the foundational theory in normal form.[9]

Under this conception, mathematics consists in the results of the completed Book of Proofs. The process of doing mathematics is subservient to the discovery of those results.

The alternative anti-realist view emphasises the epistemology over the ontology, and thinks of mathematics as a process that leads to results. The importance of results lies in their continuing the process of establishing knowledge – 'results' are not ends. They are steps in a process.

[7]Ontology is usually presupposed to be consistent. There are no impossible objects, there are no pairs of objects whose existence precludes each other. Of course, paraconsistent ontologies are a different matter (no pun intended). For our gross sketch, we need not consider this added complexity.

[8]In the last chapter of the book, I am more explicit and subtle than this. It will turn out that the pluralist is, in some respects, a type of sceptic, and he is neutral on the realist, idealist axis of debate; but this added subtlety will be introduced in due course.

[9]The idea of The Book of Proofs has a history. In the original conception, all perfect proofs were entered. There was no guarantee that there would only be one proof for each theorem, since there was no presupposition that there was only one founding theory for mathematics. However, i
assume monism in foundations, then The Book of Proofs will only have one proof per theore

Since we shall return many times to the realist and anti-realist's considerations throughout the book, rather than exhaust the issues here, let us simply use our new vocabulary to reformulate the antagonists we saw earlier. The realist in mathematics is a reformer. The goal of mathematics is to generate proofs to enter into the Book of Proofs. The language and format of the proofs will be determined by The Foundation. The content of mathematics consists in the theorems found in The Book of Proofs.

The anti-realist, and pluralist suggest that restricting mathematical activity to entering proofs into The Book of Proofs would result in a loss, or at least a very long delay, in generating some results. This is because we have results today, which were easily and efficiently generated when ignoring the constraints advocated by reformist movement.

We should be careful, for, it looks as though the argument of the anti-realist rests on a quantifier confusion. There might be some results *more easily* gained before the reformation. However, after the reformation said results are not precluded. They are not lost altogether, just postponed. Furthermore, notice that there is a net gain in efficiency, since a lot of pre-reformation results will be completely obvious (in virtue of unifying the language of mathematics). Even better, since the reformation ensures that we are discovering the truth in mathematics, we shall not be led astray and generate mathematical falsehoods! This too is a net gain in efficiency. In terms of quantity of results, there is a standoff between the antagonists. The issue is over the order in which results are derived or produced, or, which are considered obvious and which not. If the mathematics of the plethora does not include set theory, then the anti-realist suffers a net loss since there will be some results (only obtainable in set theory), which will not be obtained by in the plethora mathematics. But if we include set theory in the plethora, then the camps are equal. Therefore, the disagreement is over the importance of obtaining mathematical results in a certain order.

2.2 Parting Company with the Reformists: Pluralism Within ZF

Here is a more serious consideration against the Reformation. We mentioned that there are rival set theories, but even 'ZF' is ambiguous. The astute reader will have noticed that I never specified whether I was discussing first-order ZF (ZF1) or second-order ZF (ZF2), and I only just mentioned one natural extension of ZF made by adding the axiom of choice to ZF to make ZFC. ZF1, ZF2, ZFC1 and ZFC2 are different formal theories. They have different axioms. In ZF2, some of the axioms will contain second-order quantifiers; in ZF1 analogues of the second-order axioms will be presented as axioms schemes. Moreover, there are many operations and concepts only expressible in ZF2. For example the concept of finitude is only second-order expressible (Shapiro 1991, 226). It follows that ZF2 is not reducible to ZF1, but ZF1 is contained in ZF2. For this reason and for the purposes of giving a foundation for 'mathematics' – given the claims above about reducing mathematics (as it is practiced independent of set theory) and then proving theorems proper to set theory – we had better specify that we are considering ZF2 as our foundation. It does not matter much, but we have to pick one, and ZF2 for a foundation is *prima facie* more plausible because of its greater expressive power. We might also want the axiom of choice since it is used essentially in a lot of proofs. But let us save that discussion for later. If we choose ZF2 we clearly have a strong candidate for The Foundation.[10]

We should not take this decision lightly. This is because there remains considerable suspicion concerning second-order quantifiers, whether, for example, to interpret these as substitutional or objectual.[11] Strictly speaking this is not a problem of ZF set theory, it is an ambiguity in second-order *logic*. But in the meta-language of second-order logic, we use the language of sets quite liberally, and we naturally interpret those sets as sets in ZF, more or less. If it turned out that there was a strong reason to favour one sort of second-order quantifier over the other, then this should affect our understanding of quantifiers in ZF as well.

The substitutional interpretation has it that the second-order quantifier gives us a license for substitution, and therefore, plays a grammatical role. We do not substitute objects but names for objects. Substitution is a linguistic act. The number of names available is determined by the name forming operations of the language. There will only be countably many.

In contrast, under the more common objectual reading of second-order quantifiers, quantifiers quantify over objects. If we are quantifying over a domain with more than countably may objects, then there are more objects than there are names. In general, a second-order quantifier ranges over the powerset of the set of objects in the domain of first-order quantification. When we read the quantifiers in, say, $\forall$P or $\exists$x objectually, the second-order universal quantifier ranges over the full powerset of the domain, since to every subset of the domain there corresponds a property (treated as an object of quantification in its own right). The cardinal number of properties will be $2^{\aleph_2}$ if our domain is $\aleph_2$, for example. In contrast, under the substitutional reading, we may only replace the P by names for properties which we can define in the language. There are only countably many definitions in any countable language, because there are only countably many symbols and all strings, which make a definition, are finite.[12]

The significance of the dispute over substitutional or objectual interpretations of the second-order quantifier shows us that even if we say that The Foundation is ZF2, we still have an ambiguity. The ambiguity is philosophically significant. It bears, at least, on the notions of (1) *ontology*, (2) *methodology*, (3) *justification* and (4) *epistemology*. If there are only countably many properties, then it is not clear what we are doing when we think we are performing mathematical operations on

---

[10]If our second-order quantifiers are Henkin quantifiers, then we only appear to have greater expressive power than in first-order ZF. See for example Väänänen (2001, 504–505).

[11]For a discussion of the difficulty in interpreting Frege on this issue see Boolos (1998, 225).

[12]Under a substitutional reading, V = L. Definitions are in the next section. For those in the know: under an objectual reading, V = L is independent. So the size of the set theoretic universe is decided if we insist that ZF2 includes a substitutional reading of the quantifiers.

sets of cardinality greater than $\aleph_0$. The different interpretations of the second-order quantifiers bear on (1) *ontology* because quantifiers are supposed to be quantifying over objects. If the second-order ones only quantify over objects, or subsets of objects which we can define in a (countable) language, then only the denotations of the definitions are treated as objects, and there are strictly fewer of these than there are possible properties of an infinite set iff there are powerset many of the latter, and we give an objectual reading of the quantifiers. While the objectual reading is more prevalent, it is not determined by the theory itself. Therefore, ZF2 with a substitutional reading of the second-order quantifiers is a genuine ambiguity.

In contrast, under an ontology-committing reading of the quantifiers, the objectual interpretation is ontologically determining, and presupposes a number of metaphysical arguments or decisions. *Ceteris paribus*, a philosopher with realist inclinations will tend towards the objectual interpretation, and the anti-realist towards the substitutional interpretation, although, not necessarily (since the so-called 'objectual' quantifier does not have to be read as forcing ontological commitment).[13] The different interpretations of second-order quantifiers also concern (2) *methodology*. For, under the substitutional interpretation, the bounds of language are to be taken more seriously in the methodology of making proofs because which inferences are legitimate will sometimes depend on the number of names available for quantification. The bounds of language, in the form of number of symbols and the arrangement of symbols in finite strings, constrains (if only ideally, and not practically (since the practical constraint is present regardless)) the range of possibilities for instanciating the second-order quantifiers. (3) *Justifications*, similarly, can invoke said constraints under a substitutional interpretation, but could not be invoked under an objectual interpretation. Lastly, the (4) *epistemology* suggested by The Foundation is different under the two interpretations. Under the substitutional interpretation, the limitations of language are part of the epistemology, whereas under the objectual interpretation they are not. Or, at least, they are not on the grounds of the interpretations of the second-order quantifiers. There might, of course, be other reasons, such as practical reasons, for bringing linguistic constraints to bear on an account of ontology, methodology, justification and epistemology, but, at least *ab initio*, these can be made quite separately from mathematical or metaphysical reasons concerning the interpretation of the quantifier symbols.

It follows that decisions, such as how to interpret second-order quantifiers are not to be taken lightly. For, one of the attractions and reasons for proposing a unique Foundation is that one gets at the truth of the matter, and this truth is not relative but unique and correct.[14] These considerations show us that even if we do claim that "ZF

[13]In systems which allow empty domains, universal formulas (ones with the universal quantifier as the main operator) do not entail the existential counter-part. So $\forall x(Fx) \vdash \exists x\,(Fx)$ is not valid in such systems. Some of these constructive systems have a separate second-order predicate E, which is a sort of metaphysical constant that does indicate ontological commitment. The existential quantifier is read strictly as 'some', never as 'there exists'.

[14]For the realist, we would prefer the word 'absolute'. If the anti-realist is of the cloth marked: 'truth is epistemically constrained', then he could still very well be convinced that there is a unique such truth, but it is determined by our epistemology.

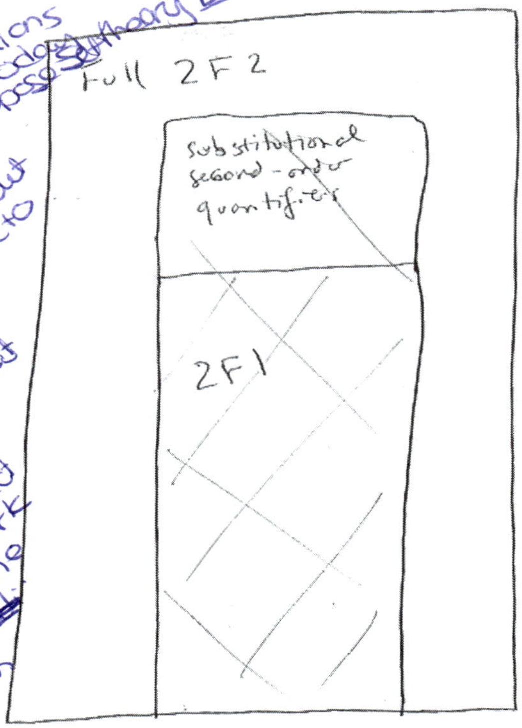

Fig. 2.2 ZF1 inside ZF2 with substitutional second-order quantifiers extending ZF1 into part of ZF2

is the orthodoxy in mathematics", we have made a highly ambiguous claim, and, unless we are willing to modify the claim to be more precise, the use of the definite article in the claim is misleading. If we are determined to be monist foundationalists, then we have to disambiguate, choose a unique foundation and be correct in our choice. Someone sceptical that it is always possible, or even desirable, to completely disambiguate and determine one foundation and confer upon it a normative role in mathematics is on the way to becoming more pluralist (Fig. 2.2).

2.3 Pluralism in Extensions of ZF: ZF as a Programme

Noting that the above problems are not resolved to this day, we can simply suspend judgment and remain agnostic (which is a form of pluralism). Under our newly found agnosticism, as a textual convention, we return to writing 'ZF' as ambiguous between ZF1, ZF2, and two readings of the second-order quantifiers in ZF2, that is, 'ZF' is ambiguous between at least four options.

The next problem to confront the reformists is that there are axioms that extend ZF. Examples include the axiom of choice, $V = L$ and the higher cardinal axioms. Explaining the terms: the axiom of choice states that for any set of sets, there is a function that will select one member of each of these sets to form a new set.[15] The continuum hypothesis, CH, implies that V = L. The continuum hypothesis is that $\aleph_1 = 2^{\aleph_0}$. V = L is a statement about the set theoretic universe, and how each stage in the hierarchy is constituted. V is Zermelo's cumulative hierarchy where at stage $\alpha + 1$, we find the union of all of the combinatorially–determined subsets of α. The notion of combinatorics is ontological. The universal quantifier in the term 'all of the combinatorially determined subsets' ranges over all subsets of objects, regardless of how we would describe these or name them. Sometimes this is referred to as 'the iterative conception' of the set theoretic hierarchy. In contrast, the set theoretic hierarchy L (developed by Gödel) is formed in the following way. At each stage $\alpha + 1$, we take all of the subsets at stage α which can be *defined* by a first-order formula in the language of set theory and whose quantifiers range over objects as they are determined (by a first-order formula) at level α (Maddy 1997, 65). The higher cardinal axioms are statements to the effect that a set exists which has a particular cardinality. Examples of such cardinal numbers include strongly inaccessible cardinals, measurable cardinals, Woodin cardinals and so on. These axioms are not reducible to ZF. So adding some of the axioms that correspond to the different cardinals makes a new theory, which extends ZF non-conservatively.[16]

What do we think of these? Many mathematicians are interested in these axioms, and the theorems that result from adding them to ZF. Most mathematicians are indifferent, since they have no bearing on their studies. Nevertheless, some mathematicians believe that some of the proposed extending axioms are true. Apart from phenomenological evidence, what we do know is that some of these are fruitful, in the sense that we use such axioms to learn about 'smaller' numbers. In further support of such axioms, we also have indirect proofs that individual axioms and some combinations of axioms are consistent with ZF. But not all pairs of such axioms are consistent with ZF. We cannot just extend ZF by adding all of them, because then (if ZF is consistent) we would have a trivial theory.[17] If we are committed to there being a unique extension of ZF, then we have to make choices, since the phenomenological evidence does not concur.

If we think that there is a correct extension of ZF, then we think of set theory as a (foundational) programme, rather than as an axiomatised or closed finished theory. Under this conception, ZF does not capture the whole of mathematics, but rather the

[15] There are different versions of the axiom, or family of axioms of choice. We shall only be as specific as we think it necessary for the purposes of the considerations being made.

[16] An extension of a theory is conservative if no new theorems can be proved, so really the 'extension' is in redundant shortcuts. A theory is non-conservatively extended if new theorems can be proven.

[17] A trivial theory in mathematics is one where every well-formed formula written in the language of the theory is true, so in particular, the negation of every formula is also true. This make the theory quite useless. To make such a theory, one would have to consider *ex contradictione quodlibet* inferences to be valid, and there would have to be a contradiction derivable in the theory from the axioms. We could then prove any formula using *ex contradictione quodlibet* arguments. We shall visit trivialism several times in this book.

core of mathematics. Some extension of ZF, probably some extension of ZFC will be a fuller core. What happens to the alternative extensions? Those, with Gödel, i.e., the monists, who are attracted to the idea of the uniqueness of a foundation, hold the belief that if[18] there is an extension, then it is unique and correct. Gödel believed in the systematicity of philosophy (Parsons 2010, 167) and the powers of the human mind. In Parson's words: Gödel's philosophical convictions were driven by a

> rationalistic optimism. This is first of all a belief, probably held before any philosophical arguments for it had been developed, of the powers of the human mind, especially in the sphere of human reason.... With respect to mathematics, he shared the conviction associated with Hilbert that for every well-formulated mathematical problem, there is in principle a solution, although his own incompleteness theorem implies that this might require introducing new axioms beyond those used in current mathematics. (Parsons 2010, 169)

Some mathematicians and some philosophers share Gödel's convictions.[19] We can call such a person a Gödelian optimist. A Gödelian optimist believes that the mathematical community will reach agreement over which is the correct extension of set theory, since she will be swayed by reasoned argument. Even more ambitious, the Gödelian optimist will have the faith that the mathematical community will be correct in their collective judgment.[20] If we project into the Gödelian optimist's future happy time of correct convergence, The Foundation will give a formal representation of some independent (from the theory or its formal representation) ontological reality. In other words, for the monist foundationalist who takes the possible extensions of ZF seriously, she thinks that today, we are still looking for The Foundation. She is an optimist because she believes that we shall eventually find said foundation (Fig. 2.3).

We part company with the Gödelian optimist if we consider that such belief and faith are not mathematical convictions but metaphysical convictions, even if they are born of mathematical experience and bear on mathematics; such beliefs and convictions cannot be assumed to be reliable, even in mathematics! When we part company with the optimist, we think that Gödel's beliefs are either unlikely to be met or we simply insist on being agnostic about what the future will bring. We are agnostic if we think that we do not have sufficient evidence to maintain that there is a unique underlying reality over which we shall eventually have full mastery through formal representation. Our agnosticism is based on the grounds that phenomenology, or gut feeling, or vividness, or mathematical experience, or even genius are unreliable guides to metaphysical truth; and this, simply on the grounds that different people, including the geniuses, have different gut feelings and have been misled.

---

[18]It is possible that no extension is correct.

[19]The rub lies in what is to count as a solution. I think it is safe to say that a naïve view of solution was assumed in these writings. That is, a solution is a definite and unique answer. With the Gödel archives being made increasingly available, this view of Gödel might be revised.

[20]Gödel was not using ZF as a foundation. He, with Bernays had developed their own set theory.

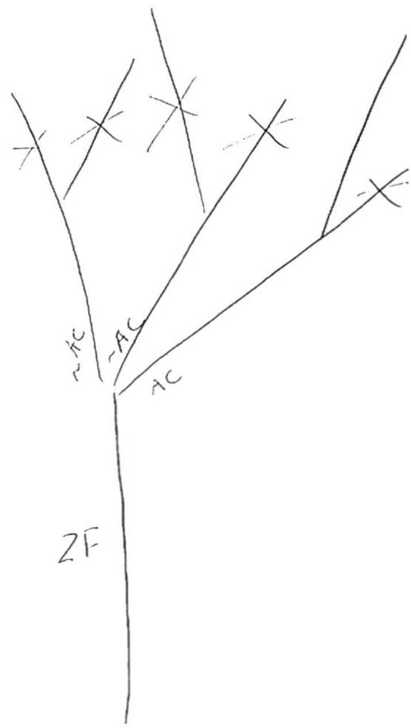

Fig. 2.3 ZF, represented as a line with extending lines representing axiom extensions. All but one of the lines is crossed out. This is the unique accepted extension

To take our first step towards pluralism we have to admit that we do not have sufficient evidence to support Gödelian optimism. The pluralist accepts that on present evidence, there are rival extensions, and ironically, we have learned from Gödel that there is no absolute mathematical way (yet)[21] to adjudicate between them (Fig. 2.4).

To adjudicate we need to make some extra-mathematical considerations. One promising route to this is to follow some work done by Maddy. Her purpose is to give formal expression to principles that would help us determine one extension of ZF, thus bringing us closer towards the desired convergence. Maddy's principles and her work will be looked at in detail in the next chapter. Suffice it to say that, for present purposes, Maddy's principles encourage convergence to one extension of ZF; and so they encourage the monist foundationalist.

However, they are *only* principles (she calls them 'maxims' but this makes no difference to the point here). We can turn the tables on the Gödelian optimist who follows Maddy's route towards convergence, and give formal expression to principles that encourage divergence as well! In the final chapter of the book I suggest extending her approach in a pluralist vein. I suggest we develop other formal principles. Some of these will run into conflict with Maddy's particular goals. Since

[21] Gödel is usually interpreted as having hoped and believed that we would eventually find some very powerful axioms that would determine the correct and unique truth for us. This interpretation of Gödel might soon be revised.

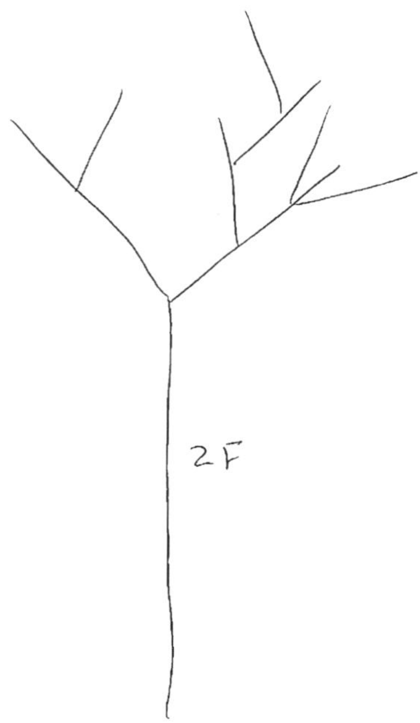

Fig. 2.4 ZF represented as a line with extending lines representing axiom extensions

we can develop alternative, nay conflicting, principles, this indicates that there is no *mathematical* reason to favour convergence towards one unique extension, this is a metaphysical preference, and is not mathematically necessary.

The pluralist is exactly someone who will at least in principle entertain the possibility that there are several extensions of ZF. The extensions are not equivalent to each other, and some pairs of extensions preclude each other – because if added together they would make set theory trivial. Moreover, the evidence is not in, to say that there will be convergence, unless it is metaphysically, or politically, forced. Here, 'forcing' presupposes a strong hope for convergence or a conviction that mathematics *has to be* unified by a single theory or programme (in the sense of one extending direction of ZF). If this hope or conviction is sufficiently strong, then it will blind the mathematical community to alternatives. The hope and conviction are applied *by force* if there is insufficient mathematical justification for them. The pluralist prefers to avoid force, and seeks a way to accommodate the present mathematical situation as he sees it.[22]

[22]Of course, if the foundational monist *is correct*, then he is right to try to force mathematicians to stay on the straight and narrow path. It is the same with religious fundamentalists. If they are correct in their beliefs, then they are doing exactly the right thing to try to force others to adhere to the correct faith. Until more evidence is in, however, such attempts at forcing are, at best, patronising.

2.4 Pluralism in Foundations

Already this is a bit heady and too pluralistic for some thinkers, but we have only gone a few steps in our journey towards pluralism. To continue our journey, we should consider alternative foundational theories. Let us begin with alternative set theories. The word 'set' is not, and has never been, fully determined by ZF.[23] 'Set' is implicitly defined by the axioms of a set theory, and even then the implicit definition might be ambiguous.[24] For example, Aczel's non well-founded set theory does not have a first smallest set from which the other sets are constructed only 'upward', so the conception of 'set' does not preclude 'negative', or 'wanting' sets, the hierarchy of which, descends downwards in mirror image to the (ZF-type) hierarchy of positive sets. The conception of 'set' we find in the theory of semi-sets developed by Vopěnka and Hájek (1972) and (Vopěnka 2013) is the 'usual (ZF) one' but we also add to this, the notion of semi-set which is related to the 'usual conception of set', so, strictly speaking 'set' is ambiguous between 'proper set' and 'semi-set'. What makes the notion of semi-set different from that of ZF set is that a semi-set is not ontologically neat in the sense of having a particular cardinal number of elements. Very roughly, we might think of a 'semi-set' as corresponding to a mass noun, as opposed to a count noun, or we can think of a semi-set as having indeterminate boundaries.

Hrbacek has also developed a foundational theory that uses multiple layers of ZF set theory. So a set in Hrbacek's set theory is not only found at a level in the set-theoretic hierarchy, but (replicas of 'it') are found at multiple such levels, each is then further contextualised by a level of magnitude (relative to which we find macro and micro levels). The level of magnitude is something determined by our present interest or application of set theory. One of the advantages of their approach is to make better sense of the notion of infinitesimal, and so make better sense of calculus than does ZF. Hrbacek's theory still makes reference to 'ZF set theory', so he uses ZF as a reference point, since he uses many copies of ZF super-imposed. The reasons for using ZF and not another set theory are partly to use a mathematical theory that is *familiar* today to mathematicians. In other words, other set theories *could* have been used to the same effect (of explaining the calculus, for example) and the choice is one governed by present day popularity, not so much a sense of *a priori* correctness!

---

[23] This point was very nicely made by Sebastik in the Logic Colloquium 2010 presentation: On Bolzano's Beyträge, Paris, 31 July 2010. There he points out that he will insist on the use of the word 'set' when translating Bolzano, but to remember that the word 'set' is ambiguous, especially in Bolzano. This is nothing new, and it is not outrageous to think of set this way. As a point of comparison, Sebastik points out that he word 'atom' too meant something very different to Democritus than it did to Bohr. So while many modern readers might see the word 'set' and think immediately of ZF sets, and therefore be confused or feel deceived when reading Bolzano, they should instead realise that ZF has no monopoly over the use of the term.

[24] See the discussion of the proposed axiom $V = L$ above.

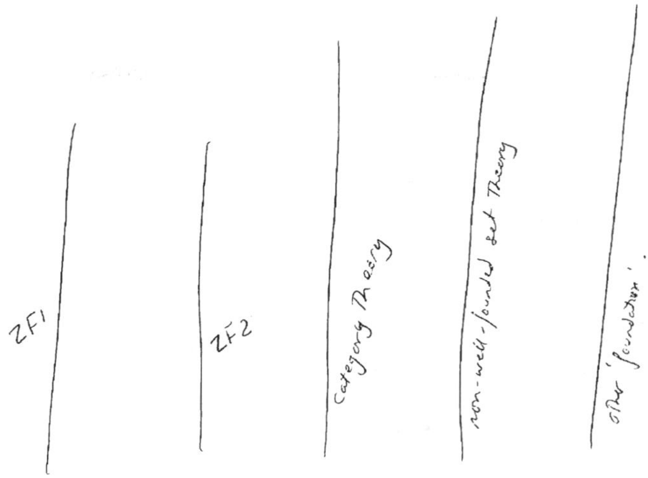

Fig. 2.5 Several foundations

The implicit definition of 'set' in each of these theories is quite different. Notice too, that in describing all of these alternatives, we mentioned ZF set theory as a starting point. These alternative set theories are not reducible to ZF set theory. But they make reference to ZF, and use ZF as a core. In all cases, ZF is embedded in them.

If we are to take these set theories seriously, then we have to tell something like the following story about ZF. ZF does not *determine* the (implicit) definition of the word 'set'. There are others. Nevertheless, we can observe in the practice of mathematics that ZF acts as a *point of reference* for many mathematicians. What does this mean? It means that we take ZF as a standard of measure. We have few absolute consistency proofs, and few verifications of mathematics against the physical world; so we require some reassurance that our mathematical theories are 'correct'. We use ZF as such a measure of reassurance.

If a proposed mathematical theory can be reduced to ZF, then this is *very* reassuring. If we can show that the theory is equi-consistent with ZF then this is *quite* reassuring. If we can show neither, then we have to look elsewhere. In other words, under this 'reassuring default' attitude, ZF is just a familiar and reassuring theory. If ZF is a point of reference, then it does not determine what counts as mathematics and what does not. ZF neither prescribes mathematical practice, nor does it circumscribe the content of 'mathematics'. It does not hold the monopoly on ontology, methodology or *epistemia*. The pluralist believes that we are not warranted in thinking that ZF is an absolute guide, it is just a point of reference (Fig. 2.5).

The next step on the journey is to consider that there are alternative foundations that not only completely absorb ZF or other set theories, but they use an essentially different language, so embedding has to go through a translation and interpretation.

These include: category theory[25] and type theory. Both of these are mathematical theories that can inform us about set theory. For example, in category theory we can form a category of sets. Then the whole of set theoretic semantics, closed under the axioms and operations of set theory form a category. We can then use this way of thinking about set theory to compare, for example, the category of sets to the category of proper classes and we use the functors of category theory to make the comparisons.

Straying further, we also have paraconsistent set theories, where we think of classical set theories as a special case. It counts as a special case in the sense that it is consistent. In a paraconsistent set theory, we can derive contradictions, and we have contradictory sets. If we are careful about how we make the paraconsistent set theory, then all of the classical results are preserved for the classical part of paraconsistent set theory, and we get more, which is the special work with contradiction.

Explaining to those unfamiliar with these notions: a contradiction is a conjunction of a formula and its negation. It takes the form: $P \,\&\, {\sim}P$, where P is a well-formed formula, '&' is conjunction and '$\sim$' is negation. In a classical theory, any formula of the language is derivable from a contradiction. We call a theory in which every wff of the language of the theory is derivable a 'trivial theory'. The paraconsistent and relevant logicians block the inference from a contradiction to any formula at all, and therefore, have different constraints on their proof theory. See appendix one for an introduction to one sort of paraconsistent logic, the logic LP (logic of paraconsistency). We shall meet paraconsistency many times in this book.

Definition The *pluralist in foundations* believes that there is insufficient evidence to think that there is a unique foundation for mathematics. Moreover, the pluralist in foundations works under the assumption that there is no reason to think that there will be a convergence to a unique theory in the future. He takes seriously the possibility that there are several, together inconsistent, foundations for mathematics.

So now we have pluralism about mathematical foundations. This is our first sort of pluralism. Pluralists in foundations understand that candidate foundational theories are not all closed completed theories. They can all be thought of as programmes in their own right, as we saw with set theory and its extensions. There are other sorts of pluralism. We turn to these to give a first taste of the wider view of pluralism.

[25] There is a heated debate between the category theorists and the set theorists on the 'Foundations of Mathematics' website. The argument is over whether set theory can say everything category theory can, so is the more fundamental theory, or whether category theory can say everything set theory can, so is the more fundamental. It seems, from the outside, that at this point in the debate, the two theories, or programmes, climb against each other. At present there is no obvious end to the debate.

2.5 Beyond Pluralism in Foundations: Pluralism in Perspectives

There is more to pluralism, than pluralism in foundations. So far, we have noticed that alternative foundational theories have been developed, and if we are pluralists about these then we intend to take them all seriously ‘as foundations’. However, in using this language we notice that we start to lose grip on the notion of foundation. A plurality of foundations is not stable, and is not itself *a* foundation. Therefore, ultimately, ‘pluralism in foundations’ is an unstable position, or one where ‘foundation’ is completely divorced from its original metaphorical meaning. Since the metaphor of ‘foundation’ has lost its grip, it would be better to think of a plurality of theories, each of which can be used to get some sort of very general perspective on the rest of mathematics. So, what were thought of as foundations, are now more *perspectives* or *vantage points*. Leaving behind pluralism in foundations, we shift to the idea of pluralism in perspectives.

Definition The *pluralist in perspectives* demurs from favouring one perspective on mathematics. Each mathematical theory, which is powerful enough to give a perspective on quite a lot of mathematics, will have its philosophical sins and virtues. There is insufficient evidence to think that there is an absolute perspective that is best either philosophically or mathematically.

To understand what is meant by ‘perspective’, forget the foundational theories or programmes. Instead characterise mathematics by giving an organisational perspective on mathematics. What distinguishes a foundation from a perspective is that the perspective is not necessarily in the form of an axiomatic theory, but looks more like a programme. Moreover, it has fewer of the metaphysical pretentions of a foundation. There might be some axioms, but the list of axioms, if there is one, is incomplete in the sense that we could add more. There might be some sense of truth and ontology, but not the familiar realist one. An excellent example of an organisational perspective with its attending philosophy is Shapiro’s structuralism. We shall see this in greater detail in the chapter on structuralism. For now, what is important is that Shapiro does not consider himself to be giving a foundation for mathematics, but rather he is working with model theory and the language of second-order logic. Model theory is not presented in axiomatic form. Yet it is used to compare mathematical structures to each other, where a structure is a domain of objects together with predicates, relations and operations pertaining to the objects. A mathematical theory can be thought of as a structure, so with model theory we can compare mathematical theories. Model theory is not a foundational theory in the philosophical sense of foundation since it lacks some of the characteristics we associated with foundations. Running through these:

(i) in Shapiro’s structuralism, model theory circumscribes the content of mathematics,[26]

[26]Shapiro does make some conciliatory remarks about being more general, and adopting alternative perspectives. Nevertheless, while he acknowledges that this is a possibility, he proceeds as tho

(ii) the ontology of his structuralism is quite different from that of set theory, since the ontology is structures, and what counts as a structure is determined by a further meta-structure, so ultimately ontology is always relative to a structure.

(iii) Shapiro's structuralist perspective does not claim a monopoly on methodology, since these are determined within a structure and can be imported from one structure to another.

But (iv) Shapiro's structuralist does claim to say quite a lot about epistemology. What we can know in and about mathematics is determined by model theory.

We know about one structure from the point of view of another (meta-)structure, and there is no other *epistemia* in mathematics.[27] Shapiro defends his choice of organisational perspective, but not against all possible alternatives. In a pluralist vein, one could develop a version of structuralism-as-a-perspective from a constructive point of view, for example. In this case, structures all have to be constructed (according to constructivist sanctioned operations), where the permitted constructions are those that respect certain constructive epistemic constraints.

It is important to note here, since this issue will re-surface, that foundationalist aspirations are neither intrinsic, nor essential, to the philosophy of mathematics. We can do a lot of good philosophical work once we adopt a perspective. The work could be local, or it could pertain to a very large part of mathematics. See, for example, the recent work by the 'maverick philosophers of mathematics',[28] represented by Larvor, Corfield, Cellucci and Rav, amongst others.

2.6 Pluralism in Methodology

Pluralism in perspective invites us to consider pluralism in methodology:

Definition The *pluralist in methodology* is tolerant towards proof techniques, methods and results being imported from one area of mathematics into another. He is also not averse to the suggestion that techniques in disciplines outside mathematics can be useful to mathematics and the philosophy of mathematics.

model theory is the only perspective. There is some tension in his writing. Whether the model theory perspective is an end point or a starting point depends on one's reading of Shapiro. I invite him to join me in becoming a pluralist in perspective, if he is not so already.

[27] *Epistemia* should not be confused with heuristics, how we learn, our private experience of ... we come to form beliefs and so on. *Epistemia* is an idealised notion of knowledge

[28] ...erick philosophers of mathematics" appears in a conference announcement for a ...e 2009 held at the university of Hertfordshire. Originally, it comes from Asprey ...3, 17).

We shall see examples of importing proof techniques from 'foreign parts' in the chapter on formalism. It is a brute observation that a lot of present day mathematical writing mixes methodologies and results from various areas. We could even say that there are two styles of mathematical work. In one, we do work purely within a mathematical theory. For example, we might produce a result in analysis using only the techniques developed in analysis, but increasingly we also see the other style of work in mathematics where results and techniques are imported from different areas. A good result in one area will be given an analogue in another and then it is used in the second area. Given that the 'areas' have very conflicting things to say, this might not always be legitimate.

Definition The *methodological monist* is someone who objects to the use of imported methodologies.

In contrast, the methodological pluralist will allow it with caution. We shall see this more closely in Chaps. 5, 9, and 14.

Pluralism in methodology does not have to be restricted to mathematical methodologies being used in mathematical contexts. If we accept that there is no sharp distinction between mathematics and philosophy, then we can also be pluralist towards philosophical and other scientific methodologies, such as those used in sociology, psychology and neuro-science. Each of these disciplines can inform both the philosopher and the mathematician. Arguing for the antecedent of the conditional, i.e., that there is no sharp distinction between mathematics and philosophy, consider first that many good mathematicians are also philosophers, and vice-versa. Examples are: Russell, Dedekind, Frege, Cantor, Vopěnca, Hilbert,[29] Brouwer and Martin-Löf. In each case, their mathematical and philosophical work inform each other. From quite a different perspective, we can say that there are mathematical ideas that are partly philosophical. For example, Church's Thesis that a function is effectively computable iff it is Turing computable[30]; where 'effectively computable' is not a concept that is absolute or constant across all interpretations. Church meant it as an intuitive concept (Folina 1998, 302). This is what makes the Thesis interesting, as opposed to tautological, or true in virtue of being a stipulation.

The conclusion to draw from these considerations is that philosophy and mathematics are not easily separable. Note that much of the meeting between philosophy and mathematics occurs with logic. Note also that there are, of course, areas of mathematics that are quite untainted with philosophy, and areas of philosophy untainted by mathematics. While philosophical argument pre-supposes some logic, the author of the argument might not have a particular formal representation of logic in mind. In fact, he might think of formal representations as distortions of a

[29]Some philosophers would not count Hilbert as a philosopher. For them, he was a mathematician, whose mathematics and suggested programme had philosophical implications. I prefer to err on the side of generosity, and allow him into the philosophical fold.

[30]There are different ways of stating the thesis. The point is that in most versions there will be an irreducibly vague or ambiguous philosophical term, otherwise we do not have an interesting thesis, but instead we have a tautology, or stipulative definition (Folina 1998).

more primitive, or pure, reasoning, and therefore disingenuous. The influence, in such cases will go the other way: a logician might be intrigued by a verbal form of argument, and might try to give that form loyal formal representation. See Chap. 7 for details. Once this is done, it enters the body formal, and is readily available to the body mathematical and the body philosophical. Thus, once we become pluralist in methodology, we see a blurring of the distinction between mathematics and philosophy and other disciplines too.

2.7 Pluralism in Measures of Success

One can be even more radical in one's pluralism. So far, we have been interested in complete mathematical theories, or programmes that are well established. We have ignored incomplete theories, not in the technical sense of 'incomplete', but rather in the sense 'uncompleted': where not all the details are yet written down, not all the ideas a fully thought through. We have also been ignoring unsuccessful mathematics. For the pluralist, ignoring incomplete or unsuccessful mathematics is a mistake. If, as philosophers, we want to explain mathematics, part of the explanation has to include an account of mistakes and error in mathematics. Moreover, errors can be small, easily corrected, or they can be disastrous (such as when we discover that our theory is trivial). But even the disastrous ones contribute to our understanding of mathematics. This is witnessed by the fact that we learn from the mistakes, and develop good mathematical theories only after having made the mistake. This point was famously made by Lakatos. We sometimes even find that we have misdiagnosed a mistake. There are many examples of this. For a start, see the work done in patching up Frege's formal system. There are several non-equivalent ways of making a good mathematical theory based on the wreckage of the original theory. The philosophical and technical importance of each is then a source of debate. There is then some sense in which Frege's work was not a waste of time. It had a measure of success, but not the one he intended. See Chap. 14 for details. We might call this sort of pluralism 'pluralism in success'.

Definition *Pluralism in success* is the view that while there are different measures of success in mathematics, and these are sometimes well accepted, an unsuccessful theory (according to the first measures) is very successful *tout court*. There might be other respects in which the theory is very successful, and exploring this is sometimes philosophically fruitful. We shall return to these issues several times in the book. This was a first wash.

2.8 Conclusion

We have now completed the first journey. Recalling the sights we visited, we saw pluralism within ZF foundationalism. This took the form of noting the ambiguity in the very idea of Zermelo-Fraenkel set theory. We then also noted that ZF is, in some sense, an uncompleted theory. There are many, together incompatible, extensions of ZF. So, we might think of ZF as a foundational programme, rather than as a foundation. We might then examine our attitudes concerning different extensions, and the pluralist's attitude is one of agnosticism concerning eventual convergence of set theory to one theory with the four philosophical characteristics: determining the scope, ontology, methodology and epistemology of mathematics. We then ventured further and recognised alternative other 'foundational' theories. This introduces us to the idea of pluralism in foundations. But this pluralism is unstable. So if we follow the pluralist this far, we have to go further to a more stable position: pluralism in perspectives. We then discovered that we can push pluralism along other dimensions. Pluralism in perspectives invites us to consider: pluralism in methodology and pluralism in success.

On the journey we have seen the pluralist as more tolerant than his counterpart. However, we might ask: is the pluralist ever intolerant, and if so, what about? The answer has been implicit throughout our journey. The pluralist is intolerant towards dogmatism and absolutism. He calls for making explicit all contexts within which philosophical claims are made, when they are in the forms of judgments such as: 'X is true', 'X is best', 'X is correct' and so on. Or at least, he urges our investigating the limitations of our claims. The pluralist is not merely admitting that dogmatic claims might be *erroneous*. Rather, there might well be no fact of the matter, and dogmatism is simply misplaced on present evidence. At least this is the pluralist's moderate face. But I warned in the introduction to this chapter that I would tell some lies. When in close company, and feeling less moderate, the pluralist will outright admit that looking for an absolute or correct point of view might be locally useful, but ultimately is probably (i.e., on present evidence) quite futile and is a mistaken way of proceeding in philosophy of mathematics. We shall see this in greater detail in the final part of the book where we put pluralism to work on a few test cases.

In his intolerance, the pluralist flirts with paradox as we might see from the question: is the pluralist always/ absolutely intolerant towards dogmatism? To see the flirtation through, the reader will have to wait for the third part of the book, where the pluralist transcends his own pluralism.

References

Arrighi, P., & Dowek, G. (2010). On the completeness of quantum computation models. In F. Ferreira, B. Löwe, E. Mayordomo, & L. M. Gomes (Eds.), *Programs, proofs and processes; 6th annual conference on computability in Europe; 6th conference on computability in Europe, Ponta Delgada, Azores, Portugal, June/July 2010, proceedings* (pp. 21–30). Berlin: Springer.

Asprey, W., & Kitcher, P. (Eds.). (1988). *History and philosophy of modern mathematics* (Minnesota studies in the philosophy of science XI). Minneapolis: University of Minnesota Press.

Balaguer, M. (1998). *Platonism and anti-Platonism in mathematics*. Oxford: Oxford University Press.

Boolos, G. (1998). Whence the contradiction. In R. Jeffrey (Ed.), *Logic, logic and logic* (pp. 220–236). Cambridge, MA: Harvard University Press.
Folina, J. (1998). Church's thesis: Prelude to a proof. *Philosophia Mathematica*, Series III, *6*(3), 302–323.
Giaquinto, M. (2002). *The search for certainty; A philosophical account of foundations of mathematics*. Oxford: Clarendon Press.
Hrbacek, K., Lessmann, O., & O'Donovan, R. (n.d.). *Analysis with ultrasmall numbers* (in prep.) (Manuscript, 2009).
Maddy, P. (1997). *Naturalism in mathematics*. Oxford: Clarendon Press.
Parsons, C. (2010). Gödel and philosophical idealism. *Philosophia Mathematica*. Series III, *18*(2), 166–192.
Potter, M. (2004). *Set theory and its philosophy*. Oxford: Oxford University Press.
Shapiro, S. (1991). *Foundations without foundationalism; A case for second-order logic* (Oxford logic guides 17). Oxford: Oxford University Press.
Väänänen, J. (2001). Second-order logic and foundations of mathematics. *The Bulletin of Symbolic Logic, 7*(4), 504–520.
Vopěnka, P. (Forthcoming, Manuscript, 2013). *The great illusion of 20th century mathematics and its new foundations*, Springer.
Vopěnka, P., & Hájek, P. (1972). *The theory of semisets*. Amsterdam: North-Holland Publishing Company.
Wright, C. (1986). *Realism, meaning and truth*. Oxford: Basil Blackwell.

Chapter 3
Motivating Maddy's Naturalist to Adopt Pluralism

Abstract In this chapter, the reader is taken on a journey from Maddy's naturalist position to a more pluralist position. The pluralist is inspired by Maddy's mathematical naturalism in the following respect. With Maddy, the pluralist is very interested in the practice of mathematics, and is quite willing to let mathematical practice delimit what is to count as 'mathematics'. The pluralist parts company with Maddy over the data concerning the philosophical inclinations of mathematicians. Given different data, the pluralist finds himself driven towards a more pluralist conception of mathematics than Maddy's.

3.1 Introduction

The idea behind this chapter is to show a reader who has naturalist sympathies how naturalism inspires and can lead to pluralism. Whether or not the arguments are convincing will depend on how seriously one takes certain evidence found in mathematical practice, what one includes in such practice, the version of naturalism one has adopted, how one interprets Quine and how courageous one is in one's naturalist convictions. Thus, the arguments are not definitive, but rather, qualitative. They are intended to give another sort of introduction to pluralism.

I begin the chapter with setting the stage. In the first sections, I discuss naturalism with reference to Quine and Maddy.[1] naturalists are wary of *a priori* metaphysics and theories which cannot be tested against the natural world. They understand 'science' to include physics, chemistry and biology as canonical examples. Traditionally, Quinean naturalists have had notorious difficulty explaining the place

---

[1]A lot of my reading of Quine is indirect and comes from Maddy. I deliberately take Maddy's Quine in order to emphasise her developments of naturalism. I am not so concerned with loyalty to Quine. There are less and more sophisticated (and flexible) interpretations of Quine. It turns out that the very sophisticated readings approach my overall point of view quite well.

M. Friend, *Pluralism in Mathematics: A New Position in Philosophy of Mathematics*, Logic, Epistemology, and the Unity of Science 32, DOI 10.1007/978-94-007-7058-4_3,

© Springer Science+Business Media Dordrecht 2014

of pure mathematics. Is it *a priori* metaphysics or science?[2] Quine's solution is to make ontological commitments to only the part of mathematics that is indispensable to science. For many pure mathematicians this is unacceptable. This is because Quine's position gives second place to pure mathematics against applied mathematics and science, and, from the mathematician's point of view, this makes little sense. Maddy wants to develop Quinean naturalism, but she makes some changes. She treats mathematics as a science in its own right. Maddy's version of naturalism is a significant improvement on Quine's, in the sense that it respects the mathematician's view of mathematics. To make the change, she draws on Quine's meta-philosophical pragmatism,[3] and gives a naturalist philosophy of *mathematics*, as opposed to a naturalist philosophy of *science*.

After discussing this background, I move to the main point of the chapter, which is to criticise Maddy for not going far enough in her naturalist inclinations. I invite her to follow them through and become a pluralist. I shall then discuss Colyvan's naturalism, but shall not give it centre stage, since he follows Quine in considering only the causal, or physically based, sciences as 'science'. He is not given centre stage because convincing Colyvan of pluralism would require quite different arguments. Within his framework – of what he counts as real mathematics – some aspects of pluralism follow very easily. But other aspects would require careful treatment, which I prefer to save for another project.

3.2 Quine's Naturalism

Quine is a naturalist, and as such, favours scientific explanations. The reason is that they involve *causal* entities and give us accurate predictions. These causal entities are 'real', 'in the world', 'independent of us and our theories', but he recognises that in theory-building, we have many possibilities open to us. We build a scientific theory with a combination of observed data and metaphysics. We restrict the number of possibilities through considerations such as: deploying Occam's razor, fit with the rest of our web of beliefs, simplicity and so on. Nevertheless, what is essential in a science is that it accords with the data, and gives us predictive power. For Quine, naturalism means favouring: scientific methodology, scientific ontology and

[2]Quine understood that some metaphysics is inevitable in science. This was his critique against the positivists (who thought at one extreme that it is possible to do all science without metaphysics). When more moderate, the positivists wanted to rid science of as much metaphysics as possible. What Quine, like Popper, was against was *a priori* metaphysics, where we construct irrefutable theories.

[3]Quine had a preference for scientific investigations, but his meta-philosophical pragmatism is what Maddy draws on. She acknowledges her preference for investigating pure mathematics, as a science in its own right. I should like to thank Zawidzki for helping me appreciate the distinction between Quine's preferences and his meta-philosophical pragmatic naturalism, which allows him to accommodate modifications to his views on the basis of different preferences.

scientifically discovered facts over other forms of enquiry, where 'science' means: physics, chemistry and biology. When Quine was writing, these were the canonical examples because they were the most successful of the sciences.

From a naturalist point of view, the 'softer' sciences should try to approach the 'hard sciences' as much as possible: rely on observed and measured data, and speculate judiciously, according to what some call the theoretical virtues: simplicity, elegance, fit with other ideas and theories. Mathematics and philosophy are soft sciences, in this sense. There is little, or no, observed data.

Since philosophy is a 'soft science', the philosophy *of* a science takes second place to that science, at least with respect to determining the scope of the subject of enquiry and the methods of enquiry. Quine calls this philosophical attitude 'science first'. He takes the attitude very seriously. In Quine (1981), he *characterises* naturalism as the "abandonment of the goal of a first philosophy." (Quine 1981, 72).

Explaining: 'first philosophy' is the opposing view to 'science first'. If, *contra* Quine, a philosopher has a 'first philosophy' attitude, then he thinks that *philosophy* should determine the parameters of a science. If, with Quine, he has a 'science first' attitude, then he lets *science*, as it is practiced, determine the parameters of what counts as 'science'. Under a science first attitude, the philosopher's task is to make a philosophy, which *accommodates* the scope of the practice. Since part of the practice is philosophical, the philosopher should also pay heed to the philosophical inclinations of practicing scientists.[4] Of course, the philosopher does this critically, since she has more training in philosophy. Maddy interprets this to mean that philosophy is relegated to second rank to science, and this interpretation of Quine will, of course, breed resentment amongst philosophers who are inclined towards a first philosophy attitude. At this stage in our considerations, we have said nothing to justify the feeling of resentment. We shall return to the resentment later, when we discuss 'the status problem', and more extensively, when we examine Maddy and Colyvan's naturalisms. Before we do this we should visit the relationship between mathematics and science, since important features are shared in the relationship between mathematics and science, on the one hand, and philosophy and science on the other.

---

[4]There are two distinct aspects to taking heed of the scientist. One is to take what practicing scientists do and discover seriously, the other is to take what they say about the philosophy of science seriously. The first is less controversial, since it will be at least a *starting point* for any philosophy of science. The second is taken seriously, especially when scientists are also philosophers. In fact, some would say that to be a good scientist, one has to be a bit of a philosopher too. But this is thought of as the exception by some philosophers who pay no attention to philosophical remarks made by scientists – since they are not qualified to make them. This dismissal relies on an easy partitioning of philosophy and science, but such partitioning is a little strained. This is plain when one considers that this sort of distinction only starts to make sense when we think of the modern education system where people are asked to specialise in one area early, and this prevents them from spending a lot of time in another. This is a very recent phenomenon. Newton and Leibniz, for example, were hardly 'trained in science' to the exclusion of other forms of 'training' or enquiry.

Turning to the issue of mathematics: on the one hand, mathematics is perfectly rigorous and is used extensively *in* science. Quine is not anti-mathematics and logic, in fact, he has made contributions to both fields, so he does not consider mathematics to be like astrology or bad metaphysics. Nevertheless, the entities of mathematics do not play a *causal* role in the world, and therefore, mathematics is somewhat suspect for Quine.[5] Esoteric, pure mathematics, involving, for example large cardinals, intentional operators, infinite spatial dimensions, hardly helps us with our predictions. As a result, all too often, the only counter-check to mathematical theories is other mathematical theories. The reconciliation between mathematics and science comes from the famous indispensability argument.

This runs as follows. First, notice that some of mathematics is indispensable to our best scientific theories. Quine's naturalist then concludes that all the parts of mathematics, which are indispensable to science, are *bona fide* 'real' mathematics. We then 'round out' the mathematics to give a completed mathematical theory. Lastly, we commit ourselves to the ontology of the whole of the minimal mathematical theory which will 'do the job' required of it by science. This is Quine's pragmatic move. For the Quinean naturalist, it follows that we are ontologically only committed to *that part of* the mathematical ontology, which is *indispensable* to the hard sciences. Thus, Quine partitions mathematics into the 'real part' (indispensable to science) and the 'recreational' part (Quine 1981).[6] What he means by 'recreational' is *not* that it is designed for amusement by amateurs; but rather, that we lack sufficient reason to be committed to the ontology of that part of mathematics, *given the present state of science*.

Of course, *probably* (by induction on the history of science) *it will turn out that* science needs more mathematics than it presently uses; as scientific results increase, we sometimes find that the 'real' part of mathematics also increases. In the future, these hitherto 'recreational' parts of mathematics will turn out to be indispensable,

[5]This is especially true of the 'early Quine'. The 'later Quine' accepted much more of mathematics and logic. This is done mainly through a fairly extensive 'rounding out' of the part of mathematics needed for science. Nevertheless, even the later Quine's starting point is science, not mathematics itself.

[6]The scope of 'recreational mathematics' is a matter of debate. Quine was in favour of first-order Zermelo-Fraenkel set theory. Colyvan argues that if we take a 'holistic approach' then pretty much all of present day practiced mathematics is in the 'real' part, since there is some link between the immediately applied (to science) parts of mathematics and the 'nether reaches' (Colyvan 2001, 107 footnote 23). Insofar as Colyvan's holistic attitude is convincing the recovery of most of mathematics is an artefact of the vast development of crosschecking in mathematics, the application of one mathematical theory, to check another. I discuss this especially in Chaps. 7, 8, 9 and 14. The difference between the pluralist and Maddy, on the one hand, and Colyvan and Quine on the other, lies in the presumed *reason* why mathematics is good science. For the pluralist and Maddy, the reason is that mathematics is rigorous, has a perfectly good methodology. For Quine and Colyvan, the physical world, prediction and causation are the ultimate reasons we can seriously engage in mathematics.

at which time those parts are promoted from 'recreational' to 'real'.[7] But, *until* the Quinean naturalist finds applications in, or links to, the physical sciences, he remains *agnostic* as to the reality of the purported ontology of 'recreational' mathematics. Arguably, at the present state of play, scientists and Quinean naturalist philosophers should make *no ontological commitment* to: proper classes, perfect spheres, categories, sets of cardinality greater than $\aleph_1$, spaces of infinite dimensions and so on.[8] Although, in part, this depends on how generous we are in our 'rounding out' and our criteria for choice of minimal mathematical theory.

Now we shall consider an objection to the indispensability argument and then turn to Maddy's solution. The objection comes from mathematicians, and we shall call it 'the status problem'. What counts as science is determined by scientists, and mathematicians are there to supply the formal machinery necessary for science. Therefore, what counts as 'real mathematics' is also determined by scientists. So, mathematicians play second-fiddle to scientists. Presently, the *recreational* part of mathematics is 'a lot' of mathematics.[9] This is alarming for the mathematician because a lot of mathematics is taken less seriously than the very small fragment of mathematics needed for science.[10]

This is the present state of play, but the future does not look much better. There is no guarantee that science will catch up with the developments in mathematics.

[7] We also have to be careful about the relationship between the rate of increase of the scope of real mathematics and the rate of increase in practiced mathematics. If the rate of increase of the first is far inferior to the second, then more and more of mathematics will be considered to be recreational by the Quinean naturalist. There is no guarantee about the relative rates of increase of 'real' and 'recreational' mathematics.

[8] Arguably, we do not need the full set of reals for science. We actually only ever (will) use a finite (and very small) number of reals. In fact, we do not need the real numbers at all. All we need is an approximation of some of them. Thus, strictly speaking, we only need the rational numbers for science, and not even all of those. Even if the mathematicians have an algorithm for producing the expansion of Π, for example, the scientists only need it to be expanded finitely, not infinitely. So they do not need the full mathematical theory of the reals. Regardless of whether or not the reader agrees, the cut off point between real and recreational mathematics is not important for the argument here. The cut off could be higher. Even if we consider a very high cardinal axiom as important for science, we would still be missing quite a lot of other high cardinal numbers, an infinite number at least!

[9] As noted, Colyvan makes a case for most, if not all, of mathematics being necessary for science. We shall look at objections to Colyvan's naturalism in Sect. 3.4 of this chapter.

[10] If we were *very keen* on positing as few mathematical objects as 'possible', and if we are willing to entertain the idea that inconsistent objects are possible, then if we first note that we only ever use a finite number of predicates in science, then "Any mathematical theory presented in first-order logic has a *finite* paraconsistent model." (Bremer 2010, 35). So, provided we agreed that a first-order language was enough for science (which Quine would agree to), and we allow the existence of inconsistent mathematical objects, then we only need a finite theory: one with a largest finite number. Since allowing inconsistent mathematical objects is not to everyone's taste, we then engage in a negotiation between our metaphysical taste and our keenness to mathematical ontology.

That is, the recreational part of mathematics might well 'increase'[11] at a greater rate than the rate of mathematics used in science. So, ever increasing amounts of mathematics will be considered to be recreational to the Quinean naturalist. Of course, things might not turn out this way, the real and the recreational part might increase at the same rate, or it could turn out that the real part increases at a faster rate, and they eventually merge, developing in tandem, for ever, or only temporarily.

Merging seems unlikely for two reasons. One is that mathematicians working in 'pure' mathematics tend to be driven by considerations internal to mathematics, and not so much to the applications of mathematics in science. This could be changed with politics, culture and economics – we simply stop having many positions in universities for research in pure mathematics. While such a cultural decision would affect the rate of development of recreational mathematics, especially if we were to make this decision in a sufficient number of countries, recreational mathematics, done in free time, as a hobby, would not for these cultural and economic reasons have to increase only at the rate of 'real' mathematics. This is not a knockdown argument, but depends on what we take to be secure cultural interests and the independence of those interests from the drive of mathematicians to produce results regardless.

The other reason it seems unlikely that the part of mathematics needed for science will catch up to the recreational part of mathematics is that science deals with the measureable. Our measuring instruments can only use rational numbers and rational lengths. Put another way, a rational length, provided it is precise beyond a certain measure, will be all that is needed for tests against 'the real world' – even if our (pure) mathematical theory tells us that there is an irrational number needed *in the theory*. So, we do not need a very developed theory of 'real numbers' (obviously real numbers are not 'real' in Quine's sense!) in order to carry out science and test our mathematics. Rational approximations will do. There are perfectly good (to serve as instruments of science) mathematical theories which can accommodate this naturalist attitude. However, mathematicians studying pure mathematics will continue to develop mathematics well beyond what is measurable by physical instruments. Moreover, the theories built upon theories of reals – all of the branches of analysis, and the mathematical theories which house pure analysis continue to be developed, and there is no mathematical reason to think that this development either has to slow down or stop (Fig. 3.1).

Depending on how the graphs develop, fewer or more mathematicians will feel that they have been given short shrift by the Quinean. Nevertheless, the objection to Quine so far is more-or-less an argument from 'political correctness', where we try not to offend too many mathematicians. But there is a deeper issue.

---

[11]The notion of 'increasing' mathematics is at least ambiguous. It is not clear if we are counting theories, theorems, objects or what. Nevertheless, at an intuitive level we can see the point. No doubt, there will be some measures of 'quantity of mathematics' where the 'rate of growth' of mathematics outstrips scientific applications significantly – say by adding sets of large cardinal numbers; and yet other measures will show a slower 'rate of growth'.

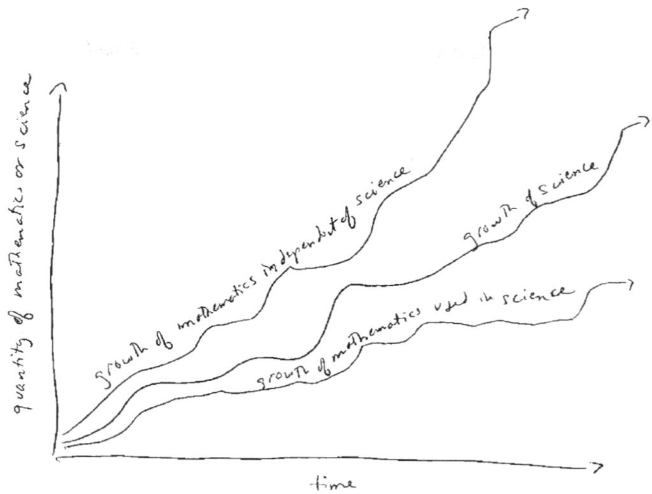

Fig. 3.1 Graph showing progress in science, growth of mathematics used in science and growth of mathematics independent of science

There is a conceptual mismatch between the Quinean and some sorts of mathematician. The mismatch is that much of the 'recreational' part of mathematics is 'foundational' from the point of view of mathematics, and therefore, it is more important, in the eyes of the mathematician. For this second reason, mathematicians object to Quine's scientific naturalism. Recall Quine's distinction between 'real' and 'recreational' mathematics. This is not a distinction recognised by working mathematicians. The closest distinction mathematicians recognise, is that between pure and applied mathematics. But the mathematician's distinction is *importantly different* from Quine's. The mathematician's distinction concerns *use* and *purpose* or *end* in developing techniques, not on justifying ontological commitment.[12] That is, some mathematics is developed to solve a particular applied problem. Ancient mathematics was developed with very practical problems in mind, and modern pragmatists could be forgiven if they surmised that ontological commitment by the Ancients was never to mathematical entities, but only to physical ones. However, those tempted to make this surmise should do well to remember the enchanted and metaphysically rich world of the Ancients. For them the pure geometrical forms

[12]The mismatch can also be identified with what Buldt and Schlimm call an Aristotelian conception of mathematics and a non-Aristotelian conception (Buldt and Schlimm 2010, 40). Roughly, twentieth century mathematics takes a more top-down, structuralist approach to mathematics, so applications are almost accidents; whereas an Aristotelian approach is one of abstraction from the observable world. If Buldt and Schlimm are correct in their diagnosis of the change in mathematical conception, then, in this respect, we might see Quine as still being entrenched in an Aristotelian conception of mathematics, (at least in this respect) whereas the practice of mathematics today is non-Aristotelian.

were quite real, and pertained perfectly to the physical world. Nevertheless, they were perfections, never realised on earth. Regardless, mathematics has changed in important ways since Antiquity, and mathematics has developed as a discipline in its own right. This development is only occasionally driven by practical problems in application. More important, the ontology of mathematical theories is not determined by application at all. It is determined by: quantifiers, models, semantics, intuition and the like. Thus, while the mathematician distinguishes applied from pure mathematics in terms of purpose, it is quite foreign to him to distinguish his ontological commitment along this dividing line. So Quine's scientific naturalism rings a false note with the mathematicians.

We can also detect some irony in Quine's position. In Quine's naturalism, it is science that determines the distinction between real and recreational mathematics. So while the distinction is one we find in mathematics (as a subject or academic discipline) it is a distinction that is neither recognised by mathematicians, nor is it a scientific distinction. It is a philosophical distinction that determines ontological commitment. It is the Quinean naturalist *philosopher* who has spoken, and decided to relegate the decision to practicing scientists as to what counts as real or recreational mathematics. Ontological commitment is a properly philosophical concern. I join Maddy in criticising Quine because of the status problem. We leave Quine to face his critics, and turn to Maddy's mathematical naturalist with her solutions.[13]

3.3 Maddy's Naturalism

As she remarks in her *Second Philosophy* (2007), the term 'naturalism' is over-used now, in 2007, and has come to mean too many different things. This is why she shifts to 'second philosophy' as the name for her position. While I agree with Maddy that 'naturalism' is over used, I shall continue to use the term here. For the purposes of discussing naturalism, I shall be largely inspired by Maddy's earlier

[13]Arguably, this criticism of Quine relies on taking too seriously Quine's *taste* for science, and attributing this to the naturalist position. Arguably, this is a distortion of Quine, and if we look at his meta-philosophical pragmatism, then we see that he would, if pressed, endorse taking mathematics seriously in just the way Maddy does, and he would simply acknowledge that she has different concerns from his. I do not mind if one takes this reading of Quine. If one does, then one says that Maddy just extends Quine's naturalist programme to include mathematics. For this reason I sometimes call her position 'mathematical naturalism'. Under this reading of Quine, Maddy is Quinean, and both resonate with the pluralist. The reason for not treating this interpretation of Quine in the main text is that Maddy sees herself as departing from Quine on this issue, and I am really concerned with introducing the reader to pluralism, not in making an accurate critique of Quine. I apologise for any clumsiness in representing Quine's position. More sophisticated interpretations of Quine approach pluralism.

work, her (1997).[14] In her (1997), Maddy disagrees with Quine's partitioning of mathematics into the dispensable and the indispensable part. She treats mathematics as a science. Despite her disagreement with Quine, she does consider herself to be a naturalist. Because Maddy's naturalist is someone who treats mathematics as a science, we could call her a *mathematical* naturalist. That is, she takes the mathematician's words and practice seriously when trying to determine what is to count as 'mathematics'. She tailors her philosophy to fit the reports of mathematicians and the practice of mathematics.

In other words, Maddy agrees with Quine that there are better and worse methodologies, forms of enquiry etc. Where she departs from Quine is over the scope of the best sort of enquiry. Departing from Quine, she does not restrict her ontological commitment to entities which can play a causal role, or rigorously inform this role (the rounding out of the mathematical theory), but widens her ontological commitment to include the entities postulated, and necessary for, the best and most encompassing part of mathematics, where that 'part' is determined, not by science, not by philosophers, but by mathematicians themselves.

This extending of the notion of 'best enquiry' solves the status problem. Maddy will not alarm the mathematicians, since, under her reading it turns out that all of set theory, for example, is now part of our 'best scientific enquiry'. The distinction between real and recreational mathematics disappears in Maddy. Ontological commitment is determined in the usual *mathematical* way, and is not hostage to application outside pure mathematics. Set theory acquires central position in mathematics, which, according to her, is in accord with how most mathematicians view mathematics.[15]

Once Maddy has solved the status problem, we find that the solution brings with it other revisions for her version of the naturalist philosophy. Maddy's mathematical naturalist will inherit some of the irony of Quine's position. The irony has to do with *using* philosophy to *argue* that philosophy takes a secondary position with respect to mathematics. This is more clearly her argument in her (2007) where she calls her position 'second philosophy'. The position is consistent (many mathematicians would agree with her here), but it is ironic.

We shall look at two problems. One is associated with what I shall call 'the topological argument', the second is more ironic and self-referential. To see the first, we should revisit the issue of ontological commitment. Concerning ontology, we learn that Maddy is a realist about the foundational part of mathematics, namely ZFC. It is *to this extent*, and *in this way*, that Maddy solves the status

---

[14]Her more recent book Second Philosophy, Maddy (2007) gives motivation for and develops the work done in Naturalism (Maddy 1997), but in it, too, Maddy does not take her own philosophical directives far enough, according to the pluralist. In some ways, for the pluralist, her more interesting work is done in her (1997). We shall see this in Chap. 14.

[15]This is not central, in the sense of what it is that most mathematicians do most of the time, but in the sense of is a recognised foundation of mathematics. Of course, one wants to specify in what respect a theory is a foundation, but that does not have to be done for these purposes.

problem. In Maddy's mathematical naturalism, it turns out that most of mathematics (everything reducible to set theory) is serious and real mathematics. All of the entities in the ZFC set theoretic universe are real objects. So, most mathematicians will be happy, indeed, all those who study a part of mathematics reducible to set theory. But remember the original concern of the scientific naturalist: that ontological commitment be justified, and in some sense be testable (against 'reality'). Applications of mathematics to the physical world will not sanctify set theory, since 'the real world' can only be used as a rather crude test for a tiny fragment of set theory,[16] if it can be used as a test at all.[17] The physical world can hardly help us choose between mathematical theories which all 'fit' the observed data. We cannot use the physical world to test the consistency of set theory in any absolute sense. All such checks are *dependent* checks for consistency: ZFC is consistent, iff any theory reducible to ZFC is consistent. But this is a circular argument for: the consistency of ZFC, the truth of ZFC or the merits of ontological commitment to the entities of ZFC. Strictly speaking, from a circular argument, we can draw no conclusions.

The most obvious *independent* checks (for consistency, truth and ontological reality) for ZFC are relative consistency checks against more powerful set theories: ZFC plus some other axioms such as higher cardinal axioms. Unfortunately, these

[16]A short anecdote: at a conference on Brouwer and Intuitionism, Martin-Löf was asked by John Thomas "how much of mathematics is really needed for science?" Martin-Löf's reply was: "a very tiny amount". As has already been remarked, how one measures 'a lot of mathematics' and 'a little mathematics' is simply not clear. Nevertheless, we can say that if we imagined physicists, chemists and biologists being asked to decide on the basis of 'usefulness for their science' which mathematicians in a mathematics department to keep in employment, the great majority of mathematicians would be fired. In fact, they might all be fired, since we already know quite a lot about the mathematics that is already used, and how to use it, and so further investigation of the mathematical theory might be thought to be quite useless. For other, philosophically quite different sources which explore what is the minimum mathematics needed for science see Field (1980) or Bendaniel (2012).

[17]A lot of philosophers would think that it is to get the order wrong to think that physical 'reality' can act as a test for mathematics. At best it can only be used to test a particular application of mathematics. Such philosophers are the ones who think that there is a hierarchy of knowledge with logic at the top, then mathematics, then physics, then chemistry then biology, then the 'softer' human sciences. The relationship between the levels of the hierarchy concern necessity or laws (of a discipline) so, for example, biology is responsible to chemistry, i.e., biology has all of the laws of logic, mathematics, physics and chemistry. Biology cannot violate those, and has a few extra laws, which do not pertain to chemistry, or any discipline above it in the hierarchy. Thus, moving up the hierarchy, chemistry contains laws. Those are all of the laws of logic, mathematics and physics. None of these can be violated by chemical reactions. Also chemistry has its own laws not found in physics or any area higher in the hierarchy (Fig. 3.2).

We are not concerned with philosophers who have this view of natural or scientific laws here, and as a reader might well suspect, the pluralist has a rather more complicated picture of laws. Nevertheless, the pluralist will maintain that holding mathematics hostage to the physical world makes no sense. What does make sense for the pluralist is to use 'physical reality' to judge the fit of a mathematical theory as a model for some part of physical reality, and nothing else.

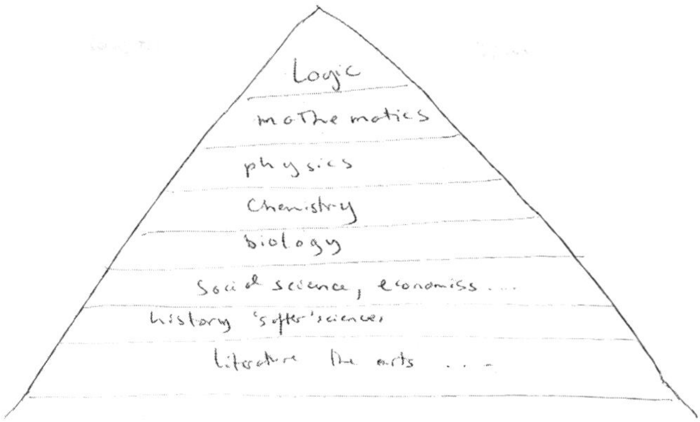

Fig. 3.2 The hierarchy of knowledge

more powerful theories are even less (epistemically) likely[18] than ZFC, to be consistent, true or capture the real mathematical ontology. At the very least, this is because extensions of ZFC have not undergone the scrutiny of ZFC, but also the new axioms are less widely understood by mathematicians, or are understood by fewer mathematicians. For really independent checks we would have to look to alternative set theories, or alternative foundations. In fact, there are a number of proofs of equi-consistency between ZFC and other theories. But these are unavailable to Maddy! Maddy can hardly say that she has good independent evidence of the reality of the ZFC ontology because of the equi-consistency of ZFC with a theory with a quite different ontology. This is because she follows what she takes to be Gödel's philosophical view of mathematics as representative of mathematicians, and so she supports (with some qualifiers) a realist philosophy of mathematics. If one is an ontological realist about set theory, then one can have no truck with alternative theories – since they are based on a fictional or false ontology. One would hardly find comfort in being told that one's theory is consistent with a false one.

We are then left with the question-begging argument. Let us widen the circle of argument, and see if we can release it from the charge of question begging. Call the wider argument: 'the topological argument'.[19] Examining the topological argument more closely, it is that:

ZFC plays such a central role in mathematics: it is very frequently used as a reference point by mathematicians, the equi-consistency results show us that if ZFC were shown to be inconsistent, then so would be most of mathematics.

<u>ZFC is very fruitful.</u>

[18] This is meant in the sense of epistemic likelihood, that is 'given what we know' or what we 'are certain of'. In these senses, ZFC+ is less likely true than ZFC.

[19] 'Topological' is chosen here just to refer to the notion of centrality and of fruitfulness together.

Therefore, it cannot be a mere mistake that ZFC plays this central role, there is an underlying reason, namely the truth of ZFC and the reality of the ontology.[20]

The argument is a mixed type of topological, inductive and inference to the best explanation argument. It is topological because appealing to the notion of centrality and anchoring of ZFC. The mathematics on the periphery can be modified, but if ZFC is central, then its consistency affects most of mathematics. It is an inductive argument, in the sense of 'ampliative'. The conclusion is made more probable, or supported by the premises, but not logically guaranteed by them. Moreover, the conclusion relies on the idea that the best explanation for convergence is truth, and ontology is what 'makes' the truth true. Thus, the conclusion very much depends on certain philosophical inclinations, some would say 'prejudices'.

Ontological realism concerning ZFC is a philosophical position; so is realism in truth. The tests for the ontology and the truth are more abstruse mathematics combined with philosophical argument. For this reason, the mathematicians who are realists (in ontology or truth) hold a philosophical stance. But Maddy adopts a mathematics-first attitude *vis-à-vis* the relationship between *mathematics* and *philosophy*. For Maddy, philosophy is not part of good scientific or mathematical enquiry. Therefore, she is no more entitled to use philosophy or endorse philosophical arguments (even if they implicitly come from the mathematicians), or philosophical conclusions than she is entitled to use or endorse arguments from astrology. Maddy (1997, 204) makes a direct comparison of astrology to mathematics, but not to philosophy. She says more about philosophy in her (2007), but her treatment there is not fully satisfying, since Maddy is not entitled to use philosophical arguments, because, strictly speaking, philosophy does not hold methodological merit, unlike science or mathematics. If she insists on drawing a line between the 'good' areas of enquiry and the 'bad', then the philosopher finds himself on the 'bad' side, and this is the problem of irony.

Recapitulating: Maddy's ironic problem is that she uses philosophical arguments to suggest a philosophical position, which (1) leaves little room for doing

[20]There is a lot to say about this argument, but it would be distracting here. One function of a footnote is to discuss issues, which would interrupt the flow of the main text. Using the footnote to this end: briefly, the first premise is only a description of a recent fact in the history of mathematics. It could be dismissed as an historical accident, having to do with the greater communication between mathematicians in the twentieth century etc. The second premise makes a lot of assumptions, and these can be questioned. For example, it holds sway if we assume that triviality is the only alternative to consistency. The third premise I like to call 'the argument from fruit', i.e., *etcetera* that if practice or assumption or theory X bears fruit, it follows that it must be true, correct etc. The argument from fruit is undermined by under-determination of truth by practice, assumption or theory. There are a number of similar arguments from other virtues: simplicity, beauty, symmetry, parsimony etc. The topological argument is not strong, when properly analysed. For some reason, it still persuades. We might *learn* more, if we give it a more sensitive treatment, as we do in the main text.

philosophy of mathematics, and (2) should not be taken seriously at all, since all arguments from philosophy are not good scientific ones. They use the wrong sort of methodology.[21]

The first consideration is ironic in the sense that she should not be writing a philosophy book, by her own lights; although to do her justice, she does 'go native' by the end of her (1997), and there she does some interesting mathematical work. The irony of (2) shows up with the question: "Should we follow, or dismiss, her philosophical arguments, since they do not employ a good methodology, by her own lights?" A good mathematics-first philosopher should dismiss her arguments. Not only should she dismiss the topological argument (which is the first problem), but she should also distrust the philosophical argument that vindicates her mathematics-first attitude (which is the second problem). As a result, Maddy's argument does not amount to more than a declaration of position. It should convince no one, not even herself.

Is this a straw-man interpretation of Maddy's argument? After all, philosophers do *use* arguments to bring readers from one position to another, in order to then *discard* the original position. This more sophisticated interpretation reads the development of Maddy's position as leading to our 'throwing away the ladder'. The final position is that philosophers of mathematics should do mathematics, but for all that they should not completely dismiss their philosophical training. The philosophy second stance is meant to set a pecking order between philosophers and mathematicians. If we follow her argument, then the conclusion is that the philosopher (*qua* philosopher) should adopt a fairly minimal role (not *no role at all*).[22] This she accepts. True to her convictions, by the end of her Naturalism book (1997), as mentioned earlier, Maddy 'goes native' and helps the mathematician to achieve his stated (set theoretic realist) goals using mathematical methods, hence demonstrating that she takes her own arguments seriously.

In effect, what we have is an implicit 'indispensability' argument for philosophy. We use the part of philosophy indispensible to the philosopher of mathematics. Given the 'minimal role' reading of Maddy's mathematical naturalism, the only role left for the philosopher is to work out what are the goals of mathematicians and help them – using mathematics. But this is a role assumed by someone who has made an anthropological or psychological observation of what mathematicians are after, and then practices mathematics to address the observed problems faced by the mathematicians. This does not leave much for the *philosopher* to do, at least *qua* philosopher, especially since mathematicians tend to assume these roles themselves!

---

[21]It is quite interesting that in her more recent book: Second Philosophy, she practices first philosophy to argue for second philosophy, and only practices second philosophy starting on page 246, in a 411 page book, excluding index, bibliography etc. And even then, she does not spend the rest of the book practicing second philosophy. Rather, she deftly moves from first to second and back. So, looking at her own practice, she cannot really object that much to first philosophy.

[22]In her book Second Philosophy, Maddy devotes a whole section to the question of the philosophical role left for the second philosopher. She is both well aware of the problem, and does answer it in a way that is similar to what I say here.

They are usually asking what it is that they want to work out, and how to do this using mathematical means! It seems obvious to leave it up to them, especially since they are better trained in mathematics.

In passing, we should note that, in general, Maddy's naturalism will encounter some resistance from the philosophical camp, especially in the philosophy of mathematics. So, now we are back to the status problem, this time for philosophers. Some philosophers[23] are *unwilling* to adopt this minimal role *vis-à-vis* mathematics, since it does not seem to be much of a *philosophical* role. Maddy's mathematical naturalism is not meant to be *philosophical* precisely because the philosopher is not allowed to *judge* the practice of mathematicians.[24] So some, less naturalistically inclined philosophers will not see much point to mathematical naturalism, but at this stage of our investigation, this is just a matter of taste. We have a draw. But we shall revisit the relationship between philosophy and mathematics in more detail later. For philosophers inclined towards realism, structuralism or formalism, they are invited to consult the other chapters in this part of the book. But for those who are still not sure, they might be interested in another form of naturalism, or rather, another way of adopting, and adapting, the lessons of Quine and providing a good philosophy of mathematics.

3.4 Colyvan's Naturalism; Colyvan, the Pluralist and Maddy

Prima facie, Colyvan's naturalism does not suffer from the problem of irony. He does not take seriously the philosophy-second attitude, since, for him, there is no hard and fast line between mathematics and philosophy. For Colyvan, the two inform each other and are indispensable to each other. All we can say about the distinction is that there are extremes: areas of philosophy, which have almost nothing to do with mathematics, and areas of mathematics, which have almost nothing to do with philosophy. Nevertheless, there is a significant and interesting vague border between them.

However, for Colyvan, there is a border between *science* on the one side and mathematics and philosophy on the other. Because of the border, Colyvan faces the status problem. The indispensability arguments are what determine the real part of mathematics. On the theme of the 'widening of scope' of what counts as 'good science', Colyvan will include all of mathematics and philosophy necessary to do science. Nevertheless, for Colyvan, it is 'science' (in Quine's sense) that sets the standard to which mathematics or philosophy should aspire.[25] He re-captures

[23]We shall encounter these again soon, so do not forget them.

[24]In this case, appearances are deceptive, as we shall see shortly.

[25]As has been mentioned, this is 'back to front' for some mathematicians and even some computer scientists. The borderline between science and mathematics, and the relationship between them is quite intricate. For example, if we consider simple scientific experiments to be algorithms (they

a lot of mathematics by appeal to the notion of 'rounding out' the mathematical theory, which is indispensable to science.[26] So, he will not, *prima facie*, offend so many mathematicians: almost all of them are doing real mathematics, according to Colyvan. Of course, there is still the issue of priority. The mathematician does not think that she needs science to sanctify her research, so she will not appreciate the idea that we begin with science, and from that recover the real part of mathematics. We note, but leave aside, the priority issue here.

One way to take issue with Colyvan is to show less generosity in the rounding out of the mathematical theory. As we saw in the first section, if we are *very* suspicious of mathematical ontology, then we only need a finite mathematical ontology:

> A mathematics which does not commit us to the infinite is a nice thing for anyone with reductionist and/ or realist [about physical entities] leanings. As far as we know the universe is finite, and if space-time is (quantum) discrete there isn't even an infinity of space-time points. The largest number may be indefinitely large. So we never get to it (e.g., given our limited resources to produce numerals by writing strokes). (Bremer 2010, 36)[27]

Similarly, it is only rather crude philosophy, which is needed to execute science. The distinction between indispensable and recreational philosophy cannot be drawn neatly along lines of quantities of ontology, nevertheless the observation holds that for science to 'progress' it needs very little philosophy. Of course, how we detect and measure progress is a fraught issue, but we leave it aside or now. It follows that most philosophers are doing recreational philosophy (since they are not helping science). For this reason, Colyvan's naturalism offends many philosophers.

For Colyvan's naturalism, the irony problem re-surfaces. He too is a philosopher, whose methodology is, therefore, dubious at best. He should, like Maddy 'go native', but do science, not mathematics. Moreover, he should not take his philosophical conclusions seriously at all. *Grosso modo*, the problems re-surface along different lines than for Maddy, but they re-surface nevertheless. So, what does the pluralist have to say about all this?

The pluralist does not single out science as marking the standard to which mathematics or philosophy should aspire, and so, parts company with Colyvan, and joins Maddy. The pluralist widens Maddy's 'good methodology' to include

---

are just finite procedures), See (Beggs et al. 2010). What we discover is that the notion of physical experiment carries not only imprecision in measurement, but also a type of uncertainty. We can use mathematical techniques to detect and measure the uncertainty of the data obtained through such simple experiments! This points to an inadequacy in physical science, *vis-à-vis* computational science, and this inadequacy cannot be *recognised* by a Quinean naturalist, since *science* sets the highest standard (for truth, measurement etc.) not mathematics.

[26]Colyvan argues that the 'rounding out' process re-captures a lot of mathematics. This is a move made to please the mathematicians who, *prima facie*, cannot recognise their mathematics when seen through the naturalist lens. But his project can be reversed. Rather than show that it is 'reasonable' for the naturalist to recapture a lot of mathematics, we could just as well stick to our suspicion of mathematical ontology. If we do this, then the 'rounding out' need capture very little, only a finite number of numbers.

[27]Careful. The finite mathematics is an inconsistent one. Discussion of paraconsistent mathematics will be resumed in Chap. 6.

(at least) philosophy, so the pluralist parts company with Maddy. For the pluralist, an important ingredient to philosophical enquiry is defining, revising and qualifying what sort of role one is taking, when, for what purpose, and how, to evaluate success in said role. Success will be evaluated in terms of communication along with standards of rigour. Rigour in argument is the guarantor of good methodology. And standards of rigour can only be evaluated philosophically, in the first instance.[28] Thus, philosophy is indispensable to philosophy, mathematics and science, and is what ensures the high standards in methodological practice of all three!

For the pluralist, there is no absolute evaluation, which says that one area of enquiry is sanctioned while another is not. Rather, the pluralist recognises that such evaluation is made relative to a set of goals, and these, in turn are revisable, more or less informative, more or less interesting, useful etc. As a result of these differences, where Maddy sees a narrow set of tasks for the philosopher, the pluralist sees a wide range of tasks.

Despite the difference in final position, the pluralist is quite delighted with some aspects of Colyvan and Maddy's work. He takes from Colyvan the lesson of not making any strong distinction between philosophy and mathematics, or between mathematics and other areas of enquiry. The pluralist takes from Maddy the commitment to take seriously what the mathematician says, and more important, will take her up actively on her suggestion concerning the positive contribution the philosopher can make to mathematics. For, despite her mathematics first attitude, Maddy's work carries philosophical import. We shall see this in the final part of this book when we widen her work and propose our own principles which are both philosophically and mathematically driven. Rather than giving 'a reduced role'[29] to the philosopher, the suggestion the pluralist takes from Maddy opens on to a vista of important projects. To see this picture clearly, see the final chapter, in Part IV. What will be salient to note here is the general idea behind the technical work Maddy does, since we shall return to it in that final chapter.

In general, Maddy starts with identifying set theoretic realism as the preferred philosophy of mathematics by mathematicians. She is chiefly inspired by Gödel. However, as she develops her naturalism, she begins to realise that present day mathematicians are not realists, in the same way as Gödel. Instead, she observes that mathematicians have a general goal to extend ZF set theory in a way that is *fruitful*, where Gödel would have had us extend formal set theory to represent mathematics as it really is. More strongly, and more precisely, if we are interested in the fruitfulness of extensions of ZFC, then we add new axioms which do not contradict, or change anything that we find in ZF. Moreover, the extension should be

[28] We could test rigour according to some fairly rigid, nay formal rules, but these, in turn would have to resonate with a pre-formal sense of rigour. Thus, at best we could engage in a dialectic between attempts to give a formal or very precise definition of rigour, and the intuitive idea.

[29] An anonymous reviewer to a paper where I develop these ideas commented that "Maddy accepts the reduced role of the philosopher." The pluralist follows her naturalist arguments, but assumes a very important philosophical role. The full pluralist does not assume a 'reduced' role at all.

maximal, in the sense of being as generous as possible (short of triviality). This goal she calls MAXIMIZE and she gives it formal expression. We shall see the formal expression in the final chapter of this book. Despite her avowed departure from realism, a vestige of it remains, in that she develops another principle, which she calls UNIFY.[30] This principle urges that we converge on *one* MAXIMAL consistent extension of ZF. The philosophically important move she makes towards pluralism is the trading of realist truth, for maximizing and unifying *principles*. This trade is a good one, according to the pluralist, and encourages other similar trades.

3.5 From Maddy's Naturalism to Pluralism

The pluralist who has followed Maddy through her reasoning and practice is inspired to take up some of her themes, but only after making the substitution of plausibility for truth. The substitution is applauded by the pluralist. This is because one of the upshots of the substitution is that ontological commitment and 'truth' (as the term is used by the working mathematician) only makes sense in the context of a theory.[31] Principles reflect aspirations, and aspirations are not taken to be independent, absolute or true concepts. They are based on experience, phenomenology, styles of learning, education and other traits that are irreducibly subjective or inter-subjective. Once expressed, they can be recognised to be shared by others, but this is no guarantee that they will lead to truth or correctness. Once we take seriously the idea that ontological commitment is internal to a theory, we have a form of ontological pluralism in mathematics.[32]

Definition The *ontological pluralist* is someone who believes that the ontology of mathematics is not unified by one semantics or one model. Rather, 'ontology' is a term that is relative to a theory.

See the chapter on Structuralism for a proper development of the concept. The ontological pluralist, rejects UNIFY, or better, restricts its import to serving some realist set theorists. Under ontological pluralism, alternative extensions of ZFC can co-habit side-by-side without conflict since there is no global, or absolute,

---

[30]Since she is careful to acknowledge that these are maxims, and not guarantors of truth or correct real ontology, we can take the liberty to call these 'principles'.

[31]We assume here the following default relationship between ontology and truth. The ontology of a theory is the 'truth-makers' of the theory. For example, what makes it true that $2+2=4$ is that in Dedekind-Peano arithmetic, the entities 2 and 4 exist, and the function $+$ and the relation $=$ all conspire to make the formula: '$2+2=4$' true. There are ways of separating truth values from ontology, but they will be ignored here. If we do separate them, then the above argument will work for at least one of truth or ontology, but not necessarily both, in which case a second argument would have to be given.

[32]The relationship between ontological pluralism and fictionalism is interesting, but will not be developed in this book. Some things, even some obvious things, have to be omitted.

mathematical ontology or methodology.[33] It might seem surprising that, Maddy (1997, 208) explicitly agrees with this, since her agreement is at odds with some of the things she says elsewhere where she identifies realism as representative of mathematician's philosophical inclinations. Indeed, she uses this realism as one of her motivations for developing the principle UNIFY. The tension can be resolved if we are careful, and admit that realism can be identified as an aspiration of some mathematicians. 'Aspiration', is being used as a technical term, so here is the definition. Inspired by Maddy's Maxims, or principles:

an *aspiration* is a general mathematical goal identified by some mathematicians.

An example might be to make all of mathematics constructive according to certain parameters on what counts as constructive. This was an explicit goal of Bishop and Bridges (1985). Another might be to unify all of mathematics under one foundation. Another might be to develop as many *incompatible* extensions of ZFC as possible. Perverse aspirations are possible! These are only examples. The pluralist thinks that aspirations in mathematical practice are very important. Identifying them, giving them an as precise as possible expression, maybe even formal representation, is an aspiration of the pluralist. Identifying aspirations in mathematicians is good pluralist and naturalist philosophy.

Moreover there is no reason Maddy should take too strict a view concerning 'mathematics-first'. As we saw above, the pluralist detects an ironic problem with the attitude. Maddy would be advised by the pluralist to take a feather from Colyvan's hat and take a more nuanced attitude towards the relationship between mathematics and philosophy, namely, to consider the relationship to be mutually informing and beneficial. In this way Maddy would widen the scope of 'best inquiry' to include not only mathematics, but also philosophy, and thereby admit her place as an important philosopher of mathematics.

Colyvan's view of a dialectical relationship between the philosopher and the mathematician is well supported by the history of mathematics and philosophy, especially if we consider mathematicians who are also philosophers such as Brouwer, Gödel, Whitehead, Russell, Dedekind, Poincaré or Frege. Adding philosophy (of mathematics) as a seamless part of mathematics to what counts as 'best enquiry' restores philosophers to the place of positive contributors to knowledge and understanding in mathematics. Whether or not this is too high a price to pay for Maddy is up to her to decide. But it is not a mere matter of taste. It is a philosophical matter of contention.

To summarize the discussion so far, if I am correct about the data, given her avowed mathematics-first attitude, Maddy should be much more of a pluralist than she concedes, on the grounds that mathematicians are pluralist.[34] If she is willing to consider more than the data she observes with some set theorists, then her

[33]The details about how this is done will be given in Part II of the book.

[34]I have not supplied evidence for this here, but there will be a lot surfacing in different parts of the book.

'native' contribution to mathematics supports pluralism in her trading of truth for plausibility. Moreover, she is free to express, or even give formal representation to, many aspirations, held by all sorts of different mathematicians. She can now discuss, and give formal definitions of 'plausibility' as it would be received by, say, constructivists, or category theorists, or people who study non-well-founded set theories, and so on. That is, she, or her followers, should take up her challenge to widen the mathematical naturalist programme to other theories of mathematics and other philosophical attitudes held by mathematicians.

The residual concern, which the naturalist might have, is to ensure that the philosophy of mathematics, and mathematics are tested against something independent of them or that they predict something. The pluralist answer is that the test for mathematics cannot lie in physical causation, and it is not based in a physical ontology. Instead it is contained in the attitude and methodology of mathematics and the philosophy of mathematics. 'Testability', in the sense of 'physical-causal' is replaced, by the pluralist, with rigour of argument, context of theory and mathematically established crosschecks between mathematical theories,[35] and perhaps some application in science. Safeguarding the rigour and plausibility of arguments, proofs, contexts and crosschecks between theories is carried out by the community of mathematicians, philosophers of mathematics and the institutions thereof.[36] I shall develop this in the second part of the book. As for prediction: mathematics has very wide scope. We use mathematics all the time, and especially in predicting. If I have eight eggs in the refrigerator, and I take two away, I can predict I'll have six left. This is based on a simple calculation. Is calculation prediction? No; not in the sense of predicting an event in removed the future. But calculation does share with prediction the feature that I could, of course, be wrong. Someone might come in and take another two eggs from the refrigerator. We can think of calculation as a limit sort of prediction. When the calculation is correct, then I have the surest sort of prediction!

References

Beggs, E. J., Costa, J. F., & Tucker, J. V. (2010). Limits to measurement in experiments governed by algorithms. *Mathematical Structures in Computer Science, 20*, 1019–1050.

Bendaniel, D. (2012, September). *Constructability and space-time*. Presentation at the first international conference of logic and relativity: Honoring István Németi's 70th birthday, Budapest.

Bishop, E., & Bridges, D. (1985). *Constructive analysis*. Heidelberg: Springer.

Bremer, M. (2010). Contradiction in mathematics. In B. Löwe & T. Muller (Eds.), *PhiMSAMP philosophy of mathematics: Aspects and mathematical practice* (Texts in philosophy 11, pp. 29–37). London: Individual Author and College Publications.

[35]Examples include embeddings, reductions, modeling or equi-consistency proofs.

[36]For a more extended discussion see: Goethe and Friend (2010).

Buldt, B., & Schlimm, D. (2010). Loss of vision: How mathematics turned blind while it learned to see more clearly. In B. Löwe & T. Muller (Eds.), *PhiMSAMP philosophy of mathematics: Aspects and mathematical practice* (Texts in philosophy 11, pp. 39–58). London: Individual Author and College Publications.
Colyvan, M. (2001). *The indispensability of mathematics*. Oxford: Oxford University Press.
Field, H. (1980). *Science without numbers: A defence of nominalism*. Oxford: Blackwell.
Goethe, N. B., & Friend, M. (2010). Confronting ideals of proof with the ways of proving of the research mathematician. *Studia Logica, 96*(2), 273–288.
Maddy, P. (1997). *Naturalism in mathematics*. Oxford: Clarendon Press.
Maddy, P. (2007). *Second philosophy*. Oxford: Oxford University Press.
Quine, W. V. (1981). Success and limits of mathematization. In W. V. Quine (Ed.), *Theories and things* (pp. 148–155). Cambridge, MA: Harvard University Press.

Chapter 4
From Structuralism to Pluralism

Abstract The reader is introduced to pluralism from the starting point of structuralism. The starting position is Shapiro's structuralist position. The pluralist is inspired by Shapiro's position, especially by his anti-foundationalism and by his self-avowed pluralism. The pluralist joins Shapiro in argument against a naïve realist position. Thus, in this chapter, the arguments of Chap. 2 are revisited and made stronger and more precise. Similarly, the pluralist joins Shapiro in endorsing the idea of there being several mathematical structures or theories each only compared to others from a meta-perspective/theory/structure, which, in turn, can only be judged, or compared, from a further meta-perspective/theory/structure. However, the pluralist pushes Shapiro's pluralism further. In particular, the pluralist will not be confined to the perspective guided and blinkered by classical second-order logic and model theory. What is in dispute, is both the classical conception of logic, especially the idea that inconsistency necessarily leads to disaster in the form of triviality, and the notion of success in mathematics. To make the last point, a distinction is drawn between the optimal and the maximal pluralist. The pluralism advocated in this book is a maximal pluralism.

4.1 Introduction

For purposes of definiteness, and precision of argument, I confine my attention to Shapiro's development of structuralism. This is because it is Shapiro's anti-foundationalism which is inspiring for the pluralist. Thus, henceforth, 'structuralism' refers to Shapiro's structuralism, unless otherwise indicated. I leave to a future project, the discussion of how other types of structuralism fit with, and differ from, pluralism.

The end of this chapter is written in a more aggressive style than the previous ones. This is simply for reasons of alleviating boredom.

M. Friend, *Pluralism in Mathematics: A New Position in Philosophy of Mathematics*, Logic, Epistemology, and the Unity of Science 32, DOI 10.1007/978-94-007-7058-4_4,
© Springer Science+Business Media Dordrecht 2014

In this chapter I discuss in what respects Shapiro's structuralism inspires pluralism, and where the pluralist parts company with structuralism. I will not present direct arguments that will necessarily convince the structuralist. Instead, in clarifying the affinities and differences between structuralism and pluralism, I present an indirect argument, where the reader can make an informed and justified choice between the positions.[1] We begin with definitions of structuralism and of structure, and then turn to the sections on affinities, followed by the sections on differences.

Definition *Structuralism* consists in the view that within the discipline of mathematics, we find a number of theories. Theories can be thought of as structures or as a set of formulas closed under some operations. All of the formulas of a theory are then true in the theory. Objects are only ever objects in a structure. The truth *of a theory* can only be judged from a meta-perspective: a meta-theory or meta-structure. Whether a theory is true, depends on the choice of meta-structure.

The *standard* meta-perspective is provided by model theory, where we prove, for example that the models which satisfy a structure will also satisfy another. But this is not always the case. We can appeal to looser principles and partial theories to occupy a meta-perspective. For example, Euclidean geometry is true, from the perspective of hyperbolic geometry, it is just a special case. But, from a naïve (object level) Euclidean perspective, hyperbolic geometry is false, since the parallel postulate of Euclidean geometry fails. The story is quite entangled since, following Beltrami, we can model hyperbolic geometry *within* a Euclidean framework (or structure) at a meta-level. Hyperbolic geometry becomes a special case of Euclidean geometry. However, Lobachevsky's view was that Euclidean geometry was a special case of hyperbolic geometry.[2] Thus, which meta-perspective one adopts makes a difference to the truth-values of theorems of a theory. We shall be encountering this example again in Chaps. 5, 9, and 14.

Definition What counts as a *structure*, and how structures relate to each other is usually determined by model theory. Roughly, a structure is a domain of objects (or class of domains unique up to isomorphism) with (maybe) some constant: objects, predicates, relations, functions and operations. Parameters on interpreting the constants, variables and other symbols in the structure are set by axioms or definitions.

For example, ring theory has the integers as a standard domain, but we can form rings using other domains. The structure also includes the 'lesser than' relation, <, and two operations that behave much like addition and multiplication. The structure of a ring is: <D, 0, ~, <, +, ×>, where, D is a domain of objects, 0 is a designated

---

[1] Taking this attitude is pluralistic. As we shall see from the temper of the writing, the pluralist ... combative spirit, rather, at the end of the day, what the pluralist hopes for is a ... positions and a deepening of understanding of positions: a sense of why one holds ... on, and what some opponents have to say.

[2] ... developed hyperbolic geometry. It was one of the first non-Euclidean geometries.

element in the domain, $\sim$ is an unary operator, not the operation of subtraction, but rather, that of negation.[3] A lot of interesting mathematics concerns the limitative results of theories, and the relationship between theories. The results are generated within model theory. For example, field theory can be thought of as ring theory with the operations of subtraction and division added to the structure of ring theory.

In concert with structuralism, the pluralist makes four arguments against a number of other philosophies of mathematics: all those that are foundationalist. In Chap. 1 we saw skeletal versions of the first three of these arguments. Here I give them more flesh. All the arguments rely on some sympathy for naturalism; a sympathy which is shared by the structuralist and the pluralist. The arguments can be numbered as follows.

1. Foundationalist philosophies make an illegitimate slide from a description of mathematics (in terms of a global foundation) to a norm for success for future mathematics.
2. Even if we did think that there was a unique theory, or structure, then our position would be highly unstable, since, whatever founding theory one has today, 'it' will grow – by adding new axioms or rules. Some new axioms are independent, and therefore, there are *no grounds* for faith in there being one, fixed, absolutely true foundation for mathematics.
3. *De dicto* and *de re*, mathematicians are very often pluralist.
4. Still in concert with structuralism, I argue that 'truth in a structure' is better received by many mathematicians today, than 'mathematical truth' *tout court*, such as might be proclaimed by a realist or Platonist. Not only are many mathematicians pluralist, at least in this respect, but they are highly aware of the context surrounding a theorem or proof, or the limitations of those theorems. It is for this reason that they are attracted to structuralism.

The plan for the chapter runs as follows. The first section of this chapter sets up the anti-foundationalist arguments, by giving the monist and dualist arguments in favour of The Foundation. Each anti-foundationalist argument is then treated as a separate section of this chapter. The fourth argument (Sect. 4.5) concerns the structuralist and pluralist, and not so much the naturalist. It is the more positive argument, in the sense of giving positive direction about how to think of truth in mathematics. It is these four arguments, which show affinity between the structuralist and the pluralist. After presenting these arguments the pluralist further develops the implications in Sect. 4.6. In the development, we surpass the structuralist and give more indication of the scope of pluralism – well beyond what was said in the first chapter. In particular, we look at, what I shall call: 'bad mathematics'. In Sect. 4.7 I want to make the notion of 'scope of interest for the pluralist' more precise. This will sharpen the difference between the structuralist and the pluralist. I distinguish between an optimal pluralist (the structuralist) and a maximal pluralist (the pluralist). Lastly, in Sect. 4.8, I address the concern that pluralism is unstable and degenerates into a sort of sociology of mathematics.

[3] Subtraction is a binary operator. Negation is an unary operator, or connective.

4.2 Setting Up the Arguments: The Monist and the Dualist Arguments for the Foundation

Pluralist philosophies (such as structuralism) take issue with most of the standard traditional philosophies of mathematics. To see this in detail, let me first give the monist argument for The Foundation.[4] This will be a presentation of a strong version of monism, i.e., one that is properly revisionist of mathematics. The argument for it is weak, so I could be accused, here, of setting up a straw man. I hope it is. Regardless, it will be rhetorically useful to set it up. I shall then point out how this differs from the dualist position, but I shall not give an argument for the dualist position. The positions will be important since I shall refer back to them when I run the arguments against foundationalism.

To fix an example, consider the proposal that ZFC[5] is The Foundation for mathematics. Presenting The Foundation has three parts: a technical part and two philosophical parts. The technical part is a result achieved through a reduction of all (or most) existing successful mathematics to The Foundation. We show, for most areas of successful mathematics, that they can be translated into the language of ZFC, and the theorems or results of the area of mathematics can be generated in ZFC too. In other words, we show that the language and proof apparatus of the reduced area of mathematics is strictly redundant with respect to ZFC.

This is not enough to convince mathematicians to cease to work in the language of the reduced theories and to use only the proof apparatus of the reducing area. For, the original language was designed to suit that area, and might be much more workable, less awkward, more suggestive *etcetera*. Nevertheless, the technical result has been achieved since all the *philosopher* needs to know is that it is *in principle* possible (if a little awkward) to do all the work of the reduced disciplines in the reducing discipline.

After we have the technical result, we make two philosophical moves. The first is to state something to the effect that mathematics is 'essentially' the reducing

---

[4]Some readers might think that such revisionist arguments are absurd. In particular, they will think so if they are not seeped in the tradition of the philosophy of mathematics. If the reader does think that the arguments are absurd, then so much the better for the pluralist. This just makes his job easier, and his position more easily accepted. I applaud the anti-revisionist attitude, and I think that it is pretty plain and obvious in the contemporary climate. Pluralism is a thoroughly modern position, in this respect. Note the reverse: pluralism is a radical position. It does break from past philosophies. It would be irresponsible of me to not warn a reader of this if she has not been seeped in the tradition.

[5]In the first chapter I use ZF as The Foundation. This is because there, I wanted to discuss certain points about the axiom of choice. ZFC is a much more plausible foundation, since the axiom of choice is rife in mathematical practice. The switch also illustrates the flexibility of the pluralist. I could have also chosen category theory, or paraconsistent set theory, as The Foundation. I chose to stay on relatively familiar territory.

discipline.[6] That is, we have captured almost all of mathematics in the reducing discipline, and therefore, all other languages, symbols and supposed ontology of reduced disciplines is strictly (philosophically/conceptually) redundant. Therefore, all we *strictly need* for most of mathematics is the apparatus of the reducing discipline. We have unified mathematics into one Foundation. This is quite a philosophical coup! It is quite remarkable to do this, and the only, or best, explanation for our success is that we have got it right. We really have identified what is essential to mathematics.

The second philosophical move is to introduce a normative element; the essence of mathematics becomes a norm for *success* in mathematics.[7] Straying away from The Foundation is dangerous. It could lead to inconsistency, falsehoods or nonsense. When we make the slide to normativity, we judge future mathematics against the backdrop of the essence capturing theory, i.e., The Foundation. If a proposed mathematical theory is not reducible to ZFC, then it is not 'properly' or not 'really' mathematics. At the very least it is not (going to be endorsed as) successful mathematics. This is because the purported mathematics does not enjoy the essence of mathematics – as identified in The Foundation.

Meta-interlude: Both philosophical arguments are rather weak in themselves. Stripped to its bear bones like this, I am giving a straw-man argument for monism. I regret this, but I cannot think of a stronger argument. The monists I have known tend either to insist that their philosophy is true by appealing to intuition or phenomenology – but these cannot be the basis of an argument, since professional's intuitions and phenomenology differ from one another. Or they appeal to the miracle of applications or to some form of convergence. But both are forced or question begging. The further weaknesses to the arguments are these. The first philosophical move relies on essentialism, or something like it, and presupposes that there has to be one essence. This forces the monist to ignore or dismiss proffered alternative foundations. The second philosophical move commits the naturalist fallacy and begs the question. Thus the structure and presuppositions of the arguments weaken them. Nevertheless, there is something to monism, if not the arguments for it. The criticisms we look at later will partly indicate the strength (and weakness) of the position. Despite appearances to the contrary, monism is taken seriously here, otherwise there would be no point in giving detailed counter-arguments. As we shall see, the counter-arguments also suffer from weakness – a reliance on naturalism. We conclude that the pluralist is willing to trade monism against naturalism.

Dualists will run a similar argument, but it will include an added complication. The technical reducing result will split mathematics into two: 'the best part' and 'the suspect part'. The 'best' part is so, in virtue of holding some desirable philosophical

[6]Many philosophers are leery of using the term 'essence', so euphemisms are used instead. Feel free to substitute in your favourite circumlocution.

[7]Some philosophers would call this move 'the naturalistic fallacy': the fallacy of inferring a norm from a description.

properties such as finiteness, definiteness or analyticity. For an example, let us consider Fregean logicism. The two parts are the arithmetic part, and the geometrical part. Similarly, there will be two 'essences' in mathematics. In the Fregean case, they will be the analytic essence and the synthetic essence, respectively. 'Success' in mathematics is judged relative to the different parts. There will be different norms, according to which part of mathematics one is operating in. For example, a proof in the analytic-arithmetic part has to be able to be turned into a gapless proof – in the language and proof system of Frege's *Begriffsschrift* and *Grundgesetze*. In contrast, for Frege, a proof in geometry may invoke intuitive gaps (which draw on our spatio-temporal intuitions).[8] The normativity in the dualist philosophy can surface in two ways: (1) we favour one part of mathematics over another, or (2) we refuse to consider the purported mathematics that lie outside these two parts (for example, modal intuitions, in the form of intentional attitudes, lie outside both arithmetic and geometry). 'Outlying parts of (purported) mathematics' are either not recognised *as mathematics* or simply not discussed. Dualists can be normative in one or both senses. Frege was normative in both.

The technical reduction results (to the two parts of mathematics) are shown in the usual way. The descriptive claim is now made, not in terms of essence, but in terms of some other philosophical property. The philosophical property will have been argued for in advance, for example, as sure ways of avoiding paradoxes, as really demonstrating the foundation of arithmetic in logic *etcetera*. The suspect part of mathematics then might either be thought of as maybe in principle reducible to the best part, or not so reducible. If it is reducible, then we have a programme for showing the reduction. If the suspect part is not thought to be reducible, then it has other philosophical characteristics, which distinguish it, and set a different standard for work with, or in, those areas of mathematics. For example, Cantor divided mathematics into the part that is amenable to mathematical manipulation and the metaphysical part. The metaphysical part does not admit of the same mathematical treatment as the best part. But it is still worth thinking about: it is very beautiful and brings us closer to God! (Hallett 1984, 13). Thus, for the dualist, the two parts have different standards and different philosophical attributes.

4.3 The First Anti-foundationalist Argument: The Slide from Description to Normativity

What is wrong with the monist and dualist foundationalist positions? Recall the naturalist insight that the philosopher is not there to set norms for success in mathematics on purely philosophical grounds. As Maddy puts it "if our philosophical account of mathematics comes into conflict with successful mathematical practice,

[8]I am ignoring Frege's later work of 1914 where he tries to found mathematics on geometry. In the above, read 'Frege' to mean the Frege of *Begriffsschrift, Grundlagen* and *Grundgesetze*.

it is the philosophy that must give." (Maddy 1997, 161). In contrast, implicitly, or explicitly, and to different degrees, foundationalist philosophies endorse the general idea that once the philosopher has developed a philosophy of mathematics, that philosophy should, amongst other things, determine the limitations, and the future development, of mathematics, and therefore, what counts as success in mathematics.

Illustrating the slide from description to setting the norm for success, Vopěnka discusses the development of set theory. When set theory was first introduced, and because it is such a powerful theory, it led to all sorts of developments in mathematics. For this reason, the proposed theory, and the reducing programme leads to a lot of insights and results. Something which is often overlooked by philosophers discussing the reduction, is that the reduction of existing mathematics to set theory was not easy, and it was not smooth. This might be one of the reasons why the reduction was surprising, remarkable and insightful. Vopěnka cites calculus as an example of a part of mathematics, which was very difficult to 'fit' into set theory.

Once the technical result was pretty much established, set theory changed its aspect from *contributing to* mathematics, to *setting limits on* our interests within mathematics. So, while set theory was proposed as a reducing discipline for giving us further insights into mathematics; once this was done, set theory became *a norm for success* in mathematics.

> Set theory opened the way to the study of an immense number of various structures and to an unprecedented growth of knowledge about them. This caused a scattering of mathematics. [It is interesting that Vopěnka does not say "unifying"!] *Moreover, most results of this kind derive their sense only from the existence of the respective structure in Cantor set theory. Mathematics based on Cantor set theory changed to mathematics* [only being recognised in terms] *of Cantor set theory*. (Vopěnka 1979, 9)

In other words, Cantorian set theory became the standard by which proposed mathematics was judged to be 'good' mathematics.[9] Today, ZFC has replaced Cantor's set theory as a point of reference.[10] Under the ZFC norm for success, much of category theory is not mathematics, nor is the ramified type theory, nor, ironically, is all of Cantorian set theory. The details of which theory is taken to be The Foundation are not important here. What is important is that Vopěnka notices a shift in the role we attribute to set theory: from description to norm for success, and that the norm precludes some potential further developments in mathematics, just because they are not recognised as *bona fide* mathematics. His development of the theory of semi-sets is an example. "This book on the theory of semisets presents an attempt to create a theory whose universe of discourse extends that of set theory; thus the new theory admits the existence of certain objects which cannot exist from the point of view of [standard] set theory." (Vopěnka and Hájek 1972, 7).

---

[9]Vopěnka is highly revisionary of mathematics too. But he proposes a different founding theory. This does not interest me here. What is important is that we should realise that the Platonist or realist proposal to found mathematics in set theory is sometimes taken to be normative of mathematics.

[10]Already this is interesting for the pluralist, since it instanciates the claim that what the mathematical community "takes to be The Foundation" changes over time.

Unfortunately for the monist, history has not born out her slide from description to norm setting. Alternatives to ZFC have been developed. Some have been proved to be equi-consistent to ZFC, (and this is much weaker than using ZFC as the norm for success). Other areas of mathematics have not been shown to be equi-consistent, but might be so in the future. Other areas might never be, and might in principle never be able to be (such as, for example, a paraconsistent set theory, where a proposed proof of equi-*consistency* would make no sense, since consistency is not a characteristic of paraconsistent set theory, equi-non-triviality might be the better option in this case). None of these would have been developed if set theory had really taken on a strong normative role in mathematics.

A prioristic monists, that is, monists who will not hear of counter-evidence to their position on grounds that, in principle, there can be no opposition; could ignore these developments, and call them all 'non-mathematics'. But then they would be begging the question against themselves. Apart from begging the question, they run the risk of rejection by the mathematical community which works in these areas and they run against the naturalist insight that philosophers should (if they are naturalists) want to observe and take seriously not only what mathematicians say about their subject, but also observe what their behaviour reveals of their *de facto* attitudes and observe what it is that they accept as part of mathematics.

Before dismissing the monist on socio-political or institutional grounds, we can be a little more careful in our deliberations. The monist is not completely off-track, since ZFC does play a very important role in mathematics. Maybe it is not a foundational role in the traditional (essence-seeking) philosophical sense. But it plays a central role. Therefore, it is worth while, as naturalists, to ask: How do mathematicians think of set theory?

It is true that a lot of present day mathematicians take it as a good verification of their work that it can be done in first-order set theory. But this does not weaken my point, since there is a difference between using ZFC as one, amongst other, means of verification, and recognising ZFC as the only means of verification.[11] I think that most mathematicians today use ZFC as a means of verification *amongst others*. It's not hard to use ZFC. It will, as a matter of fact (demonstrated by the famous reduction) be sufficient for most mathematics. However, showing (or its already being obvious) that an area of mathematics is reducible to ZFC is not *necessary* for acceptance by mathematicians.

---

[11]Model theory is used also. But my point remains. To illustrate: model theorists sometimes complain to Harizanov that she is drawing distinctions not recognised by model theory. For example, she sometimes insists on more than 'uniqueness up to isomorphism'. She insists on including certain properties used to measure complexity when she is identifying structures or patterns. Model theorists cannot recognise these properties. This illustrates that there are norms which are not recognised by model theory, but are used in complexity theory. To illustrate the second point, Harizanov's reaction is not to stop doing her mathematical work, or to consider what she is doing is not mathematics. Rather, she suggests to the model theorists that they should pay attention to more than what they can 'see' from a model theorist's perspective. The illustration comes from conversation. There is no written reference. However, Rodin (2010, 25) gives a further example.

There is a further complication which should be addressed. When Vopěnka makes his anti-foundationalist complaint, the naturalist should note that it was mathematicians, not philosophers,[12] who set said norm. Given this observation, it seems then, that, as naturalists, we should give a philosophy which advocates ZFC or Cantorian set theory, as a foundation.[13] This is, indeed, what Maddy does. According to the naturalist, if mathematicians are setting norms in this way, then the philosophers should take the norm setting seriously. But, even as naturalists, we can be more careful.

First, there is a difference between taking seriously what mathematicians say about *mathematics* and taking seriously what they say about the *philosophy* of mathematics. As philosophers, with greater training in philosophy, maybe we should be less conciliatory about mathematician's views of philosophy. But even if we want to give the mathematicians the benefit of the doubt, and try as best we can to accommodate their philosophical views about their subject matter, there is a further complication.

As we saw with the quotation from Vopěnka, not all mathematicians agree to follow the norm, immediately, since he is an example of a mathematician who does not. So then what are the *philosophers* to do about the rival norms internally set in mathematics? The pluralist has no difficulty with this.[14] He observes that the Cantorian set theory norm was temporary. This is why Vopěnka's alarm is short lived. Contemporaneous with, and subsequent to, when Vopěnka was writing the quoted passage, many developments in set theory have taken place. Furthermore, his own theory of semisets has enjoyed some success. It takes time for some mathematical ideas to be accepted – and this is not only because there is a particular foundation that is in the historical process of setting the norm for acceptance. Since his writing, a much more significant number of higher-cardinal axioms have been proposed as extensions of ZFC set theory, and the prevailing attitude (I think) is that, in the light of the rival foundations, pluralism has succeeded set theoretic monism. But what of dualism?

The dualist does not fare much better. For, he proposes a foundation for some part of mathematics, and this part will suffer from the same criticisms. The part of mathematics for which we provide a proper foundation: second-order logic for the Fregean logicist, finististic (real) mathematics for the Hilbertian, all of mathematics save 'absolutely infinite magnitudes' for the Cantorian; is good mathematics, the

---

[12]The distinction is, of course, somewhat artificial, and if we do not accept it, then we rephrase the structure of the foundationalist philosophy appropriately. Many mathematicians are also philosophers, and the same person can play both roles. I follow Colyvan in not recognising a clear distinction between philosophy and mathematics, either in terms of persons or in terms of roles. Despite my agreement with Colyvan, it will be useful for the arguments here to adopt this artificial distinction.

[13]It is exactly on these sorts of grounds that Maddy, in her earlier work forges a realist naturalist philosophy of mathematics.

[14]Oddly, Maddy was reluctant to make this observation, or take it seriously. This is one of the contentions between Maddy and the pluralist, which we saw in Chap. 3.

rest is suspect. With Hilbert, we engage in a project of trying to widen, or determine the limitations of, and the scope of, the good part of mathematics, to minimise the suspect part.

The naturalist observer of mathematics will disagree with this nuanced normative attitude. He will observe that mathematicians work in both 'good' and 'suspect' areas of mathematics, and do not always agree that the 'suspect' part of mathematics really is suspect. Take, for example, all of the work on higher cardinal axioms. A Hilbertian would find less value in this work than in 'proper' engagement in the Hilbertian programme of reducing the existing suspect part of mathematics to the good part, since, for Hilbert, mathematicians should not be *extending* the suspect part![15] Sporting my naturalist hat, I am not sure that the mathematician working on the higher-cardinals would agree to change the direction of his work! The very notions of 'good' and 'suspect' mathematics are not happily applied to the practice of mathematicians.[16] So, here, since they all share a naturalist attitude, the naturalist, structuralist and pluralist part company with the dualist. The slide, from developing a very powerful theory and showing its scope, to having the powerful theory set the norm for mathematics as a whole is illegitimate. It has not been born out by history. Note that the strength of this argument depends on the philosopher having some sympathies with naturalism.

4.4 The Second Anti-foundationalist Argument: Growing Foundations

This argument is very much a repetition of the argument we saw in the first chapter of this book. A few details are added. As was mentioned in the argument of the foundationalist, the foundationalist begins with the technical result that most of mathematics *can* be reduced to The Foundation. This is a twofold mis-description. First, the reduction is sometimes too contrived, and therefore, not successful. Second, any proposed foundation is only that: a foundation. That is, we can add more to the foundation. Whatever the founding theory is, 'it' grows. As it grows we understand the founding parts in a new light. So it is not a fixed foundation.[17]

---

[15]'Bad', of course, is an over-simplification, especially in light of Hilbert's famously stating that he was not willing to be expelled from the paradise Cantor had introduced to mathematics. Nevertheless, there is a tension in Hilbert's attitudes towards the finitistic and the ideal.

[16]Frege's logicism (paradoxes aside) is a little more subtle, since we can take logicism in a fairly neutral way. I don't think that this is loyal to Frege, but I do think it is an interesting position. Under the neutral reading, the analytic part of mathematics is not so much 'best', but just analytic. We have a description, and no norm. Then it is just philosophically interesting to know what the scope is of the analytic part. The pluralist argument against the neutral reading of Frege, concerns much more the further part of this chapter, where we consider 'bad' mathematics.

[17]Brouwer agrees with this, so in this respect, he too, parts company with the monist. The issue about where Brouwer fits in my account is quite subtle. Where Brouwer and I part company is

Vopěnka complains about the initial reduction: "Some [mathematical] disciplines pursued in pre-set-mathematics [mathematics before the development of Cantorian set theory] had to be considerably violated in order to include them in set theory." (Vopěnka 1979, 9). Vopěnka gives the calculus as an example of such a violation (Vopěnka 1979, 9). There are many other examples. Try proving that $7 + 92 = 99$ in Frege's logic (for the cheeky reader: even using the inconsistency generated from Basic Law V, and then making an *ex contradictione quodlibet* argument would be quite hard in Frege's system) or in Russell and Whitehead's type theory. It is in principle possible to 'do calculus' in set theory, but it is so awkward that no one does it. Why? This is because the proofs are too long and not explanatory, so we lose sight of what we are trying to do, and much of the proof is very mechanical, and should be skipped, since going through all of the mechanical steps is not informative, and certainly not 'doing calculus'.[18] In this way, the reductions do *not* give the 'essence' of what mathematics is about, how it is practiced, what is interesting about it. The phenomenology is wrong. Nevertheless, the reducing discipline does give some philosophical insights. For example, we might learn, with Frege, that arithmetic is really analytic, *pace* Kant. However, it is a much larger philosophical step to take to say, with a fictional successful Frege, or a neo-Fregean, that all of arithmetic and analysis are essentially second-order logic. These thoughts should recall the argument from the second chapter concerning the two communities of mathematician, one community working exclusively in set theory, and the other in mathematics with a more traditional presentation. It is not at all clear that the set theoretic community does make a net gain over the 'plethora' community, nor is it clear when, or by what path, it could make such a gain.

Apart from the artificiality of the reduction, there is the second problem of instability of the foundation. Even lovely, all-encompassing, legendary, great, mathematical theories grow. New axioms and techniques are suggested and tried. When we add new axioms, we shift the foundation, for, we change the implicit definition of the primitive concepts or elements (such as the relation of membership, or the empty set). If we endorse the naturalist attitude, then we can observe (rather than resist) co-variance between 'founding theories' and 'essences': as the founding theory changes, so the essence changes. What is worse is that it does not just grow in one direction. Like the Hydra, it grows many heads, and they do not get along. For the pluralist, this observation makes a mockery of foundationalism as essentialism. This is not a logically necessary argument, since the foundationalist could insist on

---

in his emphasis on intuition. I think that mathematical intuition is interesting, but I disagree with Brouwer that "mathematics takes place in the mind", if we interpret the omitted quantifier at the beginning of the quotation as 'all'. I save this issue for a paper.

[18]Hrbacek et al. (2009), are all working on a way of doing calculus using ZF set theory, but they add a notion of small and large relative to a frame of reference. There are many layers of ZF sitting on top of each other, so a number is very small relative to where one is sitting. They have tried teaching calculus in the classroom in this way, and found that it is much more intuitive for the students than calculus as it is normally taught! The very fact that there are mathematicians working on this shows us that the relationship between ZF and calculus is strained.

fixing the foundation, and resist extensions, but then he would beg the question against himself. Or the foundationalist could insist on a hope for convergence, but this would be to rely too heavily on shared hopes. Neither insistence can be supported.

The pluralist, however, does not fare much better. His argument relies on naturalist sympathies and is inductive: based on the history of mathematics: every successful proposed foundational theory has spawned new developments or additions to the mother theory, and fostered the development of rival theories. No sooner had Whitehead and Russell introduced their type theory, then they developed the ramified type theory. Other type theories have sprung up since, some more successful (studied more), than others. After Cantor developed his naïve set theory, rivals were forthcoming: Zermelo-Fraenkel set theory, Gödel-Bernays set theory. Moreover, additions to these were made, such as the axiom of choice, the development of class theory, higher cardinal axioms were added *etcetera*. Category theory too has seen development.

Looking more closely, we *extend* the foundational theory with new axioms which make a new theory (assuming we individuate theories by the language, plus axioms, plus inference rules). For example, we can extend ZF set theory with the axiom of choice, which gives us ZFC. Moreover, as is well documented, there is considerable dispute over the *admissibility* of new axioms, which extend ZF set theory, none more notorious than the axiom of choice (Martin-Löf 2006, 2). Admissibility (classically) requires at least consistency with the original theory, but some proposed axioms are consistent with, but independent from, the original theory; and therefore we can add the independent axiom, or an axiom which is consistent with it, or several axioms. The problem now is to arbitrate between the alternative proposed extensions, since pairs of new axioms will lead to contradictions.

To arbitrate, we have to modify our original notion of the *essence* of mathematics – since it no longer rests in the founding theory. The Foundation is strictly broader since we think it can accommodate extensions. Moreover, we have to do this in such a way as to accommodate only *one* proper subset of the proposed additional axioms. Witness the debates about $V = L$. If we choose $V = L$, then we preclude a number of other axioms. If we choose $V \neq L$ as a new axiom, then we preclude other proposed axioms. Accommodation, in the face of choices which preclude other additions, is no easy task, since the founding theory itself cannot arbitrate. Here, 'choice' relies on some underlying sense of 'the' theory – not individuated by a language, set of axioms and rules of inference – but by some (not yet formally represented) intuitions which, one hopes, will become explicit through discovery and formal representation. But these vague intuitions are not good philosophical justifications for foundationalism, since these sorts of intuition vary from one mathematician to the next. Moreover, history has shown us that mathematicians have sometimes been badly mistaken.

We might dress up the intuitions by introducing considerations of fruitfulness, simplicity, elegance *etcetera*. But these considerations alone will not do, since, remember, we are providing a foundation, not for generating 'lots' of mathematics,[19] or aesthetically pleasing mathematics, but for correct mathematics. If the foundationalist does go ahead, and opts for one extension over another, to fix the 'essence' of mathematics, then he shows a weakness in the original monist argument for his chosen foundation. For, arguing for one extension over another, is a covert admission that he did not have the full essence properly formally represented in the first place. Thus, if, as philosophers, we are going to take history seriously, then the lesson we learn is that proposing a particular theory as The Foundation for mathematics is a highly unstable position. At best, a particular theory will represent something of the core of The Foundation, where The Foundation, is now something which we attempt to represent formally, but eludes our attempts. Again, relying on naturalist sympathies, we are better off making candid observations, and accept that The Foundation is simply not in the offing. There is no convergence of opinion.

4.5 The Third Anti-foundationalist Argument: *de dicto* and *de re* Many Mathematicians Are Pluralist

An aspect of structuralism, which the pluralist adopts, is his naturalism. Shapiro is a naturalist in the sense of paying attention to what it is that mathematicians do and report about what they do (Shapiro 1997, 3). To some philosophers this attitude is quite obvious. To others it is not. So, it should just be born in mind, in reading the following, that, within the philosophy of mathematics, naturalism is not a shared trait.[20] This distinguishes the philosophy of mathematics from other areas of philosophy, especially when naturalism is associated with favouring scientific methodology *etcetera*, and where 'science' does not include mathematics, except when mathematics is indispensable to science. Set aside these more usual associations with the term 'naturalism'. Here, we are interested in Maddy and Shapiro's sense of observer of mathematician's activities, where the boundaries of the discipline of 'mathematics' is determined by mathematicians themselves. Structuralists and pluralists observe that mathematicians are rarely monist.

---

[19]We have to be very careful about quantifying over mathematical results, for, adding almost any axiom will add an effectively enumerable number of new theorems, so then we might count only 'important' new results, but how these are determined/chosen is again a problem; at least at any given time, since we might later discover that a theorem or result is important only many years later.

[20]It might turn out to be a matter of emphasis. Obviously every philosopher of mathematics, has to have been exposed to some existing mathematics, and taken that as a starting point. The difference in emphasis is over the hesitation or reluctance with which a naturalist will think he can tell the mathematician what counts as mathematics, once the technical result on his foundation has been demonstrated.

De dicto many mathematicians are anti-foundationalist. Or, more mildly, they view foundations with suspicion.

> Many working mathematicians (though by no means all) are suspicious of logicians' [and philosophers'] apparent attempt to take over their subject by stressing its foundations. ... [Moreover,] I have been persuaded by Edwin Coleman that foundationalism in mathematics should be regarded with considerable suspicion; or at least that proper 'foundations', ... would be much more complex and semiotical than twentieth century mathematical logic has attempted. In which case it would be arguable whether 'foundations' is an appropriate term. (Mortensen 1995, 4)

In conversation, Andréka, Chubb, Enayat, Harizanov, Kauffman, McLarty, Miller, Mourad, Németi, Székely and many other working mathematicians have all declared themselves to be pluralist, in some sense of 'pluralist'. I think that pluralism 'is in the air', but it has not been worked out as a whole philosophical position, only as an attitude within other positions.

Moreover, many mathematicians are not only *de dicto* pluralist, many are *de re* pluralist. That is, their behaviour at conferences, and in their written work, displays an open-mindedness and acceptance of alternative foundational theories – if not a complete disregard for the (philosophical) notion of foundation. More than this, in their proofs and methodology, mathematicians will often avail themselves of whatever hypotheses are useful and can support the desired result.[21] We shall see this in detail in the next chapter on formalism.

Consider the idea of a proof in mathematics. If we are monists, then the best proof will be one carried out in the foundational theory. Moreover, the proofs had better be pretty explicit, and rigorously carried out in the foundational theory, but this picture is distorted. According to Thurston, for mathematicians, the "*reliability* [of proof] does not need to come from mathematicians *formally* checking formal arguments [so working within one foundation]: it comes from mathematicians *thinking carefully and critically* about mathematical ideas." These ideas are not restricted to the ideas found in one foundation. The choice of which method or result to use in a proof is pragmatic, and there is a sense in which said method or result is considered to be trustworthy because it is "quite good at producing reliable theorems that can be solidly backed up." (Thurston 1994, 171 (emphasis added)). Following Cellucci, real 'mathematical' proofs are not carried out in a particular foundational theory. Instead, they are derivations. The derivations are not formal proofs as we would expect in a formal logic, such as a sequent calculus or a natural deduction system of proof, as developed by Prawitz. The derivation will mix together meta-language and object language. To make the point, I choose an arbitrary proof.

21 There are even worse cases, from a foundationalist point of view. Kauffman showed me an example of a knot. He then translated from the language of knot theory into the language of set theory. The knot then seems to be an impossible object, since it is a knot where $a \in b$ and $b \in a$, and this is set theoretically impossible. Assuming that the translation from knot theory to set theory is best possible, in some sense, then it is surprising that this makes no difference to the practice of knot theory. They do not defer to set *theory* at all, except to use the *language* on occasion.

METALEMMA. If φ, ψ are fully representable in **T** then φ & ψ is fully representable in **T**.

Demonstration. If $\mathbf{T}_1$ and $\mathbf{T}_2$ represent φ and ψ respectively w.r.t. $x_1, \ldots \ldots, x_n$ then $\mathbf{T}_1 \cap \mathbf{T}_2$ represents φ & ψ w.r.t. $x_1, \ldots, x_n$. (Vopěnka and Hájek 1972, 116)

The mixing of levels of languages, and the lack of formality in proof is part of the story.

The other part, following Cellucci, is that what a mathematician derives is not a conclusion, but a plausible hypothesis from problems. Problems are open questions (Cellucci 2008, 12). In our chosen proof above[22] the problem concerns the making of conjunctions in the language. And, more important, a hypothesis is said to be plausible if and only if it is compatible with existing data – which includes any mathematical results and notions available at the time of inquiry (Goethe and Friend 2010). In the quoted demonstration, we assume an understanding of conjunction and intersection. We are then simply told the relationship between symbols in the meta-language and in the object language. The metalemma, or conclusion is plausible because of our prior understanding. It remains a hypothesis, in the sense that it is not absolutely true. It is true relative to our background knowledge and the theory being developed.[23]

Especially this last point runs directly against the picture drawn by the monist philosophies of mathematics, but maybe the dualists are more accommodating. We might think that, as good dualists, mathematicians avail themselves of the better part of mathematics, when they can, and use the more suspect part with an uneasy conscience. For example, consider a constructive mathematician who first makes a classical (and constructively unacceptable) proof. He will do this on the grounds that the classical proof indicates 'the truth' of the conclusion. He will then, maybe in his spare time, work on giving a constructively acceptable proof of the same result. This is reported behaviour in some 'constructive' mathematicians. Of course, some constructive mathematicians will not accept this behaviour at all. They are monist constructivists. But those who do behave in the way described can be called dualist constructivists. There are *bona fide* dualist constructivists amongst mathematicians.

The problem for the monist and the dualist is that the monist or the dualist stories (of constructivist or other stripe) are not the only stories to be told, and many mathematicians completely disregard the advice of the monists and the dualists. They do not recognise a favoured or privileged part of mathematics – or, more carefully, what counts as favoured or privileged is thought of as a personal

---

[22] I really did choose this arbitrarily. The only constraint was to look for a short proof. I picked up the closest technical mathematical textbook I had to hand, and opened it to a middle page, and looked for a short proof. I do not think that it is important that the proof is of a meta-lemma, rather than simply an object-level lemma. Even the point about meta-language and object language still holds, since this lemma and proof mix meta-meta-level, meta-level and object-level languages.

[23] There is a lot more to be said about proof and the nature of proof. For a more thorough discussion, see Chap. 12.

choice, a matter of taste, or personal experience which will depend on one's particular education and temperament. We look at examples of mathematicians who are neither monist nor dualist in the next chapter. In other words, for many mathematicians, the purported distinction between good and suspect mathematics is a distinction without a difference. In the light of the *de dicto* and *de re* observations, the naturalist inclination shared by structuralism and pluralism make such a philosopher anti-foundationalist.

We have some *prima facie* evidence for pluralism from the claims and behaviour of mathematicians. However, this is simply an observation about the state of play in mathematics today. As philosophers we have to decide whether or not to take the observations seriously, or to think of them as a temporary glitch. We might excuse the observations on the grounds that the working mathematician is simply 'not a very good philosopher of mathematics and has not thought through the implications of his pluralism',[24] or is engaged in 'cognitive dissonance' or treats mathematical theories as tools and therefore his pluralism is due to a proto-instrumentalism or a complete a lack of philosophical thought about the matter. This might well be true in some cases. But, as philosophers, we should at least say more.

Pluralism motivated by naturalism does not prevent a philosopher or mathematician from working within the strictures of a philosophy, but we want to distinguish between being wedded to a theory for technical reasons, historical reasons or reasons of personal taste, on the one hand, and being wedded to a philosophical or mathematical theory for foundationalist reasons. Making use of this distinction, *the pluralist views the normative force of a philosophical position as only exercised within the philosophy*, or within the theory; the same will go for truth. For the pluralist, normativity stays internal to a monist or dualist perspective, and is limited to the scope of the foundational theory. This should not upset the traditional philosopher of mathematics too much: they have, after all, the same material at their disposal as they had before. But, at the end of the day, the pluralist asks them to admit the parameters of The Foundation and the accompanying philosophy. In fact, we should shift our vocabulary from discussing The Foundation to 'a big theory'. Here, 'big theory' just means a mathematical theory to which a lot of existing mathematics can be reduced. There are alternative big theories and there are other philosophies, and it simply is not clear that one is correct. In other words, according to the pluralist, what the foundationalist may not do is claim to give a philosophy for 'all' of mathematics.

---

[24] "... what the mathematician says [about the philosophy of mathematics] is no more reliable as a guide to the interpretation of their work than what artists say about *their* work, or musicians [about theirs]." (Potter 2004, 4), Even if we do not quite have such a strong point of view, it remains that mathematicians express very different philosophical attitudes. At the risk of being repetitive, my personal observation is that most mathematicians are pluralists.

4.6 The Fourth Anti-foundationalist Argument: On Truth in a Theory

I should make two preliminary remarks. One is that we have not used any distinctively structuralist ideas in our previous arguments against foundationalism. Here we finally do. The second preliminary remark is that what is presented in this section is not so much an argument against foundationalism as a presentation of an alternative conception of truth. Thus, we might say that in this section we look at a 'point of contention' between the foundationalist on the one side and the structuralist and the pluralist on the other side.

The contention is over truth. The monist identifies the truths of the foundation with the truths of mathematics. The rest are falsehoods. The dualist identifies the 'real', 'better' truths with the truths of the foundation, and the rest are either waiting to be reduced to the foundation or have some sort of lesser status. In contrast, the structuralist and the pluralist have a quite different conception of truth in mathematics.

"In physics, the same way as in mathematics, we do not address the question whether the axioms are true or not, we just postulate them." (Andréka et al. 2012). Together, the structuralist and the pluralist do not think that there are absolute truths in mathematics of the form: "$2+8=10$". Instead, what is true is: "In Peano Arithmetic, $2+8=10$".[25] To explain further: first allow the simplifying assumption that wffs are of the right category to be candidates for truth-bearers, as opposed to states of affairs, propositions, or what it is that propositions refer to, or supervene on *etcetera*.[26] Note that I specified that it is wffs that are the right sort of thing to qualify as candidates. This does not mean that all wffs are truth-bearers. Rather, it is a wff, when a particular theory is specified or understood, which is a truth-bearer. The reason for the qualification is that the pluralist takes an interest in mathematics as a series of theories, where each contains truths relative to that theory. But the wffs of the theory are not true, independent of the theory. Similarly a theory by itself (thought of as a conjunction of wffs closed under some axioms and rules of inference) is not true. This much is inspired by structuralism.

This is a more positive 'argument', in the sense of giving a positive proposal about how to think of truth in mathematics, if we are pluralist. Structuralism is a philosophy of mathematics where the notion of 'truth' is always qualified by 'in-a-structure'. Shapiro uses the highly expressive language of second-order logic to capture important mathematical concepts, such as 'is Dedekind infinite'[27] and

---

[25]This is because, for example, $2+8=10$ is false in arithmetic mod 8, where $2+8=2$.

[26]In later parts of the book, I shall revise the simplifying assumption. Not only *can* we revise the assumption, but *we must* revise it when moving outside the model theoretic perspective. So the assumption does not offend the structuralist, but it will, at some point, be revised by the pluralist.

[27]The definition of Dedekind infinite is that a set is Dedekind infinite iff it has a proper sub-set with which it can be placed into one-to-one correspondence. The natural numbers are Dedekind infinite, as are the integers, the rationals, the reals and so on. Finite sets have no proper sub-set

model theory to pick out structures (which, for Shapiro, are what mathematicians are interested in). For Shapiro, model theory is not a foundation, but an organisational perspective allowing for the clear individuation of mathematical theories, and for the comparison of various theories/structures from the point of view of chosen further meta-structures.[28] There is no ultimate structure, on pain of paradox. There is no absolute perspective. No structure is ultimately favoured over others (since model theory does not have one global structure). Model theory is not an axiomatised theory, and therefore has the flexibility to allow future extensions, without jeopardising stability. What is admitted as a structure will, undoubtedly, change over time, since model theory is a developing theory.

There are two types of individuation of 'theory' taking place side-by side. We can either individuate theories in terms of the language of the theory, the proof theory and axioms, which is, roughly, how the model theorist thinks of a theory, i.e., as a structure. A structure is just a set of objects together with some structure imposing relations which bear on the objects. Equally, we can individuate theories, in terms of an underlying idea which is not necessarily known to be fully captured by the formal representation of the theory. For example, if we think of model theory, then the formal representation is not yet fully achieved. In structuralism: model theory itself should be individuated in the latter way, since it is a growing theory; whereas particular structures should be individuated in the former way.[29] This is a very pluralist way of speaking, and not one that the structuralist will necessarily endorse. For the pluralist, the model theorist is able to 'see' a lot of mathematics, make sense of it, organise it within the strictures of his model theory and make contributions and offer insights. He individuates structures up to isomorphism and recognises all concepts expressible in a second-order language. Shapiro's structuralist can see quite a lot of mathematics, but not all of mathematics. As a result, Shapiro's

which can be placed into one-to-one correspondence with them. To capture the notion of Dedekind infinite, we need the expressive power of second-order logic. See Shapiro (1991, 100). The formula for set X being infinite is: INF(X): $\exists f[\forall x \forall y(fx = fy \rightarrow x = y)\ \& \rightarrow \forall x(Xx \rightarrow Xfx)\ \&\ \exists y(Xy\ \&\ \forall x(Xx \rightarrow fx \neq y))]$. This is read: There is a function which is such that if (two) of its values are identical, then the (two) arguments are equal. Moreover, the function operates on a proper subset of the set X.

[28] The title of Shapiro's first book on structuralism is: Foundations Without Foundationalism The Case for Second-Order Logic. Note the "Without Foundationalism". Foundationalism is identified with what I have been calling monism and dualism. Shapiro is anti-foundationalist in the sense that all mathematical theories which he recognizes are on a par. The 'foundation' is model theory. Model theory allows him to individuate mathematical theories (as structures). The model theory does not favour one structure as against another.

[29] This could be turned into a criticism of Shapiro's structuralism, but it could equally be launched against the pluralist. However, in the next part we shall see the pluralist defuse it. The criticism is made indirectly in Potter and Sullivan (1997). The criticism is that Shapiro makes different ontological and metaphysical claims concerning individual models, on the one hand, and model theory itself, on the other.

pluralism is restricted to what is recognized through the lens of model theory and what can be expressed in second-order logic, and this lens then prescribes what is to count as successful mathematics.

There are limitations to the arguments given against the foundationalist. All but the last of the above arguments rely on naturalist sympathies. We conclude that the pluralist trades with the monist and the dualist, truth for naturalism. There is no clear outcome to this debate. However, once the trade has been made, the pluralist pushes beyond structuralism.

4.7 Moving Beyond Shapiro's Structuralism

Where I part company with Shapiro is over the very important issues of what is to count as success in mathematics and what is of interest to the philosopher. The structuralist picks out as successful mathematics all mathematics which can be recognised by model theory. The rest is not.

In contrast, for the pluralist, success does not depend on being able to be recognised by a particular theory, open ended and generous as it might be. Instead, success is judged by reference to the community of mathematicians. Pluralism is more naturalist than structuralism. Examples of parts of mathematics which are not recognised by model theory, but that the pluralist deems perfectly successful include: intensional logics, mathematical theories, which are still in a stage of development and paraconsistent mathematics. Pluralism is more pluralist than structuralism. Or, we can think of structuralism as a very conservative form of pluralism.

To be fair, Shapiro acknowledges that model theory is *a* perspective, amongst others. Therefore, in principle, he is open to the suggestion of adopting other perspectives, and then becoming sensitive to notions in mathematics not recognised by model theory. This is so in principle. Nevertheless, as he observes: "the *prevailing* semantic theory today is a truth-value account, sometimes called "Tarskian". Model theory provides the framework." (Shapiro 1997, 3, my emphasis). For this reason, Shapiro develops his structuralism in accordance with model theory, adopting the perspective it affords on mathematics. So, it should be noted, that the target in this chapter, is not so much Shapiro, in his more conciliatory moments, but the structuralism he actually develops, in order to be able to say definite things in the philosophy of mathematics. But note, even with Shapiro's conciliation, he cannot recognise unsuccessful mathematics. To be dramatic, I shall use the term 'bad' to designate theories or proto-theories, not recognised by model theory.

Definition *Bad Mathematics* is any 'mathematics' not recognised by model theory, where 'mathematics' is not determined by model theory but by existing practice. This includes both what we called in the first chapter 'successful existing mathematics' and some unsuccessful mathematics!

Bad mathematics might just be mathematics overlooked by the model theorist. Or, bad mathematics might be quite unsuccessful, and for good reason – by mathematician's standards. But even in the unsuccessful cases, the bad mathematics can inform the 'good' mathematics, and therefore, according to the pluralist, should not be ignored in a philosophy of mathematics. Overlooked 'bad' mathematics includes (amongst other things): some intensional theories,[30] intentional theories,[31] not yet completely formally represented theories, paraconsistent mathematics and trivial theories of mathematics. It is in considering bad mathematics that the pluralist distinguishes himself from the structuralist.

For the pluralist, unsuccessful mathematics comes in two forms, the mathematics which no one (save the author) accepts, so, 'the community of mathematicians' or even a 'large enough part' of 'the community of mathematicians' does not accept the work. The other sort of unsuccessful mathematics are trivial mathematics. As far as I know, everyone agrees that trivial mathematics is unsuccessful, and should be avoided. We postpone discussion of trivialism and trivial mathematics to Chaps. 10 and 11. Here, let us discuss the part of mathematics deemed successful by the pluralist, but unsuccessful by the structuralist.

The problem the pluralist sees with rejecting 'bad' mathematics is that this offends against the naturalist insight and runs the risk of instability or of begging the question. The pluralist philosophy developed here is more stable than Shapiro's pluralism and the more traditional philosophies as well. The instability is temporal. That is, as judged by the community of mathematicians, what counts as successful mathematics changes over time. Odd theories suddenly find an application; some obscure result proves useful to a more central mathematical concern. Theories which were viewed as highly suspect come to be accepted in more main-stream mathematics, such as the study of non-standard arithmetic. More important, there are revolutions in mathematics, such as the discovery on non-Euclidean geometries, or of the incompleteness results. These revolutions radically alter our conception of mathematics, and the altering of the conceptions can take considerable time. On the reverse side, we see that some mathematics which were deemed highly successful drop out of use, or are no longer studied. Briefly: success is not even cumulative (it is not the case that once successful, always successful). Success in mathematics is temporary. Success is judged by a community of mathematicians who change their minds about what are the standards of success.

---

[30]It depends on how we single these out. If all it takes for a theory to be intensional is that it have an intentional operator, then some intensional theories are extensional, and can be recognised by model theory. An example is a modal logic where the modal operators have terms within their scope. Such a logic will be extensional (models will be unique, and identified, up to isomorphism). See Melia (2003, 2–4). In contrast, a modal logic with whole formulas within their scope will suffer from 'opacity of translation', and are, therefore, not extensional theories. Model theory is extensional, even if it has no axiom of extensionality (since model theory has no axioms at all, i.e., it is not presented axiomatically). An extensional theory cannot recognise the differences between wffs with intentional operators because of the opacity of translation.

[31]We postpone discussion of these to Chap. 6.

To remedy the instability, we can be more careful and qualify our account for our judgment of success by means of a temporal index, and maybe even a community index (such and such a theory was well accepted by Soviet mathematicians in the 1950s but not by 'Western mathematicians'...) then any 'rejection' is made relative, stable and harmless – and more accurate. This is a good start, but the pluralist is more ambitious than this. He wants to say something definite about the nature of mathematical truth and importance.

The rejection of bad mathematical theories might also beg the question, as when the reductionist foundationalist philosopher re-trenches and says that whatever fails to conform to her conception of what counts as successful mathematics is, by definition, unsuccessful. That is, she sets an *a priori* norm for success in mathematics. But the force of such an argument is limited, for it begs the question against the naturalist perspective. My diagnosis is that there is an inevitable tension between the naturalist attitude and the desire to give a philosophical account of (unqualified) 'successful mathematics'. Since, for the pluralist, there is no one foundation, and even the extension of the term 'successful mathematics' changes over time (and across communities), it follows that there are no truths of the form "the sum of the interior angles of a triangle add up to 180°". In contrast, "in Euclidean geometry the sum of the interior angles of a triangle add up to 180°" is true, or at least much more enduring and stable.

We have an idea of what sort of sentence is a truth bearer, let us be more precise. Truth can only be had within a known context (whether it be explicitly stated, or implicitly understood). The context is usually a mathematical theory. Let 's' be a well formed formula within a language of a mathematical theory. Or, it can be a sentence, which we know we can express in formal notation. "The sum of the interior angles of a triangle add up to 180°" is an example of an s. It is in English, but can be expressed formally if we so wish. Let 'T' stand for a theory. This can be given axiomatically, as in Euclidean geometry, or by setting rules, by setting principles, or it might arise out of a practice. For example, model theory is a mathematical theory, but it is not presented axiomatically or with a set of rules in the way that arithmetic is. Nevertheless the theory T sets the context in sufficient detail that the truth of the wff s can be verified within the theory (in easy cases of complete theories: by using the proof theory which is accepted within the practice of the theory). So, we are concerned with sentences of the form: Ts are candidates for truth bearers; i.e., any 's' written in the language of the theory T are candidates for truth-valuation.

Note that there is some slippage in the notion of T. This is deliberate, and is meant to reflect and encompass mathematical practice. The more precise we can be about T the more obvious it will be that we can generate s in T as a theorem or, given incompleteness, we can know it is true or unprovable. There might be some Ts combinations of which we do not know if they have a truth-value, this will be the case with 'open problems'.[32] We might not know what we have to put in to the

[32]Distinguish between not having a known truth value (now), not having a truth-value at all, and having the 'truth-value' 'unknown', or 'undetermined'. We sometimes use 'U' to indicate

T to derive s, so the T will just be a guess, and might be an open theory (to which we can add new material: axioms rules, methods, information). We might not know if the proof theory of T will experience the halting problem with respect to s. Ts might not get a truth-value because T is too ambiguous *etcetera*. In advance, we cannot say, although we might be able to guess, that a sentence of the form Ts is a truth-bearer. But we need to demonstrate its truth or its falsehood (or sometimes both, if we follow the dialetheist),[33] or demonstrate *that* it has to bear a truth-value (say by *reductio*), in order to know that it is a truth-bearer.

There are two further complications. One is that how we determine the truth of the sentence concerns what is 'acceptable', and this might come from somewhere outside the theory, or come from the semantics of the theory such as in one of the famous Gödel sentences which shows the incompleteness of a theory. The other complication is that we also want to make no pre-suppositions about sentences Ts being: true only, false only, neither or both. They can even change truth-value if we are not sufficiently specific about T or what counts as an acceptable way of detecting the truth of s in T. This will be the case when T is not a fully formalised theory. For the above reasons, we cannot be more specific about truth in mathematics than to say that sentences of the form: Ts are candidates for truth bearers, but not all of them have a truth-value.

The structuralist wants to be much more strict about the nature of T in the above statement. He also has classical inclinations with respect to truth-value assignment. So in both of these respects, the pluralist has taken his leave of the structuralist. We can sum up the difference this way: the structuralist wants to give a philosophy of successful mathematics, and is willing to miss out on some mathematics; we might say that he wants to give a philosophy of *definitely accepted* successful mathematics. In contrast, the pluralist is interested in giving a philosophy of, or bringing a philosophical approach to bear on, what it is that mathematicians do *qua* mathematician. We can then call the structuralist an optimal pluralist, where the pluralist is a maximal pluralist.

'unknown' or 'undecided' and treat it as a truth-value, and make three-valued truth-tables with T, F and U, each as admissible 'truth-values'. Strictly speaking this is sloppy. 'Unknown' or 'undetermined' are not *truth-values*. They are indefinite place holders for a truth-value. They are ambiguous between "there has to be a truth value T or F (not neither or both) but we have not worked it out yet" and "we do not even know if there is a truth-value T or F". Above, I am *not* counting lack of truth-value as a truth-value. The parameters for what counts as know*able* will depend on the resources we think we are allowed and to some extent on our theory of knowledge.

[33] A dialetheist is someone who holds that some sentences (or well-formed formulas) are both true and false. In particular they are true. We shall be introduced to the dialetheist more formally later.

4.8 Making the Differences Clear: Optimal *Versus* Maximal Pluralism

To distinguish 'success', from the 'rest' of mathematics, while remaining pluralist, let us distinguish between an *optimal* pluralist philosophy and a *maximal* pluralist philosophy. The optimal pluralist gives norms for philosophically well motivated theories, i.e., for definitely accepted 'successful' mathematics. Shapiro is an example of an optimal pluralist. There might be several competing norms. They might include: consistency, constructive considerations, definitions of validity, search for a robust ontology *etcetera*. The structuralist norm is 'can be recognised by model theory', and then by transitivity, all the norms enjoyed by model theory accrue. These include (presumed) consistency, extensionalism (models are identified uniquely up to isomorphism), proximity to ZF, *etcetera*. In contrast, the maximal pluralist is maximally descriptivist: tries to philosophically account for the whole corpus of mathematical activity. The maximal pluralist is loath to *set* or *fix* a norm for success in mathematics, and he will *accommodate*, *account for*, *or study* 'bad' theories without, himself, slipping into triviality; see Chap. 11. Under the maximalist attitude, the pluralist can, of course, *observe* the setting of norms by the mathematical community; norms, as given in the professional practice of mathematics. But the pluralist will not judge between competing norms (unlike the optimal pluralist). For this reason, the pluralist has to entertain, what were traditionally thought of as 'bad' mathematical theories.

Since, in this chapter I am advocating maximal pluralism, let us give the motivations for considering 'bad' mathematics at all. Recall that the bad mathematics include: some intensional theories, intentional theories, not yet completely formally represented theories (call these 'nascent theories'), paraconsistent mathematics and trivial theories of mathematics. I shall not discuss intentional and intensional mathematics here. We only need to discuss some of the items on the list in order to show some of the differences between structualism pluralism, what I am calling 'optimal pluralism' here, and 'maximal pluralism'. We skip to the nascent theories.

Nascent theories are theories that are still in progress. All theories go though a stage of 'construction', 'becoming' or (more platonistically) 'coming to be known' or 'coming to be formally represented'. Depending on how we individuate theories in mathematics, we might even say, with the Gödelian optimist,[34] that set theory is nascent! More specifically: if we do not individuate theories in the standard way in terms of a language, a set of axioms and rules of inference, but rather in terms of 'some theory to be discovered' or as a 'construction of the mind', or as

[34]A Gödelian optimist is someone who has faith that we shall one day, or that it is (*a priori*) in principle possible for us to find The Foundation, or the absolute truth about an axiom. In particular, the Gödelian optimist thinks that we shall eventually determine, for example, whether or not the continuum hypothesis is true. We shall do this by finding a new very powerful axiom which will help us to derive the solution.

having an 'informative semantics',[35] then many mathematical theories are nascent. A philosophy of mathematics that did not accommodate nascent theories would be found wanting by the pluralist (and the structuralist cannot fit such theories into a structure). Moreover, structuralism smells of paradox: model theory is also nascent, and therefore cannot be recognised by model theory, as a structure. In Chap. 11 we discuss this last point. Rest assured, the structuralist has some things to say about this. But the pluralist does not have to try. He recognises model theory as a theory amongst others and makes comparisons between that theory and others, and when doing this, he does not have to do anything surreptitious. For the pluralist, there is nothing remotely paradoxical here.

Let us move on to paraconsistent mathematics. There is now a corpus of literature on paraconsistent logics and paraconsistent mathematics. These are taken seriously by some mathematicians.[36] A philosophy of mathematics which does not treat of these is incomplete, and violates the naturalist attitude.[37] There is nothing more to say, except to stress that what we see with the pluralist is a more encompassing philosophy than the alternatives, even those which declare themselves to be pluralist.

Having considered most of the 'bad' theories, and finding that they all run foul of naturalism, it follows that, if we want to follow through with our naturalist attitude, then we should try to adopt a maximal pluralism, and not only an optimal pluralism. The virtue of maximal pluralism is greater inclusion, greater longevity, and in that sense greater stability. We should turn to the criticisms of maximal pluralism before giving a more detailed account of the maximal pluralist view in the next part of the

---

[35]For the distinction between an 'informative' and a merely 'technical' semantics see Priest (2006, 181). The distinction is not always clear, but roughly there are two parts to being informative: the intention behind the development of the semantics, and the 'sense' we can make of the semantics *post facto*. Intentions first: 'technical' semantics are developed with the intension of solving a problem, to provide *a* model for the syntax. In contrast, 'informative' semantics are developed in response to intuitions and ideas, which hold the formal theory responsible (we can judge the success of the formal theory by comparing it to the original intentions. For example, if my intention is to develop a temporal logic to reflect norms of reasoning over temporally indexed propositions, then my formal theory is judged with reference to the supposed norms). The *post facto* sense concerns what we make of the formal theory after is has been developed. For example, we might find that a purely technically developed semantics turns out to have an application, which makes 'sense' of the semantics. An example is quantum logic. In contrast, a technical semantics has only the intention, say, of proving consistency: if there is a model for a set of formulas, then that set is consistent. In this case, we just mechanically 'give a semantics', but we do not do so as an act of interpretation, which adds dimension to our understanding.

[36]There is plenty of sociological evidence for this. Witness publications by 'major' publishers, both as books and in journals; numbers, sizes, and sections newly contained in conferences. One telling example is the history of the world congress on paraconsistent logic.

[37]Shapiro's pluralist structuralism cannot recognise paraconsistent logics and mathematics, since they cannot have a structure, since the logic he uses is classical second-order logic, and only consistent theories have a model.

book.[38] We reserve for part III of the book the criticism, which evokes the danger of pluralism becoming trivial. A less pointed criticism, which we turn to now, is the criticism 'from the disdain for sociology'.

4.9 A Concern from the Right, or from a 'Disdain for Sociology'

It is quite normal, when confronted with a philosophical position which claims to account for a lot of things, too many things, to think that the position is not tight, is not rigorous, has degenerated into an 'anything goes' philosophy (which is no philosophy at all). I call this 'the critique from a disdain for sociology'. An imagined interlocutor might object:

> Michèle, if you give up on giving a philosophical account of successful mathematics, then you let in all sorts of abominations: trivial theories, crankish scribbles, numerology... Moreover with your moral-high-ground pluralism you are loath to judge, rate and rank rubbish-posing-as-mathematics as quite inferior to very good and fruitful mathematics. What sort of a philosophy are you hoping to give here? It might be 'stable', as you say, but it will also be empty/uninteresting. Have you lost all philosophical ambition? Have you turned Wittgenstinian (later, and only under some interpretations)? Are you not left with only doing sociology, history or historiography of mathematics since your naturalist attitude only allows description? Have you gone continental?[39]

There are a number of complaints included in the imagined quotation. The interlocutor accuses the maximal pluralist of philosophical, or logical, degeneration in the sense that whatever philosophy there was initially, threatens to 'degenerate' to the rank of sociology. This can be countered. There is, in fact, a lot of philosophical work to be done under a pluralist banner. We shall see some examples in part IV of the book. However, very briefly, we can answer the complaint in a general way.

I have no complaint against the sociology, history or historiography of mathematics. I also think that the theories of communication and meaning developed by the 'continental' philosophers are interesting and inform pluralism. I embrace all of these studies and attitudes. However, it would be a mistake to reason that, therefore, there is no philosophical work to be done, or that there is no *philosophical*

---

[38] I should like to thank Goethe and Sundholm for sustaining some of these criticisms against me in conversation. Note that they were much more delicate and kind in their tone than what is reported in the imagined quotations!

[39] The term 'continental' was used, and sometimes still is used, by some Anglo-Saxon philosophers as a term of abuse, and it generally refers to the sort of work being done by some present day French and German philosophers, for the most part. The use of the term here is meant as an ironic joke. As we shall see, some of my leanings are distinctly 'continental' since having to do with the notion of meaning, the politics of meaning and communication. My reason for not discussing the 'continental' theories in more detail is that I am not sufficiently familiar with them to feel ready to discuss them in writing. While no names are mentioned, see the discussion of meaning earlier in this chapter, for an instance of these sympathies.

family of positions called pluralism, or that there is no philosophical judgment in the philosophy of pluralism; on the contrary. *Ad esse, ad posse*. This is why I have included the fourth part of this book – to give a taste of some of the things the pluralist philosopher does when discussing mathematics. In general, the call of the pluralist is for greater qualification of claims, of lucidity, clarity, lack of ambiguity, a willingness to dig deeper into explanations, a willingness to face the limitations of a position in philosophy, or the limitations of a mathematical theory, and stresses that communication of mathematical ideas is important, and the discipline, which makes for good communication, should always be cultivated. Put plainly, the pluralist makes a call for communicating mathematics clearly, and candidly, and the same holds for philosophy.[40] For this reason, explicitness and rigour of argument become very important to the pluralist, and so does explicitness about what rigour consists in, since, as we shall see in the relevant chapter, rigour is not one absolute measure – unfortunately! If we do not take all of the nuances into account then we fool ourselves into thinking that that we are doing honest work, but instead, we are doing blinkered work – within very small confines, and we forget and ignore what we lack the courage to see. For the pluralist, this is dangerous and ultimately, intellectually dishonest.

A full picture of maximal pluralism will be presented in the first chapter of Part II of this book.

References

Andréka, H., Madarász, J., Németi, I., & Székely, G. (2012). A logical road from special relativity to general relativity. *Synthese, 186*, 633–649.

Cellucci, C. (2008). Why proof? What is proof? In G. Corsi & R. Lupacchini (Eds.), *Deduction, computation, experiment. Exploring the effectiveness of proof.* Berlin: Springer.

Goethe, N. B., & Friend, M. (2010). Confronting ideals of proof with the ways of proving of the research mathematician. *Studia Logica, 96*(2), 273–288.

Hallett, M. (1984). *Cantorian set theory and the limitation of size* (Oxford logic guides 10). Oxford: Clarendon Press.

Hrbacek, K., Lessmann, O., & O'Donovan, R. (n.d.). *Analysis with ultrasmall numbers* (in prep.) (Manuscript, 2009).

Maddy, P. (1997). *Naturalism in mathematics*. Oxford: Clarendon Press.

Martin-Löf, P. (2006). 100 years of Zermelo's axiom of choice: What was the problem with it? *The Computer Journal, 49*(3), 1–6.

Melia, J. (2003). Modality. In J. Shand (Ed.), *Central problems of philosophy*. Chesham: Acumen.

Mortensen, C. (1995). *Inconsistent mathematics* (Mathematics and its applications 312). Dordrecht: Kluwer Academic Publishers.

Potter, M. (2004). *Set theory and its philosophy*. Oxford: Oxford University Press.

Potter, M., & Sullivan, P. (1997). Hale on Caesar. *Philosophia Mathematica, 5*, 135–152.

Priest, G. (2006). *In contradiction* (2nd ed.). Oxford: Clarendon Press.

[40] I can only work towards attaining my posted ambitions.

Rodin, A. (2010). *Did Lobachevsky have a model of his "Imaginary Geometry"?* PhilSci Archive 2010 University of Pittsburgh.

Shapiro, S. (1991). *Foundations without foundationalism; A case for second-order logic* (Oxford logic guides 17). Oxford: Oxford University Press.

Shapiro, S. (1997). *Philosophy of mathematics; Structure and ontology*. Oxford: Oxford University Press.

Thurston, W. P. (1994). On proofs and progress in mathematics. *Bulletin of the American Mathematical Society, 30*(2), 161–177.

Vopěnka, P. (1979). *Mathematics in the alternative set theory*. Leipzig: Teubner – texte zur Mathematik.

Vopěnka, P., & Hájek, P. (1972). *The theory of semisets*. Amsterdam: North-Holland Publishing Company.

Chapter 5
Formalism and Pluralism

Abstract This chapter introduces the reader to pluralism from the starting point of formalism. Formalism is in some ways the closest position to pluralism. The characterisation of formalism is taken from Detlefsen. Adding support to the pluralist's argument in Chap. 3 against Maddy, about the philosophical conceptions of mathematicians not always being realist, we give support to the claim that many mathematicians see themselves as formalist. We also find support for this claim from the practice of mathematicians. We look at three test cases: the classification of finite simple groups, renormalisation and Lobachevsky's model for indefinite integrals. With this *de dicto* and *de re* evidence, we then argue that pluralism reaches beyond formalism, and better fits the *de dicto* and *de re* evidence. In particular, we argue for a pluralism in methodology which is not permitted under the structures of formalism, as we characterise it.

5.1 Introduction

Many mathematicians consider themselves to be formalists, and formalism has had considerable influence on modern mathematics, both on how we conceive of mathematics and how we present mathematical results. Moreover, formalism shares a number of characteristics with pluralism. Therefore, it is useful to compare the two positions. In particular, formalism is close to pluralism in methodology.

The content of this chapter is close to that of a paper co-written with Pedeferri (Pedeferri and Friend 2012). The hard research work of finding and developing the test cases was carried out by Pedeferri. We worked together on the conclusions we could draw from the cases. It is in this sense that the chapter is co-written with Pedeferri. Therefore, throughout the chapter 'we' is preferred to 'I'.

M. Friend, *Pluralism in Mathematics: A New Position in Philosophy of Mathematics*, Logic, Epistemology, and the Unity of Science 32, DOI 10.1007/978-94-007-7058-4_5,
© Springer Science+Business Media Dordrecht 2014

The theses of this chapter are that:

(I) While many mathematicians consider themselves to be formalists, pluralism answers to the aspirations of mathematicians better than formalism. Moreover,
(II) the modern practice of many mathematicians is better accounted for by pluralism than formalism.

In Sects. 5.1 and 5.2, we characterise formalism and give voice to the formalist mathematicians. The quotations from mathematicians confirm the thesis that some mathematicians characterise themselves as formalists, or fit the characterisation of a formalist. Call these mathematicians 'formalist mathematicians', on the grounds that they are attracted to the virtues of formalism.[1] The virtues are: freedom, creativity and the unwillingness to make commitments concerning truth or ontology in mathematics. In Sect. 5.3, we characterise pluralism. The pluralist shares the attraction towards the formalist virtues. However, he has a broader and more flexible position.

To show (II), in Sect. 5.4, we shall look at three test cases from the practice of mathematics: the classification of finite simple groups, renormalisation and Lobachevsky's model for indefinite integrals. In all of the test cases, it is clear that pluralism in methodology is better than formalism in accounting for the practice of mathematics. This is because pluralists are less restrictive in their notions of good practice. However, pluralism is not a position where 'anything goes'. The pluralist is informed by formalism: when mathematicians stray outside the bounds of what the formalists would consider to be good practice, the pluralist offers council in the form of a protocol to ensure good pluralist mathematical practice. The protocol is developed in Sect. 5.5, by reference to the test cases.

5.2 Characterisation of Formalism

Formalism is a philosophy of mathematics, which was developed in the late nineteenth century and the beginning of the twentieth. To be more precise, it is not simply one philosophy, and it is not identified only with Hilbert. Rather, we

[1]We could ask: why not call them pluralist outright? There are two reasons. One is that it would be anachronistic, in the sense that pluralism is a new position, so the word was not used to describe a position in the philosophy of mathematics, only an attitude amongst others, within a larger philosophy. So, formalism is a better default name for their position. Of course, once the mathematicians quoted learn about pluralism the hope is that they will then describe themselves as pluralists. Or, counter-factually, since some are dead, if they had been exposed to the philosophy of mathematics called 'pluralism', then they would have opted to call themselves pluralist rather than formalist. The second reason is that we do not want to beg any questions. We want to compare formalism to pluralism, in order to show the merits of pluralism over those of formalism. There are some subtle differences.

add Bourbaki, Curry and Robinson as good examples. We find that formalism has permeated mathematical practice, and influenced both how mathematicians often conceive of mathematics, and how they present their mathematical findings.

The characterisation of formalism presented here is based on the characterisation in Detlefsen (2005, 236–237). For Detlefsen, formalism is a family of positions, each member of which has the following five characteristics.[2]

1. The formalist reverses the Aristotelian tradition of favouring geometry over arithmetic as the measure of rigour and reliability in mathematics.
2. The formalist rejects "the classical conception of mathematical proof and knowledge" (Detlefsen 2005, 236).

The formalists replace this with a formal 'ideal' proof. The closest analogue today is a syntactic proof. While this is fairly orthodox, this conception of proof was not immediately accepted.

The contrasting 'classical conception' is also related to 'the genetic conception', or the 'synthetic conception'. Under this classical conception of proof, a proof counts as a proof, by virtue of the origins of the ideas in the proof. For example, Russell and Brouwer thought of proof synthetically, but in different ways. In *Principia Mathematica*, Russell and Whitehead stated that all mathematics should be accountable to logic. In this sense, the *origin* of mathematics is logic. Brouwer had quite a different view, but it was still 'synthetic'. For Brouwer, the only possible foundation for mathematics is intuition: "to exist in mathematics means: to be constructed by intuition." (Brouwer 1975, 96). A written proof, for Brouwer, is simply a poor attempt to communicate a mathematical intuition to someone else. It follows that the idea of an 'ideal proof' as a written series of statements is quite alien to Brouwer.

If we look at the history of geometry, and the development of non-Euclidean geometry in the nineteenth century, we discover that there was a lot of resistance to non-Euclidean geometry. The resistance testifies to the *genetic* conception, because when mathematicians did develop models of geometry where the parallel postulate was altered, these were treated, or dismissed, as mere games, or odd fantasies. They had nothing to do with truth, rigour of argument notwithstanding!

3. The formalist attitude to proof follows from a particular conception of rigour. For the formalist, rigour follows from an act of abstraction away from intuition.

The contrasting view is that rigour follows from an embedding of intuition into, or onto, a previously accepted theory (which is what we find in the genetic or synthetic conceptions, for example, in pre-Hilbert and Tarski presentations of geometry).

[2]While we often gloss Hilbert as a formalist, we follow Hallett's caution against calling him a formalist. It was Brouwer who first called him a formalist! (Hallett 1995, 141). Nevertheless, as we shall see, Hilbert's position is one in the family of formalist philosophies.

This is not to say that all formalists share the same conception of rigour. For example, for Hilbert, rigour consisted in using a precise step-wise or finitistic proof method based on the finite number of "strokes on a line".[3] Other formalists use different bases for rigour. For example, they might use a particular syntactic system of proofs.[4] Gentzen's work is a good example. Note that the conception of rigour places a methodological constraint on what counts as 'best practice' in formalist mathematics.

4. The fourth characteristic of formalism *safeguards* the idea that proof is a purely formal exercise. The characteristic is to advocate "a nonrepresentational role for language in mathematical reasoning" (Detlefsen 2005, 237). 'Non-representational' means, here, 'not tied to (or held responsible to) an interpretation'.

One part of the idea is to move away from intuition because this is not trustworthy, since ultimately subjective.[5] It follows that, for the formalist, mathematics should not be thought of as an art, passed on from teacher to student, where this is necessarily geographically and linguistically parochial. Instead, mathematics should be thought of as a universal and objective science. The other part is that mathematics is essentially formal, i.e., what we would think of as 'syntactic', thanks to Tarski. When we separate syntax from semantics (meaning or interpretation), the syntax (symbols, proof theory) can be re-interpreted in other contexts, and transferred to other areas of study. For Detlefsen, (4) is "perhaps [the] most distinctive component of the formalist framework" (Detlefsen 2005, 237). The separation of semantics and syntax results in our thinking of mathematics as a mere tool, applicable to any semantics we choose and find appropriate.

The formalist conception of proof, together with the distinction between semantics and syntax freed a lot of mathematical conceptions and led to a number of interesting innovations and insights in mathematics. But we should be careful. Added freedom comes with added responsibility. Like (2), (4) requires a methodological constraint to make the syntax properly responsible. One methodological constraint, which was chosen by Hilbert, was consistency[6] (in light of paraconsistent mathematics, we might today prefer 'non-triviality') but there were other constraints as well, such as 'finiteness', which today we can think of as roughly analogous to effectiveness. This hails the fifth and final characteristic, which concerns the content of mathematics for the formalist.

5. The formalist insists that her freedom in interpretation is what gives her *genuine* knowledge of mathematics.

[3]See, for instance Hilbert (1923), which represents his mature work on proof theory.

[4]See, for instance, Troelstra and Schwichtenberg (2000).

[5]'Intuition', here, just means relying on a sort of 'private mental feeling or insight'.

[6]This formalist proposal was controversial. Brower had strong objections to consistency playing any important role in mathematics: "the question whether a certain language is consistent, is not only unimportant in itself, it is also not a test for mathematical existence." (Brouwer 1975, 101).

The knowledge is genuine because it is unencumbered with history or intended interpretations or applications. Instead, it is what we might call 'pure' knowledge. The point is that mathematics becomes even *more* creative with this conception of mathematical knowledge, than under the more traditional content-constrained mathematics, not *less* creative, as many suspected; and it is creative of something: genuine and pure mathematical knowledge. With this characteristic, the formalists are defending their position against the claim that mathematics, under formalism, would become purely mechanical (because syntactic). The formalist denies this. If mathematics is free and creative, then it cannot be mechanical.

Detlefsen characterises a formalist as someone who adopts all five characteristics. We ignore the first, since we think it is eclipsed by modern developments in mathematics, see, for example, Tennant (Manuscript 2010). Formalists *differ* from each other over the details, in particular, over their interpretations of 'rigour', 'formal', 'proof', 'semantics', 'content', 'syntax' and so on. This is what makes formalism a family of positions.

Under this characterisation of formalism, we should play it out, and ask ourselves what the practice of mathematics looks like for the formalist. A related question is: what makes for 'bad' formalist mathematics?

At the time when formalism was being developed, the paradoxes were in the air, and non-Euclidean geometry was much better accepted. So formalists knew that intuition could lead us astray, either by leading us to paradox, or by restricting us to unnecessary intended interpretations restricting our choice in theories. Therefore, they proposed that inferential processes (proof procedures) used in mathematics should be completely formal, i.e., there should be no natural language in the proof itself. Proofs should be thorough. Each step should rigorously follow from the previous steps. For example, for Hilbert and Bourbaki, a proof takes place within a theory.[7] The theory contains axioms and rules of inference. The rules account for each step in a proof. Ideally, the proof is entirely carried out in a formal language, and it is finite. Thus, we have the development of proof theories, and an investigation into the logic of mathematics.

Hilbert, and others, knew very well that it was too much to demand that all mathematical proofs be written out in this way, so they distinguished between an 'ideal proof' and a more heuristic proof. The thought was that, *in principle*, any heuristic proof could be turned into an ideal proof. An ideal proof, then, consists in stating some axioms, using only allowed rules of inference of the formal system of proof of that theory, and coming to a conclusion. Natural deduction proofs, from logic, are perfect examples.

---

[7]Hilbert (1996) claims that "the development of mathematical science as a whole takes place in two ways that constantly alternate: on the one hand we drive new provable formulae from the axioms by formal inferences; on the other, we adjoin new axioms and prove their consistency by contentual inference".

We deviate from such a proof when we:

1. fail to specify which theory we are working in,
2. import foreign axioms,
3. use rules of inference not in the proof theory,
4. fail to completely formalize our proofs (or fail to show that we could do this in principle) or
5. leave unexplained gaps in our reasoning.

The importance of sticking to the strict methodology (or proving that we could generate such a proof) is that if we have proved the theory to be consistent (or equi-consistent with another theory) then, by following the proof theory – the given methodology – we ensure continued consistency. Losing consistency is a real danger, because, as we know, rival[8] formal mathematical theories contradict each other.

5.3 The Voices of Formalism

Many mathematicians today call themselves formalists. A large number of working mathematicians have endorsed (sometimes implicitly) formalism as their way of thinking about mathematics.

Bourbaki (1991, 4) praises Aristotle for making it known to us that "it is possible to reduce all correct reasoning to the systematic application of a small number of immutable rules, which are independent of the particular nature of the objects in question."[9] On the notion of truth in mathematics, Bourbaki (1991, 11) writes:

> Mathematicians have always been sure that they prove "truths" or "true propositions"; such a conviction can obviously only be sentimental or metaphysical, and it is not by getting on to mathematical ground that it can be justified, nor even given a meaning that does not make it a tautology. The history of the concept of truth in mathematics is the concern therefore of the history of philosophy and not of that of mathematics.

Discussing Hilbert's arithmetisation of geometry, where he treats all the geometries as varieties, and where axioms are treated as hypotheses, Bourbaki writes:

> Hilbert classifies these axioms [of the different geometries] into different groups of different types, and sets himself to determine the exact area of influence of each of these groups of axioms, not only in developing the logical consequences of each of them in isolation, but also in discussing the different "geometries" obtained when one omits or modifies certain of

[8]We use here the word 'rival' in the sense Beall and Restall (2006, 36). For them, rivalry between logics can only occur in our choices for applications.

[9]Bourbaki then regrets that Aristotle confines his attention to logic and rhetoric in this respect, and does not extend it to mathematics. We might regret Bourbaki's attributing this view to Aristotle. But this is not the issue. Rather, we are simply concerned with Bourbaki identifying virtues in Aristotle, and the fact that he considers these to be virtues, and not vices.

> these axioms . . . he thus puts clearly in the picture, in an area considered until then as one of those nearest the reality of the senses, the *freedom* of which the mathematician disposes in his choice of postulates. (Bourbaki 1991, 17)

Robinson held an unabashed formalist position. He wrote about the foundations of mathematics that:

> My position concerning the foundations of mathematics is based on the following two main points or principles:
>
> i) infinite totalities do not exist in any sense of the word (i.e., either really or ideally). More precisely, any mention, or purported mention, of infinite totalities is, literally, meaningless.
>
> ii) Nevertheless, we should continue the business of Mathematics "as usual", i.e., we should act as if infinite totalities really existed. (Robinson 1965, 232)

That is, in the practice of mathematics we disregard 'reality', or 'meaningfulness' in the sense of referring to reality or some Platonic ideal.

Along these lines Nelson gives a passionate 'apology for formalism':

> What we devote our lives to is seeking for proofs; if a proof follows the formal rules, it is correct; if it does not, it is not a proof and is worthless unless it suggests a way to find a proof. No other field of human endeavour has maintained such a consensus over such a vast extent of space and time.
>
> [. . .]
>
> Formalism denies the relevance of truth to mathematics. But, one might object, mathematics works – the evidence is all around us. Does this not imply that there is truth in mathematics? Not in the slightest.
>
> [. . .]
>
> In mathematics, reality lies in the symbolic expressions themselves, not in any abstract entities they are thought to denote. The symbol ∃ is simply a backwards E. If we conclude that a certain entity exists just because we have derived in a certain formal system a formula beginning with ∃, we do so at our peril. The dwelling place of meaning is syntax; semantics is the home of illusion. (Nelson 1997, 3)

In other words, formalism fits very well with many mathematician's reported conceptions of present day mathematics, where 'present day' means twentieth and twenty-first century. But realism lingers, along with attachment to the genetic conception of proof. Otherwise there would be no need for Nelson's warnings. Hilbert makes fun of the genetic conception (Hilbert 1923, 184).

> . . . old objections which we supposed long abandoned still reappear in different forms. For example, the following recently appeared: Although it may be possible to introduce a concept without risk, i.e., without generating contradictions, and even though one can prove that its introduction causes no contradictions to arise, still the introduction of the concept is not thereby justified. [Hilbert comments:] Is not this exactly the same objection which was brought against complex imaginary numbers when it was said: "True their use doesn't lead to contradictions. Nevertheless their introduction is unwarranted. For imaginary magnitudes do not exist"?

We might still hear such objections more recently. This is not our concern here, for, we only make the claim that formalism is an important trend in modern mathematical thinking. Nevertheless, the 'old objections', and old conceptions still hold some sway. For example, some mathematicians hold a schizophrenic position between traditional realism and formalism. The schizophrenia is described by Moschovakis:

> Nevertheless, most attempts to turn these strong [realist] feelings into a coherent foundation of mathematics invariably lead to vague discussions of 'existence of abstract notions' which are quite repugnant to a mathematician. Contrast this with the relative ease with which formalism can be explained in a precise, elegant and self-consistent manner and you will have the main reason why most mathematicians claim to be formalists (when pressed) while they spend their working hours behaving as if they were completely unabashed realists. (Moschovakis 1980, 320)

To whom Dales replies with a reversed schizophrenia:

> It seems to me that most mathematicians really are formalists for all the days of the week. It is of course very useful when seeking proofs within the formal system to have a 'realistic picture' in one's mind, and so it is temporarily convenient, during the week, to be a realist, but it is the realism that the mathematician does not really believe in. (Dales 1998, 185)

Now that we have read some testimonies from mathematicians, we should turn to the practice of mathematics.[10] The mathematicians we quoted describe themselves as formalists – with some realist leanings, but maybe the behaviour of their fellow mathematicians tells another story. We shall turn to the test cases shortly.

The test cases we shall examine are not the work of the mathematicians quoted above. So, one could object that we have no evidence that the test cases would be accepted by the formalist. Indeed not, in fact this is the point. The test cases are not *good* formalist projects. But for all that they are not good realist, intuitionist, fictionalist, structuralist or naturalist projects either; nor do they fit well with any other orthodox position in the philosophy of mathematics.

The structure of our argument is that, *of* the traditional philosophical positions, formalism fits much of modern mathematics best. This is clear if we bear in mind that the formalist cherishes creativity in mathematics and is seeking 'genuine' mathematical knowledge. He does not have a classical conception of proof, but a formal one, and this is his key to the gate to freedom and knowledge. Moreover, success is what ultimately sanctifies mathematical ideas. Hilbert cites 'success' as the "supreme court to whose decisions everyone submits." (Hilbert 1923, 184). There is nothing else on which to hang one's trust in a new mathematical notion.

[10] Of course, we could dismiss the mathematician's avowals as philosophically naïve, since they seem to think that there are only two philosophical positions: formalism and realism. This impression is artificially created and is a misimpression, since it is hostage to the particular quotations we chose. But there is a deeper point. The mathematicians are not completely wrong to think that there are only two positions, or better: to short circuit discussion of alternative positions, since their point is to admonish claims which hail metaphysical enquiry. Such claims are most obvious in realism, but can be found, in one form or other, in other traditional philosophies as well, even if they are given a negative treatment, as is done in fictionalism.

Nevertheless, our contention is that formalism short-changes the practice of mathematics. We think that pluralism is a better description for the practice. Therefore, in order for the reader to make a judicious judgment of our claims, we characterise pluralism in methodology before turning to the test cases.

5.4 Characterisation of Methodological Pluralism

Let us now turn to the pluralist in methodology.

The pluralist in methodology shows a tolerance towards different methodologies in mathematics.

We see this in the form of using techniques developed in one area of mathematics in an area 'otherwise foreign' to it. We do this in order to prove a theorem. As we would expect, methodological pluralism closely resembles formalism. In particular, if we understand the idea of 'areas of mathematics being foreign to one another' in virtue of the content, intention in developing the area, application or in terms of semantics (as separated from syntax), i.e., 'classically', then we have a version of formalism.

Nevertheless, there are four differences between the pluralist and the formalist. One is that the pluralist in methodology is not forced to understand 'being foreign' classically. An 'area' of mathematics might be 'foreign' to another in the sense of 'formally inconsistent with'. Of course, here we start to flirt with triviality, and therefore we exercise caution.

The second difference is that, unlike Bourbaki, the pluralist does not insist on uniformity in presentation of a mathematical theory (in terms of language, axioms, and rules of inference), and this is a subtle point. On the one hand, the pluralist agrees that uniformity in presentation (say, in terms of language, axioms, rules *etcetera*) allows for easy comparison between theories, but it can also be distorting, and obscure important points. This will be explored in more detail in the final chapter of the book when we look at Lobachevsky's development of hyperbolic geometry. For now, it is enough to appreciate that the pluralist is sympathetic to the idea of uniformity of presentation of mathematical theories, but he is not wedded to it, in fact, he is a little suspicious of it.

A third difference is that the pluralist is not wedded to uniformity in syntax or logic. Two proof theories might contradict each other, but it might still be possible to 'borrow' some aspects of a theory, in order to construct a proof. There are perfectly rigorous ways of analysing a proof to ensure consistency, or non-triviality, such as the method called 'chunk and permeate' (Brown and Priest 2004). We shall see this deployed in Chap. 9.

So the pluralist is also willing to break with the proof-theory constraints of the formalist, where these constraints are to use one unique proof theory in any one proof, or series of proofs. Of course, when we break with this, the mixing and matching of proof theories, or syntax, has to be handled very carefully, and we shall witness this in the test cases.[11]

The fourth difference between the pluralist and the formalist concerns the conceptions of content or meaning or the ontology and truth of mathematics. Again, the pluralist has a wider conception than the formalist. The pluralist is not wedded to the orthodox schools of philosophy where content is identified with ontology, and meaning with truth (or truth-conditions). These issues and the ways in which they are discussed in the philosophical literature are hostage to the very long and interesting debate between the realists and the anti-realists – in mathematics and outside mathematics. The pluralist finds the debate interesting and recognises the passions that drive it and the agility and beauty of the arguments which have been mustered on both sides, but the pluralist is not, for all that, committed to one side of the debate, or to thinking of meaning in terms of use, or in terms of truth-conditions, which are then guaranteed by an ontology *etcetera*. Instead, the pluralist recognises that not all mathematicians are driven by these considerations. Or rather, that thinking of 'content' in terms of ontology or semantics is sometimes unhappy. (Recall the quotation of Nelson, in the previous section). In these respects, the pluralist is in complete agreement with the formalist.

However, the pluralist recognises the schizophrenia of some mathematicians, and recognises the value of a 'realist' picture. The pluralist carves a third way of thinking about meaning in mathematics. In some circumstances, and for some practicing mathematicians, 'meaning', might better be thought of as embedded in practice or as fluctuating with new discoveries. For example, as we learn more about numbers, as we develop sub-systems of Peano arithmetic, as we develop the number system in set theory, as we consider non-standard models, the meaning of the very word 'three' changes. It deepens and alters. It is not fixed, except *within* a formal system.[12] Similar remarks can be made concerning ontology. It might reflect the phenomenology of practicing mathematics, or be useful, to think of the ontology of mathematics as fixed and 'independent' of us. Or, it might be useful to think of the ontology as contained 'in the mind' and only existing when under the consideration of a mind. For the pluralist, both approaches are legitimate, and lead to developments in mathematics, or notation, or in interpretation, and neither conception should be ruled out of court, unless there are definitive arguments. This is an important point, and we shall have occasion to return to these themes several times in the book. While these themes develop other types, or aspects of pluralism, in this chapter we shall restrict ourselves to the pluralist in methodology. Here, we shall focus on the second and third difference between the formalist and the pluralist: the uniformity of presentation and the fixed syntax and semantics.

---

[11]We also have several models for doing this in the literature on paraconsistency. One is the adaptive logic approach, another is the 'chunk and permeate' approach (Brown and Priest 2004). We shall look at the later in more detail in Chap. 9.

[12]We leave this as a working hypothesis for now. This too will be revised later, but we can do only one thing at a time!

5.5 Three Test Cases

5.5.1 *The Classification of Finite Simple Groups*

The first case concerns 'big projects'.[13] In these, mathematicians divide the main goal into different sub-goals each of which is again divided into other sub-goals, or 'cells'. The cell-structure allows mathematicians (and computers) to work in parallel. Each cell works on specific problems (that are not always directly connected with the main goal, but are necessary for its success). The success of the project depends on the success of the cells. Therefore 'success' consists in finding the solution to a problem, such as classifying a mathematical theory.

Roughly speaking, the task of people working in a cell is to prove theorems. Because success is important, the mathematicians and computers working in a cell avail themselves of whatever it takes to prove the theorem of that cell. Moreover, they will have limited amounts of contact the other cells, since the group of cells is simply too big to have contact with all of them, and it is not deemed to be important to do so.

The classification of finite simple groups started more than a century ago and ended in 1983. It has been a collective work, resulting in thousands of pages in books, articles and manuscripts written by many different mathematicians. We can think of the collection of work as one long 'proof', resulting in one long 'theorem': the table of classification (which could be written out as a conjunction of characterisations of the classes of finite simple groups). The 'proof' is fragmented into many sub-proofs. It is a collection of a very large number of different proofs made with different techniques on different topics. The general proof is, therefore, "unsurveyable by a single human being" (Otte 1990, 61).

Is there a problem with mixing methodologies or types of proof? There is some controversy concerning the classification of the quasi-thin groups. Serre showed how this could be regarded as a gap in the larger proof of 'the classification theorem' (Raussen and Skau 2003). The 'gap' is due to the length and the structural complexity of the proof of the characterisation of quasi-thin groups. Serre's criticism addresses the fact that the dishomogeneity of the general proof for finite simple groups does not prevent *further* gaps arising that have not yet been discovered and fixed.

This is not a side issue. The classification of quasi-thin groups is a key step for the main goal of the classification of finite simple groups. The final stages in the history

[13]These are increasingly popular, and are fostered by some academic infrastructures. They are 'fostered' in the sense that mathematicians, who want to head such a project, apply for very large grants of money. They involve large numbers of people: faculty members, post-doctoral students and graduate students. Projects might span several universities, and take a number of years to complete. And, attracting such grants is taken to be a measure of prestige for the academic institution. The case we examine is not one which follows a large grant donation, but since such projects are becoming increasingly popular, it behoves the philosopher to take account of this.

of the project draw out the problem. The first announcement of the classification of quasi-thin groups was made in the early 1980s by Mason. But there was a problem. In Mason's proof, critical gaps were left, due to the proliferation of unexpected groups (which is the problem identified by Serre). Only in 2004 Ashbacher and Smith gave a complete proof in two volumes, running to more than 1,200 pages.

What is surprising, if we remember the emphasis in Sect. 5.2 on rigour of proof and working within a declared theory and using a unique proof theory is that these gaps seem to have brought no discredit to the results of 'the theorem'! This runs directly against formalist constraints, while the project could not have been envisaged without a sense of the freedom cherished by the formalist.

We might think that the break with formalism is due to the proof being unsurveyable. But this is not the major problem, since 'surveyable', 'finitist' and 'demonstrable' can be flexibly interpreted to fit this case. In terms of surveyability, we could fit the case to a liberal version of formalism, where each part is 'surveyed' separately.[14] We need not insist that one person be able to survey a proof, or that the person be able to survey a proof in a particular amount of time. The material is, after all, gathered in a two-volume work. What is more damning is that, in the proof, we see examples of mathematicians deviating from a strict and unique axiomatic system or proof theory.

We claimed above that many mathematicians today consider themselves to be formalists, and we also claimed that regardless of the self-perceptions of mathematicians, formalism fits mathematical practice better than other orthodox philosophies. Yet, we witness mathematicians disregarding some formalist prescriptions. If mathematicians are formalists, then, at the very least, they are not strong formalists, at worst they do not follow their own criteria for acceptable and correct proof. They exercise their freedom and creativity in an irresponsible manner, according to the formalist. In fact, even in this project, the mathematicians almost never give what a formalist would count as a proof. This is not just a question of shortening proofs to make them more perspicuous, but, here, the mathematicians use 'illegitimate' (by formalist lights) techniques in proof. In our test case (and in many other instances as well) we can find what we shall call 'deviant' proofs. Careful. Here we mean 'deviant' from the point of view of the formalist only.

Formalistically deviant proofs are 'proofs' where mathematicians use steps which deviate from the rigorous set of rules, methodologies and axioms agreed to 'in advance' and that fit formalist precepts.

[14]We have to be careful, since every part of every written proof is 'surveyed' in some sense, at the very least by the author. In the case of computer generated proofs, there is a program, which a human surveyed (by writing). In the case of computer generated proofs where the computer came up with the proof procedure, then there is some program which was installed at some point by a human, and that was surveyed. However, none of these will quite do. How to formulate the right balance which gives enough attention to proofs is a delicate matter, but it need not prevent us for continuing with the argument.

Of course, shortcuts can be useful to speed up a proof without any danger of inconsistency, or we can change our minds about the methodology or proof theory we are using. So, 'in advance' is not taken in the temporal sense, but rather, in a conceptual priority sense – at the meta-level we decide on the proof theory. Here we are interested in something else. Deviant shortcuts or detours can help to circumvent an impasse which *could not* be overridden with the standard steps agreed upon 'in advance'. We have to be using a different proof theory, one which is inconsistent with the first. Nevertheless, many mathematicians consider that these deviant methods provide 'correct' (enough) results. So they are successful! We cannot help but conclude that either said mathematicians are not formalists or they are formalists in bad faith (since the realist might be less concerned with the method of proof since she is interested in the truth). Some practicing mathematicians insist on the attractive aspects of formalism while ignoring the constraints.

There is a rebuttal against the above argument. One might think that the argument has mis-fired. After all, the classification of all finite simple groups is hardly 'a theorem'. It is a classification. The project is not to prove a theorem, but to give a meta-level result. It was never meant to be 'carried out within a formal theory', with axioms and rules of inference given in advance. Thus, we have relied on a metaphor for our argument to show that mathematicians are not really formalists, but the metaphor does not carry.

Here is our counter-argument to the rebuttal. It is correct to say that we have stretched the ideas of 'proof' and 'theorem' by saying that the classification is the 'theorem' we are trying to 'prove'. It is also correct that no umbrella formal theory was agreed upon 'in advance' for 'proving' the 'theorem'. However, even if the work is to seek a 'meta-theoretic result', this should not entail that all standards are dropped! The classification does require careful definitions, it does require proofs – that a particular group or class of groups falls under a particular classification. The danger in mixing methodologies is that under one we might classify a group a different way than under another methodology. These proofs – even if they are carried out at the meta-level – are still proofs, which are checked for correctness and so on. An incorrect table of results is useless. The results of these proofs are gathered together as a unified result – about one subject: finite simple groups. If we are to disregard the genesis of the idea of a finite simple group, and still think of these as forming one concept, which can be classified, then we had better be using the same means for proof in making our classification! Thus, the metaphor does hold sufficiently for us to make our argument. So, the formalist is scandalised by such practice. In contrast, the pluralist is untouched.

The pluralist would urge us to think very carefully about the standards of proof at the meta-level, and whether or why they might be permitted to be different from the standards at the object level. There are pluralist answers to this, but no convincing formalist answers, since the formalist makes a strict trade between freedom of interpretation against high standards of proof methodology united under one theory with one type of (axiomatic) presentation. The pluralist judges the formalist to be too conservative.

If this counter-argument is not fully convincing, then it is up to us to find a better example. There are several 'big' mathematical projects being carried out today, and they show the features we are interested in – a lack of adherence to one 'method' of proof, and therefore run the risk of inconsistency in methodologies. The conclusion we wish to draw from the example is that the formalist theory is too strict in respect of proof to reflect current mathematical practice, and what happens with proofs at the meta-level is not sufficiently well worked out. The pluralist diagnosis is that the phenomenon of 'big projects' in mathematics requires a more subtle treatment to explain the success and the acceptance of the results of such projects, especially when they use formalistically deviant proofs.

As we see from the above example, the mathematical practice displays pluralism in methodology, which runs directly against the formalist conception. This is for the very good reason that pluralism in methodology might generate an inconsistency. The pluralist agrees that this is a possible danger in importing different methodologies. In light of these concerns, the pluralist believes that we should retain a sense of ideal, formalistically acceptable proof within a mathematical theory; that deviation should be flagged and carefully scrutinized; but not that it should be banned. The scrutiny urged by the pluralist is systematic.

Here is an example of a protocol the pluralist could urge on the big projects.

First, declare the allowed methodologies, theories, proof procedures and so on for the project. Call the latter 'the methodology agreed upon in advance'.

If we find gaps, we prove (using the methodology agreed upon in advance) that they can be filled with said methodology.

Second, if we are importing foreign methodologies, axioms, rules *etcetera*, then we need to determine whether adding these to the methodology agreed upon in advance will create an inconsistent theory.

There is no danger when the foreign and native methodologies all belong to a larger theory. In this case, we ought to have agreed to the larger theory in advance, and we simply revise what we declare as our 'allowed methodologies, theories and proof procedures'. Again, 'in advance' is not meant in the temporal sense, but in the sense of 'conceptually prior'. But this is not always possible, as we shall see in the next two examples. The protocol will be further developed after we have looked at these.

5.5.2 *Renormalisation*

Another problematic case is the mathematical procedure of renormalisation.[15] It is a procedure used in physics, especially in quantum electrodynamics, to

[15] Renormalisation has been picked up in the philosophical community as an interesting case of applied mathematics. See for example Maddy (2007).

eliminate infinite quantities during certain types of calculations. In these physical theories, integrals represent observable physical quantities, which diverge towards some specific limits. To avoid these divergences being infinite – and therefore incalculable – renormalisation is used as an adjustment to the theory, which allows us to eliminate these divergences. Basically, the procedure cuts off the divergences at a calculated number and this allows us to obtain finite (rational) values. The new values fit with our observations. So, renormalisation has a strong practical justification, since the mathematical divergences do not seem to affect the physical results. That is, after renormalisation we retain perfect accordance with the measured data and predictions. Nonetheless, there is a problem with the *explanatory power* of the procedure. The particular cut-off points decided upon by the physicist are *mathematically ad hoc*, even if they are not physically *ad hoc*. For this reason, they are mathematically unjustified. But the situation is worse than this. Renormalisation leads to conceptual and mathematical inconsistency.

For example, strictly speaking, when we renormalize, we subtract infinities from infinites in order to get a particular finite non-zero result. This is mathematically inconsistent, we should either get zero, or an infinite number. Nevertheless, this is the mathematical 'process' at the base of renormalisation. The procedure is deviant, for a formalist, because there is a back-and-forth play between physics and mathematical calculation, and physical data are foreign to formal mathematics. We might say that renormalisation is a process that 'launders' the inconsistency of the results from mathematics with the 'magic soap' of application provided by physics. The physical data 'tells us' to fix the pure mathematical results so that we never deal with infinite values. The infinite values have to be expunged since infinite values are not permissible in observation statements, we can hardly report to have observed an infinite number of something, at least in physics. Freedom and creativity in interpretation are at play, but adherence to formalist constraints about consistency are lacking. What we see is either a pluralist approach to methodology (allowing constraints from physics to alter mathematical calculations), or we see very dubious scientific practice, according to the formalist. Once again, the formalist is offended by such practice.

The pluralist is not offended, but insists on vigilance and care in these situations. It is quite right that the physical world, to which we apply the mathematical theory, sets parameters on the possible mathematical theories we can effectively use, and the sorts of results of calculations we are permitted. But reality only sets parameters. For this reason we should be very careful in trusting our particular mathematical theory to completely and closely model reality. For this reason, we have to be particularly vigilant when letting the mathematics predict physical outcomes when these are not verified by the application. However, in this case, they are so verified, and this is what makes the practice acceptable to the pluralist. To return to the metaphor, the soap is not magic, it is an independent check. Indeed, since in the case of renormalisation, we are using a compromised mathematics, we should properly modify the mathematics so it is not *ad hoc*. That is, the original mathematical theory is not perturbed by the presence of infinite values. It is the *application* that is perturbed. It 'tells us' that infinite values are impossible, and so we modify the mathematical results. These adjustments to the original mathematics are done systematically (with respect to the physical theory): whenever a 'physical' value is calculated to be infinite replace the value with an appropriate finite number (by subtracting an infinite number form the mathematically calculated infinite number). This sort of 'adjustment' is symptomatic of the systemic and pervasive nature of problems in physics of dealing with infinite and infinitesimal quantities. It is a problem that permeates mathematical applications. So, we should accept the practice with caution. In this case, the caution is met by the correlation with observation. But we could do better.

While we temporarily accept the gap between perfect mathematics and measured physical quantities, we can also work to identify and overcome the problem. Here the problem is not with inconsistency within the physical world, or within the pure mathematical theory. Rather, the 'inconsistency' has to do with a mismatch between the pure mathematical theory and what we metaphysically suppose, i.e., what we suppose we can measure of the physical reality. It is a problem of fit between theory and application. Here, the methodological pluralist will not caution against inconsistency, but will endorse more careful use of mathematics, and seek a mathematical justification for what looks *ad hoc*. Moreover, we can do this quite systematically. One suggestion is to look to the work of Hrbacek et al. (2009), since they do away with the inconsistencies of applied calculus using a modified set theory.[16] The other reference work is that of Vopěnka (forthcoming, Manuscript 2013). In this work, Vopěnka uses the notion of a semiset.

> A class is understood to mean any collection of given objects (its elements) that we interpret as being an autonomous entity or a single object.
>
> A set is understood to mean a class that is sharply defined.1 Moreover (in accordance with our former decision), every set is finite from the classical point of view.
>
> A semiset is understood to mean an unsharply defined class which is part (i.e. a subclass) of a set. Vopěnka (forthcoming, Manuscript 2013, 39)

We suspect that the quantities used and measured in electrodynamic quantum theory much better fit the idea of measurement of a semiset than that of a regular measurement on a set. But at this stage, this is only a suspicion. Other suggestions as to what to do to philosophically tidy the renormalisation procedure come from the literature on paraconsistency. Adaptive logics could play a role here, as could the chunk and permeate method of analysis.[17] We save this for a future project. The pluralist is not passive. He does not *simply* accept the practice of renormalisation. He accepts it temporarily, recognising that it 'works'; but also he recognises that there is something metaphysically and mathematically incorrect about the practice. Because of the latter, he casts around for better methods, for a better practice.

---

[16]They actually use layers of ZF set theory, so that from one perspective, at a level, a quantity will be infinite, but from another level, it will be finite. They report that their approach is quite intuitive for engineers, and other students who are more inclined towards applied mathematics.

[17]In fact, in Brown and Priest (2004), they suggest applying chunk and permeate to renormalisation (Brown and Priest 2004, 386).

A formalist could object that the example or renormalisation is one of applications, and formalism concerns pure mathematics. This might well reflect the view of some formalists. Here, the pluralist is more ambitious than the formalist. He tries to account, not only for pure mathematics, but also for applications. This has the added advantage that the distinction between pure and applied mathematics need not be drawn. Nevertheless, since renormalisation is a problem of 'application', and so, could be dismissed, let us turn to a last case, which is more in keeping with the formalist paradigm.[18]

5.5.3 *Lobachevsky's Model for Indefinite Integrals*

Another example, which is less modern, provides us with a similar case. In dealing with the problem of finding the exact solutions for indefinite integrals, Lobachevsky thought to apply to the calculus his imaginary (non Euclidean, hyperbolic) geometry.[19] The method of Lobachesvky mirrored the usual technique of using geometry as a model for this kind of operation. However he used a hyperbolic model instead of a Euclidian model. In his "Application of the imaginary geometry to some integrals", he applied what we call today, Hyperbolic Trigonometry, to calculate complex integrals. These are the only geometric equivalents of indefinite integrals. He was then able to find a solution to the indefinite integrals.

This was possible because "... the limiting surface sides and angles of triangles hold the same relations as in the usual geometry." (Lobachevsky 1914, 34). It is therefore possible "to develop the Hyperbolic trigonometry on the basis of the usual (Euclidean) trigonometry," (Rodin 2008, 19). And this can, in turn, be used to solve certain integrals "which earlier were not given any geometrical sense." (Rodin 2008, 11). That is, Lobachevsky recognized how we can 'translate' from one world to another, and then

> As far as we are (sic!) found the equations which represent relations between sides and angles of triangle (sic!) [...] Geometry turns into Analytics, where calculations are necessarily coherent and one cannot discover anything what (sic!) is not already present in the basic equations. It is then impossible to arrive at contradiction, which would oblige us to refute first principles, unless this contradiction is hidden in those basic equations themselves. But one observes that the replacement of sides a, b, c by a*i*, b*i*, c*i* [*i* is the imaginary number: square root of negative 1] turn these equations into equations of

[18] We shall see later, in Chap. 9 and in the final chapter, that fitting Lobachevsky's work into the formalist framework is an artefact of our modern conception of mathematics, which is heavily influenced by formalism. Rodin (2008) is sensitive to this, and alerts us to the dangers of giving a formalist reading of Lobachevsky.

[19] 'Hyperbolic geometry' is the modern name for the geometry developed by Lobachevsky. 'Hyperbolic trigonometry' is the trigonometric part of the theory. 'Imaginary geometry' is the name Lobachevsky used, because of the imaginary numbers present in the trigonometry. I mention all this to dispel confusion in reading the quotations.

> Spherical Trigonometry. Since relations between lines in the Usual and Spherical geometry *are always the same*, the new geometry and Trigonometry will be *always in accordance with each other*. (Lobachevsky 1914, 34) (Italics added)

Lobachevsky explains how he avoids inconsistency (or where to find inconsistency if it is there). Thus, the mixing of methods does not necessarily lead to inconsistency. This is just one danger.

The formalist is too restrictive when he insists that all proofs use only one proof theory or methodology. The pluralist in methodology allows for together inconsistent proof theories or axioms provided we are careful to avoid inconsistency, and this is exactly what Lobachevsky does. As Kagan puts it:

> He [Lobachevsky] considered the given integral as a value of length of a certain curve in a hyperbolic plane, as the area of a certain figure in a plane or any other surface, as the volume or mass of a certain solid, and since these were metrical values in hyperbolic space, the consideration he deduced on the basis of imaginary geometry indicated how to find the value of the considered integral. And when this value was found it was frequently possible to find also analytical ways [mechanical calculations] which led to the same goal. The congruency of the results obtained Lobachevsky regarded as confirmation of the correctness of hyperbolic geometry. (Kagan 1957, 59)

Nevertheless, the formalist criteria have been violated. Moreover, the deviance smacks of circularity: to use a new tool to produce 'correct results' which themselves are to be regarded as a proof of the correctness of the tool. Nonetheless, by using this method we do actually find the 'right results' confirmed by congruence. Therefore, we have what Hilbert calls 'success'.

This test case seems to confirm the fact that the mathematical practice of proof does not rigidly follow the rules of the logical system supposed to underlie it (the axioms, in our case). As Corcoran points out, in mathematical practice it is common to find

> sentences beginning with "for purposes of reasoning suppose that". Here suppositions other than axioms are being introduced not as main premises but merely to begin a subsidiary deduction. [...] The myth that a proof is simply a sequence of (declarative) formulas has its usefulness but truth cannot be claimed for it.[20] (Corcoran 1973, 32)

The supposition does not just introduce an idealization, but might introduce something quite foreign to the theory. If we were to reconstruct Lobachevsky's proof using this language, we might introduce some of the suppositions of hyperbolic trigonometry into our calculations. Therefore, following these examples we maintain that mathematicians are not as formalist as some declare, and it seems that the actual mathematical practice is closer to pluralism than to formalism.

---

[20] The remark about truth is interesting and I deliberately left it in. It smacks of realism. However, we could use a circumlocution which would be in better accord with the discussion here. For example, we could paraphrase with: "The myth that a proof is simply a sequence of (declarative) formulas made in one axiomatic theory has its usefulness but quite often we deviate from our original theory when making a proof."

Developing to the protocol we started in Sect. 4.1:

If we cannot, or do not see how to, fit all proofs into one methodology, then we should check the cases for non-triviality.

In the given case, Lobachevsky did check this. He ascertains that "since relations between lines in the Usual and Spherical geometry are always the same, the new geometry and Trigonometry will be always in accordance with each other." Therefore, not only is triviality avoided, but so is inconsistency. As Kagan writes, after Lobochevsky had obtained these results: "when this value was found it was frequently possible to find also analytical ways which led to the same goal." Thus, the next step in the protocol is to

try to find a third, independent (or seemingly independent) way to re-justify, or independently justify the first set of calculations.

This re-confirming is the buttress the pluralist adds to ensure that we do not prove nonsense, and is what allows freedom in mixing and matching formal proof procedures. Rodin does this in his analysis of Lobachevsky, and does so with an eye to preserving the insights of Lobachevsky's original proof. We shall do something similar in detail in Chap. 9, but we shall only be checking for consistency, not for preserving Lobachevsky's insights. Moreover, we shall do this in a way which is neutral with respect to Euclidean or other geometries. The neutrality of the methodology I employ is what makes it a candidates for 'independent' re-justification. It is a *re*-justification because, of course, we have accepted Lobachevsky's proof ever since Riemann and von Helmholtz introduced the general notion of a geometrical manifold in 1854 and 1868 respectively (Katz 1998, 767) and Beltrami was able to model hyperbolic geometry in *Euclidean* geometry in 1868 (Rodin 2008, 1). Thus, the pluralist not only makes a recommendation, but discovers that mathematicians and philosophers actually follow it, not because the pluralist told them about it, but because they feel an unease about the mixing of proof procedures. According to the pluralist, they are correct to feel unease, and they are correct not to abandon the mathematical results for all that.

5.6 Diagnosis and Recommendation

So why did the formalist misfire? The reasons the mathematicians were attracted to formalism were that it (1) allows for creativity in re-interpretation and freedom from the classical conception of mathematics. (2) Formalism avoids heavy foundational philosophical disputes about ontology and truth. These attractive ideas are what led to deviance. But the deviance is deviant only to the formalist.

Whereas the formalist allows freedom only in interpretation, the pluralist also allows freedom in methodology. However, the pluralist advocates vigilance when 'foreign' elements are introduced, and can suggest a protocol to counter-check such cases against triviality. When pluralism in methodology is being practiced, he makes two recommendations:

1. Know what counts as a strict proof within a mathematical theory.

Axiomatisation is fine as a starting point, but axiomatisation is just about being explicit as to one's justification, it is not to provide an ultimate justification. This is important because the gaps in an otherwise strict proof will signal deviation. The pluralist does allow deviance in proof. The trick is that we have to know when we are being deviant and when we are not. As explained in the last section, the vigilance can be made quite systematic, it is not just a vague call to caution.

2. We should bear in mind that when revising, correcting or being critical of a result we should look first to the deviant steps.

Then, as with Lobachevsky, we might be able to show that there is no inconsistency.

However, the analysis of the pluralist does not stop here. We might need to re-evaluate the original theory. For example, if our mathematical model of some part of physics (the application of mathematics to 'reality') predicted a certain outcome, and we found that the outcome was not what was predicted by the mathematics, then we look first for an error in calculation, second, we look at any deviant elements in the making of the faulty prediction, but thirdly, we might look at revising the whole theory – making a new one, adding axioms, adding rules of inference, modifying or eliminating existing ones, *etcetera*. This is where the pluralist goes beyond the formalist. This was the recommendation made in the second test case.

In conclusion, methodological pluralism better describes mathematicians' practice than does formalism. We do not need a unified presentation of theories, and we do not need methodological rigidity (to adhere to a particular proof theory) to guarantee consistency when we can use 'reality', physical theory or another mathematical theory such as hyperbolic geometry, to sanction the methodological deviance in proof.[21] But when we use deviant methods, we continue our search for mathematical justification, holding the temporary 'result' (deviantly obtained) in abeyance.

The pluralist shares with the formalist, the celebration of creativity in mathematics, and agrees that the freedom can be enjoyed responsibly or irresponsibly. Ideally, the symbols and their manipulation is held rigid, and to a high standard of rigour. See the chapter on rigour for details. But whereas the formalist stops with a unique proof system as a measure of good mathematical practice, the pluralist is more liberal. Nevertheless, he offers council when we are deviant in our methods of proof. Having an ideal standard is not binding, for the pluralist, but rather, it allows us to recognise when we are deviant. We check if we are just making a short-cut, in

[21] A pluralist does not insist on consistency, since, e.g., he endorses paraconsistent theories. However, he endorses and encourages crosschecking with other theories.

which case we are convinced that we can fill the gap with a proof of a lemma, or even refer directly to the proof of the lemma. If we are not certain we can fill the gap, then we should check we are not flirting with triviality. In this case, again, we have a protocol. In other words, rather than hold mathematicians to a rigid standard of rigour, we use the standard of rigour to make us aware of deviance. This is simply practical, and reflects current practice in mathematics.

References

Beall, J. C., & Restall, G. (2006). *Logical pluralism*. Oxford: Clarendon Press.

Bourbaki, N. (1991). *Elements of the history of mathematics*. (J. Meldrum, Trans.). Berlin: Springer.

Brouwer, L. E. J. (1975). In A. Heyting (Ed.), *Collected works. Vol. 1: Philosophy and foundations of mathematics*. Amsterdam/New York: North-Holland/American Elsevier.

Brown, B., & Priest, G. (2004). Chunk and permeate, a paraconsistent inference strategy. Part I: The infinitesimal calculus. *Journal of Philosophical Logic, 33*(4), 379–388.

Corcoran, J. (1973). Gaps between logical theories and mathematical practice. In M. Bunge (Ed.), *The methodological unity of science* (pp. 23–50). Dordrecht: Reidel.

Dales, H. C. (1998). The mathematician as a formalist. In H. C. Dales & G. Olivieri (Eds.), *Truth in mathematics*. Oxford: Clarendon Press.

Detlefsen, M. (2005). Formalism. In S. Stewart (Ed.), *The Oxford handbook of philosophy of mathematics and logic* (pp. 236–317). Oxford: Oxford University Press.

Hallett, M. (1995). Hilbert and logic. In M. Marion & R. Cohen (Eds.), *Québec studies in the philosophy of science, Part 1: Logic, mathematics, physics and the history of science* (Boston studies in the philosophy of science 177, pp. 135–187). Dordrecht: Kluwer Publishing Company.

Hilbert, D. (1923). The logical foundations of mathematics. In W.B. Edwald (Ed.). (1996). *From Kant to Hilbert: A source book in the foundations of mathematics*. II (pp 134–148). Oxford: Clarendon Press.

Hilbert, D. (1996). The logical foundations of mathematics. In W. B. Edwald (Ed.), *From Kant to Hilbert: A source book in the foundations of mathematics* (Vol. 2, pp. 1134–1148). Oxford: Clarendon Press.

Hrbacek, K., Lessmann, O., & O'Donovan, R. (n.d.). *Analysis with ultrasmall numbers* (in prep.) (Manuscript, 2009).

Kagan, V. F. (1957). *N. Lobachevsky and his contribution to science*. Moscow: Foreign Languages Publishing House.

Katz, V. J. (1998). *A history of mathematics; an introduction* (2nd ed.). Reading: Addison-Wesley.

Lobachevsky, N. I. (1914). *Geometrical researches on the theory of parallels*. Chicago: Open Court Publishing Company.

Maddy, P. (2007). *Second philosophy*. Oxford: Oxford University Press.

Moschovakis, Y. N. (1980). *Descriptive set theory*. Providence: American Mathematical Society.

Nelson, E. (1997). 'Confessioni d'un matematico apostata'. Nuova Civiltà del le Macchine, 243–247. Published online in English. http://www.math.princeton.edu/nelson/papers.html

Otte, M. (1990). Intuition and formalism in mathematical proof. *Interchange, 21*(1), 59–64.

Pedeferri, A., & Friend, M. (2012). Are mathematicians better described as formalists or pluralists?. In *L&PS – Logic and philosophy of science; Selected proceedings of the SILFS 2010 international congress* (Vol. IX, No. 1. Section 2). http://www2.units.it/episteme/

Raussen, M., & Skau, C. (2003). Interview with Jean-Pierre Serre. *European Mathematical Society Newsletter, 9*, 18–20.

Robinson, A. (1965). Formalism 64. In *Logic, methodology and philosophy of science. Proceedings of the 1964 international congress* (pp 228–246). Amsterdam: North-Holland.

Rodin, A. (2008). *Did Lobachevsky have a model of his imaginary Geometry?*. http://philsci-archive.pitt.edu/4099/1/loba8.pdf

Tennant, N. (2010). *Why arithmetize the reals? Why not geometrize them?* Manuscript.

Troelstra, A., & Schwichtenberg, H. (2000). *Basic proof theory*. Cambridge: Cambridge University Press.

Vopĕnka, P. (Forthcoming, Manuscript, 2013). *The great illusion of 20th century mathematics and its new foundations*. Springer.

Part II
Initial Presentation of Pluralism

Chapter 6
Philosophical Presentation of Pluralism

Abstract In part I of the book I gave motivations for adopting pluralism, from the starting point of several well-known philosophies of mathematics. I drew inspiration from them and rejected some of their claims as unwarranted. But we still do not have a clear picture of what pluralism is as a philosophical position in its own right. In this part, I give an initial presentation of the position over the course of four chapters. In this chapter I answer some pressing questions. I begin with the notion of tolerance. This invites discussion on three issues. One is normativity, the second is general organisation of the types of philosophical issues addressed by the pluralist, and the third is restrictions. In particular, I open the issue of how it is that the pluralist will cope with contradiction, and in what respect a paraconsistent logic can help the pluralist.

6.1 Introduction

A pluralist in the philosophy of mathematics is someone who places pluralism as the chief virtue in his philosophy of mathematics. He brings the attitude to bear on mathematical theories and on different philosophies of mathematics. Pluralism is founded on the conviction that we do not have the necessary evidence to think that mathematics is one unified body of truths, or is reducible to one mathematical theory (foundation).

The pluralist is simply agnostic on this issue, but, for all that, does not think that we have to give up trying to do philosophy of mathematics. *Au contraire*, it is in the light of this agnostic attitude that pluralism is developed. Thus, the position can be characterised as a non-radical scepticism. By this light, the pluralist is free to take an interest in mathematics as a series of theories, where each contains truths-relative-to-that-theory. Or, the pluralist is free to think of mathematics as a process, as opposed to concentrating on mathematics as a unified body of truths.

M. Friend, *Pluralism in Mathematics: A New Position in Philosophy of Mathematics*, Logic, Epistemology, and the Unity of Science 32, DOI 10.1007/978-94-007-7058-4_6,
© Springer Science+Business Media Dordrecht 2014

Most of this has already been said in earlier chapters of this book. What I have mentioned but not emphasized is the normative element in pluralism, and I have only made a start on conveying the scope of the pluralist's interests. I have also said very little about how, or why, pluralism works. So, the topics of this chapter are normativity, scope and the structure of the family of pluralisms. I use the plural because pluralism is pluralist about pluralism, but we shall see this in more detail only in Chap. 11.

This chapter will proceed as follows. In the first section, I discuss pluralism and normativity. We shall see that the pluralist is normative, not only about mathematical practice, but more importantly, about other philosophies of mathematics. We discover that the pluralist is pretty liberal in his endorsing of mathematical practice, and therefore we return to the idea of 'bad' mathematical theories in the interest of exploring the scope of pluralism in the second section. Bad mathematical theories were mentioned in Chap. 4. The discussion continues here. We look at intensional theories and why these are important for the philosopher of mathematics. In Sect. 6.3 we then turn to the inner structure of the family of positions which are versions of pluralism. We introduce a notion of levels of discussion in both mathematics and philosophy. We discover that, for example, most traditional philosophies of mathematics occupy the second level, whereas maximal pluralism – as introduced in Chap. 4 – occupies the third level. This is why the pluralist is able to discuss most other philosophies of mathematics without being directly threatened by their positions; but more will be said on this matter in Chap. 10, when we address the paradoxes of pluralism. In the last section we underpin maximal pluralism, or what we can also call 'third-level pluralism', with a standard, and well-tested, paraconsistent logic, LP.

6.2 Normativity for the Pluralist

The pluralist is 'softly' normative about specification, precision and communication.

'Soft' normativity is simply encouragement, which comes from an aspiration (to make statements as clear as possible); as opposed to *setting* a norm, and holding oneself and others to that standard.

The pluralist is normative (tout court) about some truths within a theory[1]

The normativity of the pluralist affects the scope of his interest. Because the pluralist is not normative in the sense of fastening on a particular mathematical theory to set a norm, he widens the scope of interest beyond that of more traditional

[1]I write 'some' because we have to make room for undecidable statements. One way to do this, of course, is to give such statements the truth-value 'U', for 'undecidable', however, if we are being careful, 'U' is not a truth-value – it is not a 'value' at all. It is the absence of value.

philosophies. He is not only interested in successful and what is considered to be 'good' mathematics, but also in what is considered to be 'bad' mathematics, and how the 'bad' parts inform the 'good' parts. Note that 'good' and 'bad' are in quotation marks; this is because these value judgments are not *made* by the pluralist. Rather, they are *observed* by the pluralist to be made by mathematicians and non-pluralist philosophers.[2] They will not be endorsed, except as technical abbreviations for a quite precise qualification about how the judgment was arrived at. That is, for the pluralist, there is no 'good' mathematical theory as such, there is only a theory, which is good according to: the community of mathematicians or some qualitative measure, which has been specified – or is in the process of being specified.

This attitude towards the use of the terms 'good' and 'bad' instanciates the pluralist's soft normativity. By calling for qualification of the terms 'good' and 'bad' we are forced to be more specific, and therefore, usually clearer. The act of qualifying terms like 'good' defuse the normativity of such terms and turns them into a *description*. Of course, there is a threshold beyond which specifying becomes useless, since a reader or interlocutor cannot hold very much information, but this is a practical problem about communication and its effective limits. We treat of this issue in Chap. 8. Regardless of our limitations, if we avoid the labels 'good' and 'bad', or if we qualify them, then we are liberated from the psychological normative effects of the labels. It can be philosophically useful to drop, or question, the normative words, since this allows us to make calm observations. For example, we can observe that it turns out that some mistakes in mathematics (errors in proof, in a conception, falsehoods of a theory) are very fruitful! Even mathematical falsehoods (which are usually dismissed as 'bad') can be fruitful, and this is acknowledged by Byers.

> Mathematics is so commonly identified with its formal structure that it seems peculiar to assert that an idea [in mathematics] is neither true nor false. What I [Byers] mean by this is similar to what David Bohm means when he says "theories are insights which are neither true nor false, but, rather, clear in certain domains, and unclear when extended beyond those domains." [(Bohm 1980, 4)]. Classifying ideas as true or false [or as good or bad] is just not the best way of thinking about them. Ideas may be fecund; they may be deep; they may be subtle; they may be trivial. These are the kinds of attributes we should ascribe to ideas. Prematurely characterising an idea as true or false rigidifies the mathematical environment. Even a "false" idea can be valuable. For example, Goro Shimura once said of his late colleague Yutaka Taniyama, "He was gifted with the special capability of making many mistakes, mostly in the right direction. I envied him for this and tried in vain to imitate him, but found it quite difficult to make good mistakes" [(Singh 1997, 174)]. A mistake is "good" precisely because it carries within it a legitimate mathematical idea. (Byers 2007, 256–257)

[2] 'Observe' is ambiguous between obeying, as in "Paolo observes the laws of his country" and "make an observation". I mean 'observe' in the latter sense.

If fruitfulness is 'good',[3] then some mistakes in mathematics are also 'good'.[4] For the pluralist 'good' and 'bad', as they are usually used, are empty terms, or rather, they are place-holders. They cry out for further qualification or explanation. In thinking this way about the terms, the pluralist is softly normative about the use of 'good' and 'bad' in mathematical or philosophical writing. As a result: 'faulty' theories, errors or 'wrong' proofs, which have been ignored in the past by philosophers,[5] are not so easily brushed aside by the pluralist. Revisiting a value judgment is always an option.

6.3 Scope: Intensional Mathematics

As we saw in the chapter on structuralism, in general, 'bad' mathematics include: intensional theories. Exactly what these are is controversial. In order to avoid the controversy, we shall simply stipulate that what we are considering here to be intensional theories are[6] formal (usually logical) systems with intentional operators,[7] where the operators have a whole wff in their scope, not just a term. This makes the inter-substitutivity of terms in wffs of the theory 'opaque'.[8] Opacity of this sort is an indicator of what we shall call an 'intensional logic'.[9]

---

[3]This is not meant as a silly point. Quite often philosophers will defend a part of mathematics, or science, on the ground that it is fruitful. In light of the fruitfulness of some mistakes, it behoves philosophers who value fruitfulness as a property of a theory to study fruitful mistakes in mathematics, as Lakatos suggested.

[4]Like 'good' and 'bad', 'true' and 'false' can hold normative sway. They do not when they are treated mechanically, or 'syntactically', such as when we talk of assigning truth-values exhaustively, or randomly to wffs. This artificial and mechanical use of 'true' is quite different from the use of 'true' when we say, for example, "this foundation is the true foundation of mathematics".

[5]Lakatos is a notable exception.

[6]Here, we shall restrict our discussion to these. The relationship between intensionality, intentionality and extensionality is complex and discussed with little consensus in the literature. I shall not settle any disputes here. Instead, I shall just indicate that some logical theories ignored by mainstream mathematical theories are still worthy of being thought of as mathematics. Moreover, considering intensionality gives us the opportunity to say a little about one of the thresholds of communication. This is important for the pluralist, since the communication of mathematical ideas is part of mathematics.

[7]Model theory is extensionalist, and only individuates structures and objects in those structures 'up to isomorphism', only recognizing certain properties (predicates, relations, functions) as 'counting' for mathematics. But we find, in mathematical practice, that considerations, not recognized by model theory, are also pertinent to mathematics. See Rodin (2008, 25) for a carefully discussed example.

[8]They are opaque in respect of reference to objects since terms refer to objects or sets of objects. Whole sentences, or wffs represent states of affairs, which might or might not obtain.

[9]In the preface, I made it clear that the pluralist is not interested in drawing a firm distinction between logic and mathematics. Here logical theories – formal theories (of logic) – are considered to be mathematical theories just as a particular geometrical theory might be.

Definition An *intensional logic* is one that includes intentional operators that take a whole wff as their scope (as opposed to just a term).

Definition An *intentional operator* is a logical operator that is meant to express an intention, or attitude, such as: doubt, belief, fear or *de dicto* possibility.[10]

Sometimes the last are called 'propositional attitudes' because they have a proposition in their scope (as opposed to a term).

Elaborating on this notion of the scope of an operator: in general, the scope of an operator can be terms or wffs. Different formal systems of logic have the operators range over different scopes. The differences will capture conceptual nuances. For example, it is quite possible to fear the devil (in which case the intention is directed towards a presumed object, referred to by the singular term 'the devil'). It is also quite possible to fear *that* the devil has possessed the dog. In the second example, the intension has a proposition in its scope, rather than a term. The person fears that the dog is possessed. She does not immediately fear the devil (according to the sentence). The person might fear the devil also, but this is an inference from the devil's ability to possess the dog, and our 'common sense knowledge' about devils, such as it might be. What is important is that intensional logics are not always recognised as mathematical because they are not extensional – *terms* are not identified with the isomorphism class of their extensions, because the term is hidden inside the proposition and cannot be picked out and substituted for.[11]

An unreflective supposition that mathematics is essentially extensional, sometimes rests either on thinking that some set theory, which comes equipped with an axiom of extensionality is the foundation for mathematics, or that model theory, which is implicitly extensional, is the way to determine what counts as mathematics. This type of extensionality precludes intensional logics, as defined above. Thus, using an extensional theory as a ground for supposing that mathematics is extensional just begs the question against including intensional theories in mathematics, since we defer to an extensional theory to determine the scope of 'mathematics'.[12]

A more sophisticated (non-question-begging) position is that of the more reflective extensionalist.

---

[10]'Truth' could also be thought of as such an operator, but to discuss this here would only muddy the waters unnecessarily.

[11]Given our definitions, extensional theories and intensional theories are contradictories. If we had a broader definition of intensional – to include intentional operators with whatever scope then extensional would be the sub-contrary of intensional (all theories have to be one or the other, but can also be both).

[12]This question-begging argument rests on two suppositions. One is that a theory with an axiom of extensionality (or equivalently, a theory which is implicitly extensional) is an extensional theory, and two, that extensional theories in this sense cannot also be intensional. More concisely: 'extensional' and 'intensional' are contraries. The definition I give of 'intensional theory' does some simplifying work here. In Chap. 13, we revisit this issue, adding more nuances.

An extensionalist is *reluctant* to include intensional theories in mathematics.[13]

> A holder of this position thinks that mathematics or logic *should* be made as extensional as possible, so extensionality is a regulative ideal.

This makes the extensionalist a dualist. There are the good extensional mathematical theories, and the leftover intensional theories that we try to make clean by re-representing them in an extensional theory. This view is supported by an idea of historical *progress* being made in logic and mathematics. By surveying the history of logic we notice that logic has become more 'pure', specific, clear and transparent because newer theories are increasingly extensional. This supposition (Bar-Am 2008) rests on the idea that progress is identified with increasing clarity, which in turn, is identified with formal representation, in aide of mechanical manipulation, which in turn, depends on formal theories being extensional.[14] As a general regulative ideal, the pluralist has no qualms about this form of extensionalism. Where the pluralist takes issue with the extensionalist is with the *identifying* clarity with extensional formal theories.

The reason the pluralist takes issue is that he recognises that there is a clarity/extensionality threshold, and both sides of the threshold are worth exploring. There might well be circumstances where forcing extensionality is too artificial, where a particular precisification is made at the expense of a more fruitful concept. For example, implication, is, arguably, not an extensional notion (Priest 2006b, 73). But in classical, and other, presentations of formal systems '$\rightarrow$' is introduced as the symbol for implication and is inter-substitutable with other symbols in the language, and so is supposed to be extensional. It is *not* distinguished from the conditional, and, thanks to the theorem of deduction, is also taken to be conceptually equivalent to entailment (the difference is that they occupy a different level of language and relate wffs, in the case of $\rightarrow$, and relate premises to a conclusion in the case of entailment). Arguably, this blurring of distinctions between implication, the conditional and entailment, in the name of extensionalism,[15] has led to a loss in our understanding of logic and our misuse of logic (Sundholm 1998, 184). We shall visit this issue specifically in Chap. 13.

What we should retain here, is that there are *different ways* of making *clear* what we say, sometimes it is useful to given an extensional definition, and sometimes it is not. Sometimes an extensional logic, or mathematical theory is clear, but sometimes not. Sometimes we seem to have greater transparency, since we have completely effective rules of manipulation; but sometimes this is just an illusion. We have easy manipulation rules, and transparency, but not necessarily *clarity* or

---

[13]It is not really a whole philosophy. It is an attitude: a striving towards increased extensionality in mathematics and logic. Quine, and more recently Bar-Am defend extensionalism. For reasons of definiteness, I shall concentrate on Bar-Am's defence of extensionalism.

[14]I should add that Bar-Am does concede that it is not at all clear, for him, that either mathematics or logic can be made fully extensional. To discuss this further we would have to work the definitions in much greater detail than we have here.

[15]Or so the extensionalist will claim.

greater understanding of the original notions, which might have been irreducibly intensional (especially if we loosen the constraints I place on the definition of an intensional theory to include *any* theory with intensional operators, including the relevant conditional). That is, there is a threshold we cross when we try too hard to represent a notion formally in an extensional theory. The threshold is where we have a trade of loyal representation of the *original* concept against transparency in the form of a symbol, which we can easily understand and manipulate. For the pluralist, the trade might good, if the original concept was hopelessly confused, and needed disentangling. The trade might be bad if we have lost concepts and understanding. How we evaluate, and weight the balance is not always an easy matter.

Not only are there different ways of clarifying, and thresholds beyond which no more clarification can be had, but there are also different degrees of precision, or explanation of background. Rav describes the following process:

> If some reader wants or needs more details, as for instance concerning modular arithmetic [in a proof to show that $1+1=0$ (mod 2)] it can be provided by giving further explanations, as is done in teaching unprepared students. In principle, though, one could go through the whole development of Peano Arithmetic, develop modular arithmetic and what not. How far one has to go back in one's justification of an inference is a pragmatic question; *there is no theoretical upper bound* on the number of *interpolations* necessary for an absolute justification (whatever that would mean). (Rav 2007, 313–314)

Suffice it to say, here, that for the pluralist, the process of clarification is not just one thing. The extensionalist is aware of this too, by the way. This is why he has a sophisticated position. Where he and the pluralist part company is over whether or not an extensional formal theory is *always* to be preferred to an intensional one. We shall see similar subtleties with the notion of rigour. Moreover, it is frustrating, but important to take note of the variability of the process of clarification, since, if we ignore this subtlety we run the risk of making formal theories extensional to the detriment of understanding. Why is this important for mathematics? If, as Priest argues, the conditional, implication or entailment, or any one of these is intensional, then the intensional is present in, and systemic throughout, mathematics. To not include a proper treatment of intensionality in mathematics or in a philosophy of mathematics, would be an oversight. How can the pluralist afford so much scope in his philosophy? We now turn to the structure of the family of pluralist positions.

6.4 Maximal Pluralism, or, Third-Level Pluralism

In Chap. 4 we argued for maximal pluralism over optimal pluralism. Here, I fill out the view of the maximal pluralist. In order to be a maximal pluralist, we should distinguish three levels of philosophical activity and three corresponding levels of mathematical activity. Collecting the philosophical and the mathematical, we shall call these 'levels of enquiry'. The first level of both mathematics and philosophy concerns particular results in mathematics. Examples on the mathematical side are particular theorems, lemmas, definitions and proofs. On the philosophical side

Table 6.1 The levels of pluralism in mathematics and philosophy of mathematics

Level	Mathematics	Philosophy
1	Particular results: proofs, theorems, definitions . . .	Particular claims made within a philosophical position
2	Full mathematical theories, for example: Euclidean geometry, modal logic S4, ZF set theory, Topos theory . . .	Full traditional philosophical positions, for example: naturalism, logicism, constructivism
3	Meta-discussions about 'mathematics'	Structuralism (as optimal pluralism), maximal pluralism

we have discussions concerning particular results. For example, we might discuss: theorems, definitions or the completeness of a theory, a compactness result or a proof in a theory. These might include discussions about limitative results, since these results are given within a particular mathematical theory. Thus, we might include the proof of equi-consistency of two theories at this first level if we are looking at the proof, since it occurs *in* a theory.

At the second level, we have full mathematical theories, such as: Euclidean geometry, first-order arithmetic, modal logic S4, Zermelo-Fraenkel set theory and topos theory. The larger of these mathematical theories are theories *within which* we make mathematical comparisons between other (smaller) theories.[16] For example, we might show the reduction of one theory to another, we might give an equi-consistency proof between two smaller theories, we might show embeddings, we might show how one theory differs from another in virtue of one axiom and so on. The larger whole theories are often thought of, by philosophers, as 'foundational' and are sometimes accompanied by a philosophy. For reasons stated in Chap. 2, we call these 'big' theories. Accompanying big theories at this level, on the philosophical side we have the more traditional philosophies of mathematics, such as: set theoretic realism, Maddy's set theoretic naturalism, the constructive philosophies, logicism and so on. In fact, most positions in the philosophy of mathematics are found at this level. We can make pluralist investigations at this level, provided we bear in mind the third level.

In *concreto*, we can tell apart pluralist from non-pluralist investigations at this level by remarks made about 'truth', 'foundation', 'ontology', 'correctness' and so on. The non-pluralist makes no qualifying comment when using such terms. The pluralist does. He entertains the 'non-pluralist' position temporarily, for the sake of argument, or as a hypothesis. At the end of the day, he is a principled agnostic about the sorts of terms listed – unless he has been convinced and abandoned his pluralism. Structuralism is a separate case. The structuralist occupies the third level, but only as an optimal pluralist. See Table 6.1.

The levels are not hard and fast. In order to allay misunderstanding, it might be worth making a comparison of the pluralist 'levels of enquiry' with

[16]The terms 'smaller' and 'larger' refer to the expressive power of a theory. Roughly, the more theories can be reduced to, or embedded in a theory, the more expressive power the theory has.

Tarski's semantic levels. Tarski's levels of object language, meta-language, meta-meta-language and up, were postulated partly as a result of observation – we do talk *about* language, and when we do this we need the resources to refer to everything in the object language talked about *etcetera*. The other reason, maybe the principle reason, was to block a number of the semantic paradoxes. Tarski thought it very important to avoid paradoxes, since they "force us to say falsehoods".[17] For this reason, and since he was a classical logician, he needed a very rigid structure of languages.[18] Each level collects everything at the lower level, and might add more material. All reference to linguistic or semantic entities is made downwards. It is (supposed to be) always clear which level one is talking in, or writing in.

This is not so for the levels of enquiry of the pluralist. The notion of levels is comparatively lax and flexible. There might be cases where we cannot decide if we are at level one or two, for example. We also might not be able to decide, when we are having a philosophical or a mathematical discussion, whether our problem, or issue, can be resolved purely mathematically or will require some philosophy and so on. The fuzziness of the level concept is not always important, and if need be, we can make a decision, or tackle a problem at different levels and use different approaches. Problems and answers are not always known to be of a particular level or type, but we *can* fix one for convenience. Admittedly, the convenience will be temporary, and we should bear in mind that we should revisit the parameters set, at a future date. See Chap. 10 for more details.

It is also worth noticing that the potential for paradoxes arising from the lack of strictness in regimentation of the levels is not to be thought of as a calamitous drawback. This is because of our making reference to a paraconsistent logic, so, in particular, we do not adhere to the dogma[19] that contradiction entails triviality. The pluralist is also not as disturbed as Tarski is, about uttering falsehoods. We do so all the time, at the very least because of inaccuracy. Falsity might also be a driving force behind development in both mathematics and in the philosophy of mathematics. As pluralists, we usually want to avoid uttering and writing falsehoods, but we accept that we do write them. In fact, as we saw in the passage quoted from Byers, sometimes mistakes are very fruitful. That is why, what we claim is always thought of as highly revisable. Put another way, we are softly normative about true and false claims.[20] We strive towards justified statements, we even strive towards very stable justified statements: ones which it would take a lot of work to seriously revise. In this way, proofs help us to stabilise a theorem. Proofs made in a stable

[17] I do not have the written reference for this. Wolenski mentioned it to me in conversation at the Logica conference in 2005.

[18] While Tarski was a classical logician, he was well aware, and quite sensitive to constructivist concerns.

[19] Priest argues for this at length in Priest (2006a, b).

[20] Here 'true' and 'false' are not meant in the sense of 'theorem of a theory' (ignoring incompleteness) or 'sentence inconsistent with a theory', but rather, as a value judgment made by philosophers of mathematics in general, at any level. That is, we are using the terms 'true' and 'false' loosely here.

theory, such as Peano arithmetic, are about as stable as possible. But we admit, that we cannot know in advance that any particular claim is completely stable or very stable. At best, we might have our suspicions, and they might, in turn, be justified or not.[21]

To summarise: the maximal pluralist occupies the third-level of enquiry. The maximal pluralist looks at first and second-level normative and unqualified statements as highly revisable, and aspires to make his statements as stable as possible through careful rigorous justification and explicitness. The second-level pluralist joins the foray competing against other philosophical positions. Insofar as he is agnostic and sceptical about his own position he is also mindful of third-level pluralism. Second level pluralism is unstable without third level pluralism.[22] So, for example, the claim that "ZF is the orthodoxy of present day mathematics" is made with emphasis on the 'present-day' (so this might very well change in the future), and 'orthodoxy' is taken to be socially, or institutionally indicated, and so varies, across societies: different schools and associations, and varies more obviously if we consider past institutional or social groups of mathematicians. The second-level pluralist is agnostic about the future, and again, he holds his pluralism on account of being aware of third level pluralism.[23]

6.5 Third-Level Pluralism and Paraconsistency

6.5.1 Third-Level Pluralism

At the third level, we have maximal pluralism: a philosophy that is pluralist towards the activity which takes place at the first and second level. We also have a corresponding *mathematical* activity at this level. The latter comes in two guises. One is when we use mathematics from different areas to 'solve a problem' so we help ourselves to tools and mathematical ideas from different mathematical theories. There is discussion of this phenomenon in Chap. 5, and more in Chap. 9. The other guise is when mathematicians identify transcendent (common to many big theories) ideas, or when they compare big theories and have a discussion at the meta-level about all mathematical theories. The discussion can be mathematically or logically informed. Logic is sometimes thought of as the study of universal laws of

[21] This should be taken with a pinch of salt. In fact there is a lot of stability in mathematics – probably more than in any other area of enquiry. However, the pluralist claims that the stability is not ensured by truth or ontology. Rather it is ensured by the cross-checking nature and process of mathematics. This is why proofs are so important!

[22] The instability is discussed in Part II.

[23] Naturalism and structuralism could arguably be thought of as pluralist at the second level. The argument would engage such phrases as "model theory is the prevalent theory today for ...". With the right sort of emphasis, Maddy and Shapiro can both be construed as pluralists.

reasoning, so ones which would transcend all mathematics. Even if we do not insist on the universality of logic, we can still identify notions which are common to many mathematical theories. But logic is not the only sort of generalisation which allows us to identify commonalities between theories. For example, a category theorist might compare the category of the universe of the iterative hierarchy of sets with the universe of Aczel's non-well founded universe of sets.

Pluralism towards maximal, or third-level, pluralism is pluralism at the fourth level, and is the subject of Chap. 11. In this chapter, we restrict ourselves to pluralism towards the first and second levels. It is what we called maximal pluralism in Chap. 4. But now, we can re-define it in terms of the levels.

Definition *Third-level pluralism* is pluralism towards at least:

(i) mathematical activity at the (first level) of working within a mathematical theory, or working with several mathematical theories to prove or verify purported theorems,
(ii) mathematical activity at the (second level) of developing whole mathematical or logical theories, or working within a theory to compare 'smaller' theories to each other,
(iii) philosophical work concerning particular results or notions in mathematics, such as work on the notion of compactness, with, or without, having any particular philosophical tradition informing the work, and
(iv) philosophical work at the (second) level of developing a foundational philosophy of mathematics.

Third-level pluralism includes a set of attitudes, amongst which, we find an avoidance of dogmatism, in favour of qualification and clarification. One by one, dogmatic claims are replaced by careful explanation that justifies (and shows the limitations of) what was stated as a dogmatic claim.[24] Not any explanation will do. These too, in turn, can be qualified and justified – as the need arises. This is discussed in detail in Chaps. 8 and 9. Before turning to these, there is a more immediate problem.

Clearly, in both mathematics and philosophy, conflicts will arise between claims. Some are easily resolved by appeal to clarification. Others are not. Some conflicts are just brute entrenchments. Or, more mildly, there is no reason to suppose in advance, that through a process of clarification, disambiguation or qualification, that we can *always*, even *eventually* reach an agreement. Since the pluralist is (publicly) agnostic about the outcome of the process of debate, deliberation and clarification, he has to make some sort of accommodation for sentences of the form α and not α, where α is a claim made at the first or second level. Let us call any sentence of the form 'α and not α' a contradiction.

[24]The dogmatism of exactly this assertion will be discussed in Chap. 10.

Definition A *contradiction* is a sentence of the form 'α and not α'.[25]

Here, we are particularly interested in contradictions arising at second or third level. Much to the chagrin of realists and other monists, mathematics has developed piecemeal. Paraphrasing Priest:

> ... language [we paraphrase with 'mathematics'] and the principles that govern it have developed piecemeal and under no central direction. As logicians know, inconsistency is the natural outcome of spontaneity. Consistency has to be fought for. Therefore *prima facie*, it would be surprising if our [mathematical] concepts were internally and mutually consistent. (Priest 2006b, 5)

Pluralists are aware of many contradictory remarks one can legitimately make in mathematics and in the philosophy of mathematics.

Let us take an example. We have several times made the point that the pluralist combines anti-foundationalism with an interest in foundations – as good mathematical theories in their own right, and as accompanied by philosophies of mathematics, which affect the development of mathematics. A pluralist who took an interest in 'foundations' might look at, say, axioms that are independent of a proposed particular big mathematical theory, such as the higher-cardinal axioms. The pluralist observes the bifurcations of set theories with the addition of different sets of axioms. The pluralist will not need to favour one extension over another. If asked to pick one, or favour one, he demurs, unless he is given a clear criterion by which he can favour one. Note that this demurring is not only due to a personal or collective 'lack of knowledge', but, rather, it is due to an acceptance that at the present state of play in mathematics, there simply is no definitive mathematical way to arbitrate between theories. To varying degrees, the pluralist accepts that there is no unique absolute perspective.[26] Moreover, an exercise in clarifying and justifying will not *necessarily* end in one side 'winning'. There is no reason to think *in advance*, that there will be such an outcome, even *in the long run*. We might, therefore, have statements of the form: "(set theory is best extended by adding axioms X and Y) and it is not the case that (set theory is best extended by adding axioms X and Y)" – where 'set theory' and 'best' are the same in both conjuncts. To make sense of the contradiction, 'set theory' and 'best' might not be specified sufficiently precisely to determine which conjunct is correct, or, what is more interesting, it might not be clear that we can just explain away the contradiction. Or, we might think that 'best' and 'set theory' are unstable over time. One theory might appear to be best at one time, but that judgment might shift back to the other theory at another time. Settling

[25] Such claims made at third level will be discussed in Chap. 11.

[26] It might be instructive to compare this attitude to Gödelian optimism, which is the thought that in the end, given an open problem, we shall discover a technique to make an absolute decision about that problem. Tennant has several good discussions about the Gödelian optimist (Tennant 1997). In contrast, here, we have the agnostic, who demurs. This character is either a pessimist (the demurring is then based on an inductive argument, and the pessimism might be reversed in a particular instance), or the character is a principled agnostic. It is the principled agnostic position that is explored in this chapter.

on one conjunct over the other might be thought of as artificial, and not capturing the 'deep' philosophical dilemma. All we need is the *possibility* of this persistent contradictory situation for our appeal to a paraconsistent logic.

6.5.2 Paraconsistency

Since there might be persistent contradictions, pluralism requires a paraconsistent logic at the third level.

Definition A logic is *paraconsistent* iff it is non-trivial and blocks *ex contradictione quodlibet* inferences.

An example of a paraconsistent logic is a relevant logic, which insists that there be some traceable 'relevant' connection between premises and conclusion. There are many relevant logics, each deploying a different strategy for blocking *ex contradictione quodlibet*. See Appendix 3 for two *ex contradictione quodlibet* proofs and indications of where they would be formally blocked by relevant logicians.

Definition A proof is an instance of *ex contradictione quodlibet* if, in it, from a contradiction an unrelated conclusion is drawn. $p \& {\sim}p \vdash q$, is a valid deduction in a classical or intuitionist proof system. It is not valid in a relevant or paraconsistent logic.

The reason we need to block *ex contradictione quodlibet* is that what it says is: "from a contradiction, anything (grammatical in the language) follows". The pluralist has admitted that persistent contradictions are possible, of the form 'α and not α' at first and second level. Referring back to the discussion in Chap. 4 on truth: for the pluralist, α might be of the form s or Ts. (As a reminder: s is any wff in a language, 'Ts' specifies the theory T, so s is true in T.) We shall have to consider the candidate cases very carefully, since some are only apparent contradictions.

When the α is of the form s, then it is usually quite easy to resolve the conflict, and we see that s is a theorem in one theory, but not in another. For example, $2+8=10$ is true in Peano arithmetic, but it is not true in arithmetic mod 8. So in this case, where α is of the form s and is: $2+8=10$, we can fully form the contradiction: $(2+8=10)$ & not $(2+8=10)$. Once we add the T: 'Peano arithmetic' to the left side of the conjunction, and the T' 'arithmetic mod 8' to the right side, the contradiction is shown to be only apparent since we have Ts & $T's$, and this is not a contradiction since $T \neq T'$. A little qualification and clarification is all we needed. Consistency, where it can be had, should be striven towards, but this does not imply that it should be striven towards come what may – to the point of dogmatism or artificiality – in the form of disallowing contradictions to stand at all (even temporarily).

Since the pluralist eschews the dogmatism concerning contradiction, he thinks that it is not necessarily the case that all (apparent, or *prima facie*) contradictions can be so resolved, or resolved at all. He is agnostic on this issue. This is not only for reasons of lack of knowledge. We might have available all possible information, which bears on α and its negation or denial, but there might be other reasons why the contradiction is persistent, or even permanent. We cannot appeal to logic forbidding contradictions, since this would beg the question against the paraconsistent logician. That is, in the light of rigorous non-trivial formal systems of logic where there are contradictions from which we can deduce some formulas, but not all formulas, appeal to the law of non-contradiction as governing formal systems of logic, is simply false or question begging. Of course, it would be audacious in most or all cases to claim that a contradiction were permanent, it is wiser to demure again.[27]

Instances of persistent contradictions are the semantic paradoxes or the set theoretic paradoxes.[28] In the case of paradoxes, we have a contradiction of the form Ts & not Ts. If the theory is classical, then by *ex contradictione quodlibet*, the theory is trivial. An example is Frege's famous formal theory of logic. From Basic Law V it is possible to derive a contradiction. Frege's theory is classical, so does not block *ex contradictione quodlibet*, and therefore, Frege's formal theory is trivial.

Paradoxes do not always entail triviality. In particular, they do not in the context of a paraconsistent theory. But they do not in philosophy either! When we discuss, and entertain, paradoxes in the meta-language English we do not lapse into triviality.[29] Most of the time, such discussions are serious, rule governed and statements are subject to correction – which would not be the case in a trivial

[27]This is definitely the case when we cite inductive reasons: "we have not solved the problem yet, it has been around for a long time, therefore it cannot in principle be solved". This is a poor argument.

[28]These might, or might not, be usefully separated (Priest 2006b, 10).

[29]This is a delicate issue. I do not know why, but for some reason, this seems to be overlooked or ignored by the great majority of philosophers who discuss the paradoxes. It seems that it is only the paraconsistent logician or mathematician who can face the following fact, and accept it for what it is: we discuss paradoxes as a part of our successful communication, and not only in the 'mention' position/mode in a phrase or sentence.

Slater (2010) would probably disagree with this, so it is worth investigating further, but it seems *prima facie*, to me at least, that we use the paradoxes too. We *use* them to *justify* conclusions, such as: "Frege's formal system is trivial". Moreover, when we deploy the paradoxes, we do not *then* jump into triviality. Slater would probably say that we mention the use of them to justify conclusions, say, about Frege's formal theory. If Slater is right, then the question arises as to why it is that, or is there a non-question begging explanation for, whenever we only mention paradoxical sentence, we are in a classical setting and whenever we use paradoxes, or whenever they enter our theory, we must *per force* be in a trivial situation. If we relax this last idea, and use his emphasis on the use/mention distinction to dissolve many (apparent) and therefore sloppy uses of the paradoxes, then I am all for following his careful advice in adhering to the use/mention distinction. But I do not see a good argument (where an inductive argument is not good enough) for claiming in advance that all possible purported uses of the paradoxes are mentions and all apparent uses can be cleared away with careful respect of the use/mention distinction. Slater could counter-attack with a similar question about trivialism. Do we use *use* trivialism to justify some sensible statements – and yet not lapse into trivialism when we do so? How do we know we have not done so, and do we know this is not a danger in advance, are we not just relying again on the use/mention distinction, this time to save us from triviality? I think that the only answer to this is that we have to accept that we

discussion – where all arbitrary sentences are allowed, and agreed upon, including their negations. Anything (which is grammatical) is acceptable, and true, in a trivial context.

Since the development of paraconsistent logic has separated the concepts of contradiction from triviality (it is possible to have the first without the second) we have licence to claim that we can sensibly discuss contradictions, without degenerating into triviality.[30] We *know* that there are many ways of sensibly discussing contradictions, because there are many paraconsistent logics. Let us look at one that has interesting implications for the pluralist. The logic is LP (the logic of paraconsistency). We discuss alternatives in Chap. 11.

6.5.3 LP

I shall not give a full exposition of LP here for three reasons. One is that it is amply presented in a number of sources, which are readily available. Two, we shall be interested in some upshots of considering Priest's LP, not in mounting a direct defence of the formal logic and, three, we are interested in some comparisons of this theory with other paraconsistent theories and approaches, but in a very general way. For those who are interested in more details, I indicate the semantics of the system in Appendix 1. As a result, it will be better to simply highlight details as necessary, rather than present a number of details, to which we make no future reference. Again, detailed expositions are amply available in the literature. See Priest (2006b, 74–81, 223–228).

LP is a paraconsistent logic.[31] Since we accept the possibility, but prefer to avoid inconsistency, LP is useful because it does not affect classical inferences.[32] We can assume that we are working in a 'consistent setting' until we learn otherwise (Batens 1986). In other words, we can carry on with classical reasoning, provided we have no reason to think that we are reasoning over a liar sentence, or other

might be in a trivial world! But it does not appear to be so. It might then be interesting to see if we can jump in and out of such a world (and know that we are doing so, or attribute this to a third party). See Chaps. 11 and 14.

[30]The argument structure is a bit odd. We observe phenomenon x as actual, we then show that we have a model for x, and therefore that it is possible that x. The model gives us licence to carry on with the actual phenomenon. It's quicker to reason from actual to actual, than actual to possible (modelled) to licence recognising what is actual, but this structure of argument has found popularity with some recent work in the philosophy of mathematics. It does make more sense if we consider that we doubt that we should continue in practice x.

[31]Priest presents the formal system in several places. I shall be using the presentation in chapters five, six and sixteen of Priest (2006b).

[32]This idea is attributed to Batens. He has developed a number of other paraconsistent logics: adaptive logics, which give strategies for coping with contradictions when they arise. Most of the strategies are attempts to dissolve apparent contradictions.

sort of contradiction-forming sentence, or that we are in a contradictory situation (Priest 2006b, 223). Another way to put this is to say that what motivates the logic, is the conviction that there are few[33] inconsistencies in reality, so there should not be many in the theory. Inconsistency is exceptional. The logic also has the advantage, for those who are unfamiliar with paraconsistent logics, that all of the classical reasoning remains in place, all classical reasoning is consistent. In semantic terms we say that for any combination of wffs with only the truth value T or only the truth-value F, the reasoning is just as in classical logic. Where we see a difference is with wffs which carry both truth values (called 'truth-value gluts'), i.e., such wffs are both T and F. A wff p will take both values T and F, just in case it is a liar sentence, such as the perfectly grammatically acceptable proposition: "this proposition is false".[34] Since logic is supposed to help us reason over *any* situation, or any non-metaphorical, disambiguated, grammatical sentence, it had better be able to help us reason over liar sentences too. LP meets these considerations.

Let us look at a few more details, to get some feel for the logic. In LP, formulas can have one of three truth-value assignments: T, F and both T and F. Use ₸[35] to indicate both truth values. We should then look at the definitions of the connectives. It will not be enough to define conjunction as "A conjunction is given the truth-value T, just in case both conjuncts are true". For, we do not know what to do if one of the conjuncts is also false, i.e., is ₸. So we also have to say something about when a conjunction is false. In LP, a conjunction is false if either conjunct is false. It follows that if at least one of the conjuncts is ₸ then the conjunction is also ₸. If a proposition is both true and false, it is, in particular, true. So, a conjunction with one true conjunct and a paradoxical sentence as the other conjunct, is true, since both conjuncts are true. But the conjunction is also false, since one conjunct is false. *Mutatis mutandis* if both conjuncts are paradoxical. Therefore, both: conjunctions with one conjunct true, but the other true and false is itself both true and false; and a conjunction with two conjuncts both true and false is both true and false. "This sentence is false and number theorists study relations between numbers" is a true conjunction. It is also a false conjunction, since one conjunct is false (and true). Negation changes truth to falsity and falsity to truth. A negated contradictory formula stays both true and false. The other connectives are defined in terms of conjunction and negation in the classical way, thus ensuring classical inference in classical settings, i.e., in setting where no formula, or sub-formula, is ₸. This gives us a quick idea of the semantics for LP at the propositional level. LP is extended to include quantifiers, but we need not visit these details here.

---

[33]Counting inconsistencies is not really what we are talking about. We use this to order theories according to contradictions that can be proved in one theory but not the other (Priest 2006b, 224). The ordering will only work if we are comparing like to like.

[34]'Proposition', as it is used here, just means a unit of the language which is truth-apt.

[35]This is a symbol independently and spontaneously invented by students filling in a truth-table, and unsure whether to write T or F. They put both T and F together in the hopes that the professor will assume that the student had in mind the correct answer, and their pen slipped. In their honour, I have adopted the symbol, but make it signify a truth-value glut.

What about the syntax? From a contradiction we can deduce its negation, or the negation of the negation, or the negation of the negation of the negation, *etcetera*. But we cannot deduce an unrelated proposition symbolised by the letter q. For cases where p has both truth values, *ex contradictione quodlibet* fails. This is because it is possible for the premises to be true (and false) and the conclusion to be false (or both true and false). Since there are several ways of proving *ex contradictione quodlibet* in a classical system, the strategy for blocking this will have to spread across several rules, governing different connectives. Put better: our definitions of the rules of inference have to be worked out in connection with each other. See Appendix 3 for more details.

Why do we invoke such a logic? First note that we are not interested in a logic of propositions *per se*. Nor are we interested in extending this to a first-order theory with quantifiers, predicates *etcetera*. We are interested in the phenomenon of understanding what to do with, and how best to cope with, contradictory *theories* of mathematics, as well as contradictory statements in mathematics which belong to quite different theories. It will not be enough to invoke a first-order version of LP, since we are interested in being able to give the whole theory T as a qualifier of a statement s where T might not be a wff, or term, it might not be given any formal representation at all (except for our calling it T here). What LP tells us, when we come across a persistent contradiction is that we can treat it as persistent (it continues to enjoy both truth values, as does its negation) and that classical reasoning no longer works. We also learn that there is some reasoning which does work, that is, we are not immediately plunged into a trivial setting. Thus, if anything, we are making reference to the existence of such a logic in order to support the claim that contradictions, in speech, writing, or in a theory, or between theories, do not necessarily entail triviality.

The very existence of formal paraconsistent logics that are not trivial, is enough to support this claim. We know that they are not trivial because there is a wff in the language of the theory which is only false.[36] We can then refer to the logic as a guide to our reasoning (especially in cases where T is not given formal representation). Further exploration of this 'metaphorical' use of a logic is discussed in the next chapter.

References

Bar-Am, N. (2008). *Extensionalism; The revolution in logic*. Berlin: Springer.

Batens, D. (1986). Dialectical dynamics within formal logics. *Logique et Analyse, 114*, 161–173.

Bohm, D. (1980). *Wholeness and the implicate order*. London: Routledge.

Byers, W. (2007). *How mathematicians think; Using ambiguity, contradiction, and paradox to create mathematics*. Princeton/Oxford: Princeton University Press.

Priest, G. (2006a). *Doubt truth to be a liar*. Oxford: Oxford University Press.

[36]We shall refer to this as Post-non-triviality in Chap. 10.

Priest, G. (2006b). *In contradiction* (2nd ed.). Oxford: Clarendon Press.

Rav, Y. (2007). A critique of a formalist-mechanist version of the justification of arguments in mathematicians' proof practices. *Philosophia Mathematica*, Series III, *15*(2), 291–320.

Rodin, A. (2008). *Did Lobachevsky have a model of his imaginary Geometry?*. http://philsci-archive.pitt.edu/4099/1/loba8.pdf

Singh, S. (1997). *Fermat's enigma: The epic quest to solve the world's greatest mathematical problem*. New York: Penguin Books.

Slater, H. (2010). What priest (amongst many others) has been missing. *Ratio, 23*(2), 184–198.

Sundholm, G. (1998). Inference, consequence implication; A constructivist's approach. *Philosophia Mathematica, VI*, 178–194.

Tennant, N. (1997). *The taming of the true*. Oxford: Clarendon Press.

Chapter 7
Using a Formal Theory of Logic Metaphorically

Abstract At the end of the last chapter I invoked the idea that a formal system of logic, such as LP, is used metaphorically by the pluralist. It is essential to the pluralist position, and possibly to many other positions, that we should be able to make sense of this, and say something quite definite about it. Otherwise, our claims about appealing to formal systems of logic are empty. I look at three ways in which the pluralist makes use of a formal logical system. The first is in direct appeal to a rule or axiom to justify a move in an argument. The second is when the pluralist uses a formal theory in order to reconstruct another theory. This is done to understand the theory from another perspective. The third use is dialectical. In invoking or developing a formal theory to represent a form of reasoning, we bring some features of that reasoning into relief, and we obscure other features. We can evaluate the fit between the formal theory and the informal one. In the evaluation, we might well consider alternative formal representations. Thus, we enter a dialectic. Lastly, in order to remind us that pluralists are not the only ones who use formal logic informally, I look at how it is that mathematicians use formal logical theories.

7.1 Introduction: What Is a Metaphorical Use of Logic?

In this chapter, I am interested in how the third level pluralist makes metaphorical use of formal logics. In particular, I am interested in his use of formal logics when discussing trivial theories; or pairs of theories, which, when put together, result in a trivial theory. In the pairing case: if at least one theory allows *ex contradictione quodlibet* inferences, and the pair have rules or axioms which contradict each other, then we have a simple case of triviality-generating pairings. For example, if I add the axioms of Gödel-Bernays set theory to those of Zermelo-Fraenkel set theory, then I will generate contradictions regarding numbers of subsets of ordinal numbers, and since both are classical, I can then make proofs of any well-formed formula in

M. Friend, *Pluralism in Mathematics: A New Position in Philosophy of Mathematics*, Logic, Epistemology, and the Unity of Science 32, DOI 10.1007/978-94-007-7058-4_7,

© Springer Science+Business Media Dordrecht 2014

the language (of the combined set theories). Therefore, it is particularly in such cases that the metaphorical use by the philosopher is delicate. I shall also extend the idea to mathematical contexts. In particular, in this section, I shall leave aside paraconsistent logic, and discuss only the *metaphorical* use by mathematicians of formal logic. We shall develop this theme in more detail in subsequent chapters, where we work with examples. This is a first glimpse at the idea of a metaphorical use of logic, not the final word.

Philosophers make ample metaphorical use of formal systems of logic, which I shall henceforth abbreviate as ‘logics’. To appreciate this point, let us just pause to think of the contrasting scenario, that is, what it would be for philosophers to make a thorough and direct use of a formal logic in their arguments. If a philosopher were to use logic directly, she would write out a natural deduction proof for the conclusion she is trying to argue for, and then write it out effectively back in a natural language. Philosophers almost never do this. Instead, they typically use logic *metaphorically*. To fix the language, I shall stipulate that

philosophers use logics ‘metaphorically’ whenever they do not make formal logical arguments for the whole of their philosophical arguments.[1]

Why not? Or, put it another way: why do philosophers make a metaphorical use, as opposed to a direct use? After all, most of us are supposedly trained in formal reasoning.

This is so for three reasons. The reasons are: background knowledge, degree of sophistication of the argument and whether one is pluralist or monist about logic. The first reason belongs to those with little background in formal logic. To my chagrin, many philosophers are not aware that there are *a number* of different formal logical systems, or, at best they are acquainted with two classical systems. They are brought up on a diet of classical propositional and first-order logic, and that is all. If they have heard mention of alternative formal systems, they will not have studied them enough to make use of them in an argument. Thus, they are logical monists in their practice, because of their education, not for deep philosophical reasons. Or rather, the philosophical reasons are presumed to exist and be vindicated by tradition and the cannon. But see, for example (Moore 1988). They *could* write out their argument in propositional or first-order logic, but they find this tedious and unnecessary. If they have given the matter any thought, they would also realise that classical logics are inadequate for some arguments. This might tempt them to re-consider their monism, but for practical reasons, since discussing logic would make them digress too much from the issue they are concerned with, they are not

[1]Some of us tell our students that we do, but this is usually a cheap ruse. I have tried writing out the formal version of an philosophical argument, but found that either my argument in English sounded too trivial, or that I could not represent the ideas properly in a formal language. Of course small inference moves are easy to represent, but more involved arguments are much harder. Nevertheless, we have something like the logical structure of the argument in mind when we argue. So part of what this chapter is about is the relationship between formal logic and philosophical reasoning.

interested in pursuing this line of thought further. This reason is mentioned, to add completeness to our list of reasons, but it is not interesting for us here, so we ignore it, (the interest is, at best, historical and socio-political).

The second and third reasons assume that the philosopher in question knows that there are several formal logics, and that these differ over what counts as a valid argument. They are also sufficiently adept at some of these that they could, in principle, translate an argument in philosophy into a formal argument. Why do they not do this?

The second reason is that they suspect (rightly or wrongly) that their philosophical argument is too sophisticated to be accurately represented in a formal language, or 'the' formal language if they are monists. After all, formal languages, sophisticated as they are today, are not sufficiently sophisticated to represent a whole philosophical idea (of any interest), since such ideas often include several modes, which we would represent by operators in a formal language,[2] and are sensitive to context or situation. Combining several different sorts of mode (belief, possibility and morally good, for example), *and* contexts and situations in a logic is still a topic of research. So, even if they took the trouble to represent their argument formally, it would turn out to be formally invalid, for lack of sophistication in the representation. If a philosopher makes an argument, presumably she thinks that it is valid. Since she believes that her argument is valid, the problem lies with the translation (adequacy of the formal language) and the formal apparatus (axioms and syntactical rules). Or if the argument relies on induction, then it is valid modulo induction, but cannot be represented accurately enough in a formal language to demonstrate the validity of the argument, modulo induction.[3]

The third reason assumes logical pluralism. A logical pluralist might lack certainty about which formal system best fits the argument she is trying to make. She might not want to decide *in advance* which formal system best represents her reasoning. For example, a philosopher might not know if she really wants disjunctive syllogism to be allowed in a particular instance of reasoning. She might not be sure whether her statement using 'if... then...' is better represented as a strict conditional, a material conditional, a causal conditional, and so on. "Hadn't she better find out?" You might ask. Not necessarily. She might prefer to defer making a choice indefinitely, or until more of the argument has been fleshed out. That is, it is in the *process* of arguing and exchange that we discover ambiguities, or discover that a coherent reading of the argument requires that we, for example, must have read the conditional in a particular statement to be strict conditional. Moreover, this

---

[2]There do exist some formal systems which combine several modal operators, (which are not duals of each other), but there are not many, and they are often not satisfactory (in their axioms or rules of inference). Occasionally, a philosophical argument can be elegantly expressed in a formal language and the inferences can be accounted for. It is very satisfying when we can do this.

[3]What I mean by 'valid modulo induction' is that the non-inductive part of the argument is valid, provided the induction move is carried out correctly, according to some axiom or rule of induction.

deferring is not necessarily a preference due to laziness. Dialectically, she *need not* choose a formal logic until an opponent takes issue with the logic of her argument. That is, *until* an error is suspected (by herself, or someone else), there is no reason to choose between formal systems, over continuing to elaborate her position.[4] This is a delicate matter, flirting with: the notion of burden of proof, laziness, consistency, triviality and practical calculations about where to focus one's energies.

These were three reasons for not translating a philosophical argument into a formal logic. But, since philosophers do argue, they must do so metaphorically (according to the stipulation above).

What more do I mean by the 'metaphorical *use* of a formal logic'? I shall explore three metaphorical uses here.

1. One is to justify particular moves in an argument written in a natural language, such as when a philosopher writes: "... and this just follows by *modus ponens*." We then understand that the conditional in the sentence is to be taken as material, or as some other conditional which allows *modus ponens* as a rule of inference.[5] This can be done well, or poorly. To do this well, we should remain consistent (with respect to background logic) within an argument. To return to our example, we had better not count *modus ponens* as invalid later on, at least, not without a good explanation! Logical monists should remain consistent (with respect to their logic) in this metaphorical use for *all* of their arguments. Logical pluralists can change background logic from one argument to the next, since they find it appropriate to switch logics with different subjects of argument. But when they do this, they should be ready to explain or account for all switches.

The second two uses assume logical pluralism.

2. The second sort of metaphorical use is more interesting. It is when we appeal to a formal logic as a possible model for a certain sort of reasoning, for example if a philosopher appeals to a paraconsistent logic to justify continuing to reason in the face of contradictions. It is not enough, in these cases, to simply mention the existence of paraconsistent logics. Rather, we also use the logic in our argument. Thus, the formal logic is a model for reasoning, or a type of 'rational reconstruction'. We shall see an example of this type of metaphorical use in Chap. 9.
3. There is a third sense of 'metaphorical' we should address, and this is the dialectical role played by formal logics. Once we have learned some formal concepts, they are used in philosophical arguments, with the idea that the arguments could be made more precise if we were to formally represent them. Formal representation is then treated as an exercise in clarification and in

[4]The point is made in Sundholm (2012), but there, he makes the point for mathematical and logical arguments, not for philosophical arguments. However, I think that the point applies to philosophical arguments just as well. I shall return to this issue.

[5]There are formal logical theories where *modus ponens* does not hold. These are some of the relevant logics.

precision: to expose the structure of the argument so we can also question the structure (or the representation). Since the added precision might involve re-enforcing (by choosing and working with), or inventing new, formal techniques, we have a dialogue between the formal logical systems and our philosophical reasoning.

In this chapter, I shall discuss all three in turn. In the final section, I shall discuss how the metaphorical use of a logic is also used in mathematical arguments, or proofs. By the end of the chapter, we shall have a sharpened sensitivity to the pluralist's metaphorical use of logic.

7.2 Direct Appeal to Formal Rules or Axioms

In a philosophical argument, we sometimes appeal to logical rules.[6] To give a 'concrete' example, take contraposition. To fix the example, let P = "we have free will" and Q = "we genuinely make choices". A philosopher would not be logically at fault to write: "if we have free will then we genuinely make choices. Therefore, contra-posing, if we do not make choices, then we lack free will." When we disagree with such an argument we first check that we agree with the premise, and the translation into a formal language. In this case, the formal representation is something like: $P \rightarrow Q : \sim Q \rightarrow \sim P$. P and Q are our propositional constants. $\rightarrow$ is the conditional. $\sim$ is negation.

Assuming we agree with the translation, then if we still disagree with the conclusion, then our only recourse is to argue about the logic of the argument. We check that the rules of the logic have been applied correctly. The correct use of the rule of contraposition is determined by reference to a particular formal logic (and a correct proof therein), or a class of logics (all those endorsing contra-position of the conditional) (and the correct proofs therein).

A sustained argument can then take different turns depending on our background assumptions, which are revealed through the process of argument. What are the different turns?

1. If the proponent of the argument is a logical monist and believes that he knows which formal theory represents best reasoning, then he defers to that. An opponent who is monist and upholds the same logic is then defeated.
2. If the opponent is a monist, but holds another formal system as representing logic, and his logic does not allow contraposition, then the argument shifts to

[6] We might think that this is a 'border-line case', that is, that in the case of appealing to formal rules in mid-argument is hardly *metaphorical*. It is just direct use of a logic. However, it is not quite, since we should distinguish between the formal system of logic and the informal use made of it. The informal use is, arguably, a metaphorical use of the *formal* logic. I do not much mind where we fall on this. Without compromising the thrust of the argument, we can ignore this stretch in the term 'metaphorical' without great loss.

one about which is the correct logic. More precisely, neither has to be a monist at this stage, since what is being argued over is the validity of contraposition.[7]

3. The argument can become more interesting if we oppose a logical monist to a logical pluralist, since now, the debate turns over logical monism and pluralism (assuming everything else is agreed upon).
4. Logical pluralists have to address the issue on two fronts.

4a. one is to see if they agree on contraposition. For example, it might be common to all of the formal systems endorsed by both, in this case the opponent looses.

4b. If one thinks that contraposition is valid in all legitimate formal logics, and the opponent disagrees, then part of the debate can focus on contraposition or can shift to the issue of fit in the application of a particular formal logic to this instance. That is, they turn their attention to this particular case. They ask the question: "Is the issue of free will and choice one where contraposition is valid?" There will then be further permutations for the argument.

But we can see already that sophisticated arguments like this invite a dialectic between the informal argument and the formal representation of the argument, and the presumed characteristics of the formal systems supporting the formal representation. We also see that even in a direct use of a logic in a philosophical argument, we hide a number of assumptions, which can in turn be questioned.

Note that, in the mapping out of possible debates, we have already appealed to logic metaphorically, in the senses of: not constructing an entire argument formally, yet arguing rationally, and implicitly being willing to be guided by formal logic (once we agree on the parameters for what counts as a legitimate formal system). This sort of idealised and sophisticated argument is one form of using formal logic metaphorically in a philosophical argument.

7.3 Using Logic for Rational Reconstruction

A second sort of metaphorical appeal to formal logical theories has two stages.

(I) One is to appeal to a formal theory to *suggest* that there is a *possible*[8] coherent move or series of moves.

(II) The other is to *demonstrate* that the move, or series of moves, is coherent by giving a 'rational reconstruction'. The 'rational reconstruction' in this case is a formal model of the reasoning.

The difference between (I) and (II) is that the first has the mode of possibility, the second calls for an instance of the possibility. For example, someone might

[7]This might even turn into an opportunity to revise one's monism, but I leave this sort of possibility aside, for the sake of simplicity.

[8]The possibility is an epistemic possibility – "for all I know".

remark: "there are contradictions in the UN resolutions". More carefully: "There are pairs of resolutions, or pairs of claims within single resolutions, which together, can logically lead to a contradiction; assuming good translation, correct deployment of the logical rules, and a legitimate class of formal theories." A too casual retort might be "but there are formal logics which deal with contradictions, so we should not worry about these in the UN resolutions." This would be an unfortunate, and almost vacuous retort.[9] But it is not entirely vacuous since it suggests the next move; and this is to make a rational reconstruction of the reasoning we should engage in, in the face of particular contradictions amongst the resolutions.

Developing this second stage, we settle on a translation of the resolutions into a formal language. We settle on what the problem with contradiction is, in this instance. For example, it might be a fear of triviality, or that the purported contradiction leads to inaction due to a decision loop, *etcetera*. We then choose a formal logical theory which we think will vindicate the better reasoning that avoids the problem. We thereby use the formal reasoning to demonstrate that the problem *can* be avoided. If this is not possible, or after a concerted effort, we cannot reconstruct the reasoning using our chosen formal theory, then we might try to develop, or 'design', a new formal theory which does.

If we do this we should exercise caution. For, we can even *guarantee* that we can succeed, and this is worrying! We can 'succeed' quite perversely, by working backwards. We design a formal system *around* the few sentences we are concerned about and *block* the inference to whatever it is we are worried about. $P \mathbin{\&} {\sim}P$ is the contradictory pair of resolutions. Q is the consequence we do not like, so $P \mathbin{\&} {\sim}P \vdash {\sim}Q$ is a rule (or better, the only rule) in our logic. P and Q are constants. They refer to particular UN resolutions. From the contradictory pair we may only infer ${\sim}Q$ (there are no other rules, so we cannot make an *ex contradictione quodlibet* argument), which is what we wanted, since we wanted specifically to avoid Q! But this is a trivial and vacuous reconstruction or recommendation. Indeed, it is vacuous, unless the reconstruction or recommendation has wider scope than the specific problem being addressed.

The wider scope is reached by avoiding constants as much as possible, and by using rules or axioms which are familiar from other formal theories. It is an art to make an interesting reconstruction, since it remains that there is no specific, identifiable cut-off point between when a reconstruction is interesting or vacuous. Ultimately, whether a reconstruction is interesting or vacuous will be a judgment, exercised by a community, and the judgment can be revisited in the future.

[9] Batens appeal to something he calls a 'zero logic'. This is a degenerate logic where there are no rules of inference. As a result, no contradiction could be derived, since no derivations can be made. This is a degenerate logic, and extreme enough that we might be carried to suppose that it is not a logic at all. This is not important for the point here. If we can develop such a 'logic' then we can develop formal logics with very few rules of inference, and we could develop one where a contradiction in the form of a conjunction of two opposite formulas, could never be derived (just because the formal theory lacks a rule for forming conjunctions, even though there is a symbol for conjunction in the language (but then it is a 'dummy' connective)).

The existence of some borderline cases does not preclude there being clear cases. There are interesting reconstructions. What do we learn from them? Not that the writers of the original construction were thinking in the way we formally demonstrate, or reconstruct. We cannot even learn that this is how we *should* reason in this, and similar instances, since, if there is one reconstruction, then there will be alternative reconstructions, i.e., alternative formal theories to which we can appeal in order to model the reasoning. One lesson we learn is that it is *technically* possible to avoid the danger. If the 'danger' is triviality, then we have not learned much. But this is not the end of the story. We can now, again, deepen our understanding and use the reconstruction as an invitation to discuss the applications of the formal theories, examine the breadth of application, and so on. This next turn in the investigation involves the dialectic between formal logical theories and their applications; and we now focus on this.

7.4 The Dialectical Use of a Formal Logic

Let us start with the easiest case. We start with an argument in a natural language. We find it is controversial, so we choose a formal language to represent and check the reasoning. To do this we translate the natural language sentences into formal formulas. When we give a formal rendition of a natural language argument, we obscure the non-logical content, and bring into relief the 'logical', or mathematical, structure. We can flag contentious translations, as potential weaknesses. This exercise is already quite sophisticated, and we learn from the very act of translation, since, the act of translating requires that we distinguish logical structure from content. We are then made to reflect upon the deductive moves we are modelling, or representing. We would normally default to a formal logical theory with which we are familiar, or better, with which many people are familiar. We then demonstrate the reasoning in the formal setting, and disputes can be easily cleared up, since the deductive moves are as transparent as possible. Not only is the natural language argument brought out in all clarity, since the arguing is meant to be indifferent to content, but the formal logic is given added confirmation in its own turn. This two-way interaction is important in the light of alternative logics or alternative mathematical theories.

Now turn to a more problematic case. We are not satisfied with the easy steps we took. Either the conclusion does not follow according to the formal theory, or it does, but there seems to be something fishy, or unsatisfying in the formally represented reasoning. Where might things have gone wrong? They might have gone wrong in the translation. We did have to make a choice of formal language. Maybe we need a more sophisticated language, especially if the reasoning does not go through in the formal setting, or can be shown to be invalid. So, we have to think carefully about the choice of formal language. This is a question of 'fit' between a formal language and an application. We then separately have to think about the axioms, definitions and rules of inference of the formal theory, and discuss whether these are good axioms, definitions and rules. It stands as obvious that if we argue for these by appeal to a formal theory, we can raise the same question again as to the merit of the axioms, definitions and rules of *that* theory. We are then engaged in a regress. Or, we can engage in a less formal discussion and use informal reasoning. Either way, we are engaged in cross-checking our formal representation against, what we take to be, independent ideas, notions or theories. We might even be pressed to develop, or design a new formal logic, or develop a new rule, or operator, in order to give formal expression to what we have in mind. This does not happen often, but it does occasionally, and there is no bar on this sort of exploration. Such a development could become interesting in its own right, as we know from a number of mathematical developments. This will happen if the formal theory's scope is wide. We might find re-applications in areas we had not thought of when we developed the new theory.

In this case, 'dialectic' is too narrow a term, since we do not just have a back-and-forth exchange, but rather a blossoming forth, and spreading exchange. If they do not reach a dead end, then these displaced, or unintended, applications can feed back to the original problem to inform us as to the scope of the model of reasoning. We have then learned a lot about our original reasoning. Such investigation takes a community of thinkers spread over time and place. The exploration may take several generations. We can see examples of such in the history of logic and in mathematics, and this is the general lesson we can learn from that history. The dialectic is not restricted to logics, but can also include mathematical theories. But mathematicians also make metaphorical use of logic in their arguments. We shall see this later in Sect. 7.4. First, let us pause and look at a toy example of the dialectical use of a class of formal systems of logic.

7.4.1 An Example of the Dialectical Use of Logic

Let us look at a toy example. Consider the sentence, uttered by me in a state of inebriation: "either all philosophers are liars or I'm not a philosopher!" Assume I am in the company of charitable logicians or philosophers who want to make sense of what I just said. It sounds plausible, after all, philosophers say a lot of outrageous things, they all contradict each other, they might all be liars. Take the second disjunct. It might well be the case that I am not a philosopher, but we do a little investigation, and discover that there are institutional indicators that I am. Therefore, the second disjunct is false (provided we agree that the institutional indicators are sufficient to conclude that I am a philosopher).

We are then left with the second disjunct by disjunctive syllogism. The first disjunct must be true (and only true) if we are classical logicians. But the disjunct is uttered by a philosopher, i.e., by a liar, and therefore, is false (in classical reasoning). But then it is true, in classical reasoning, and I find myself in the land of triviality. Everything I say from now on is true, and I must agree to everything anyone says, for, all grammatical sentences are true. This is what classical reasoning commits me to. Once I have entered the land of triviality, my interlocutors will soon discover that they can attach no meaning to what I say, and I too will probably sense the same thing, go mad, and cease to reason at all.

However, this cannot be right, since we utter contradictions (as professional philosophers) and do not end up in the frightful mess described above. So let us try to wriggle out. Let us start with the use/mention distinction. Did I use "all philosophers are liars", or did I just mention it? I seemed to use it, especially if we are trying to make sense of it and reason from the sentence, after all, we used disjunctive syllogism on the sentence to conclude that all philosophers are liars. So, we are committed to the truth of that utterance. It closely resembles the Cretan version of the liar paradox, and this was used by a Cretan, and more important, was interpreted to qualify everything he said. Thus, we cannot wriggle out so easily here.

We could, for example, appeal to a distinction drawn by DaCosta between apparent and classical contradictions. Classical contradictions are of the sort described above, they are contradictions in a classical setting and so lead to triviality. Apparent contradictions are ones that look like a contradiction, but can be dissolved with the right sorts of qualifiers. For example, we might say that in the phrase "all philosophers are liars" there is a quantifier ambiguity over lies. Did I mean that philosophers only utter lies, or that some of their utterances are lies. The second is more plausible, and it just might be the case that in this particular utterance I was telling the truth! We might be convinced by this. Notice what I did to wriggle out. I appealed metaphorically to a class of paraconsistent logics (those developed by DaCosta and his followers), or more specifically to a distinction common to that class, and since they are formal rigorous systems of reasoning I have rescued myself from the fate of the trivialist.

But hold on! Maybe we are not convinced that we can wriggle out so easily. After all, in this particular context I might have been telling the truth, but I was lying about the second disjunct. So maybe I was lying in both disjuncts! That is, we cannot know that I was telling the truth, or, therefore, how to interpret the quantifier ambiguity over this possible lie. Therefore, by classical reasoning, we have a classical contradiction, and we step through the gates into the land of triviality again.

Not so fast. How could I possibly be sure that I can separate off the disjuncts so neatly using disjunctive syllogism. The two disjuncts are joined (in content or intent), and so a classical treatment will not do them justice. Maybe we do not have a classical contradiction (through classical disjunctive syllogism) but rather a disjunction with a contradiction as one disjunct. Maybe it is a true contradiction, and we appeal to LP. In the logic of LP, disjunctive syllogism is invalid. Here is why: one of the disjuncts in the syllogism could be both true and false, symbolised $\overline{\top}$. This is enough to make the disjunction true. The other disjunct could be just false. The disjunction is still true, because one of the disjuncts is (in particular) true (and also false). But the conclusion is false (but also true). So the reasoning is invalid. Here is the relevant part of the LP truth table for the disjunctive syllogism reasoning (Table 7.1):

Table 7.1 Relevant line in the table for disjunctive syllogism in LP

p	q	p V q	~p	q
₸	**F**	**T**	₸	**F**

So what can we conclude in LP from P V ~Q, and Q? Not much. The reasoning is invalid. But we cannot conclude that I am a trivialist or that I have lost my mind either. If we opt for the truth-value glut hypothesis, then we can engage in the philosophical work of the dialetheist. (I will leave out those details here, but we shall see them later.) However, if we take the dialetheic option as a *reductio* argument, then it turns out that the first disjunct is not a dialetheia. If we also agree that I am a philosopher then we should accept disjunctive syllogism and return to DaCosta's distinction, and the dis-ambiguation we used before. Anyhow, we enter a dialectic between our natural language statement and the different formal representations in order to understand what we can understand in this toy situation. This was an easy example. If I am the only philosopher, and I am not lying, it is not a paradox. Let us now turn to the more mathematical case of the metaphorical use of logic.

7.5 When Mathematicians Use Formal Logic Metaphorically

Some of this material anticipates Chap. 12, but let me give a foretaste. A casual look at a modern mathematics journal, book or Ph.D. thesis will reveal that not all of the proofs are done in the form of a natural deduction proof. In fact, only some moves in the proof are made explicit.[10] This is because, it is only in the face of anticipated doubt, or in the spirit of 'proof as explanation' that we write out explicit steps in a proof.

The relationship between doubt and explanation is tight. We call for explanation when we are in doubt, not otherwise. The doubt might be mild, and take the form of curiosity. We are only curious about things which interest us, or intrigue us, and doubt accompanies that interest. No proofs are needed for obvious truths, or what are taken to be obvious truths. The norm is to not require formal proof! It is unusual, or only in particular instances, that we are called upon to give a formal proof. We give a proof if we think someone might reasonably suspect we are wrong. "Being wrong is a concrete, particular issue, whereas being right is universal freedom from that concrete particularity." (Sundholm 2012, 3). We are usually right, or close enough not to elicit doubt. So we are *called upon* to fill in gaps in our reasoning, or to offer

[10]In fact, in journal articles we have a lot of proofs, and this is because we are meant to be on the forefront of knowledge. The finding and conclusion should be surprising, curious or dubious (otherwise we would not publish the finding). It is for this reason that we see so much formal proof in advanced texts and journals. In less advanced material proofs are less necessary. They are given in textbooks in order to convince the student to whom the material is new, and to accustom the student to the language of proof.

explanation and justification for conclusions. We do this only inasmuch as we need to. How much will depend on our interlocutors, or imagined interlocutors. What is appropriate: whether to fill in the gaps, or give a proof that the gaps can be filled, is also sensitive to the context, mix of interlocutors or particular circumstance.

When we fill in those gaps, we might do so by appeal to a formal theory, we might appeal to a proof in a meta-theory which shows that we could fill in the gaps if called upon to do so. Or more casually, we might appeal to the formal theory metaphorically. Again, *particular* circumstances will dictate[11] which is possible and which is appropriate. In sophisticated, delicate, subtle contexts, a dialectic will start between the formal representation and the original ideas being represented. That is, there are cases where we formally represent an argument, but discover that the formal representation reveals something surprising, or that we cannot understand it fully. An example is the discussion of Lobachevsky's argument in Chap. 5, Sect. 5.5, and we shall see more of the dialectic being meted out in Chap. 9. The development of a formal theory can lead to surprises. The dialectic, or blossoming forth, constitutes the development of the content of mathematics. It is a cross-referring process, an incomplete process, and as I mentioned earlier, there is some re-enforcing of a logic when we deploy a logic, or engage with it. We confirm its use in another application. We confirm it in our own minds, and it becomes better accepted by more people in the community.

7.6 Conclusion

In conclusion, we anticipate the next chapters. These are on the notion of rigour and the notion of fixtures in mathematics. Starting with rigour: to anchor confirmation, acceptance and understanding we use proofs. We communicate across time, culture, education and traditions with proofs. Nevertheless, how they are written out in a particular instance of intended communication is sensitive to the background suppositions of the intended audience. We try to overcome these boundaries by developing a notion of rigorous proof. The notion is an ideal, or rather, it is several ideals. The ideals have parameters. The difference between ideals depends on philosophical inclinations. A completely rigorous proof should be accepted by all the mathematicians who share the philosophical inclinations relevant to the proof.

Since the notion of rigorous proof does not have a unique extension, we should account for this and fend off the concern that 'anything goes' in mathematics. This is not at all the case, as we all know. The philosophical task is to account for this, in light of the variability of the ideal of rigorous proof. Remember that *prima facie* the pluralist wants to recognise and accommodate different orientations and different

---

[11] Circumstances are individuated by the people receiving the information, and their knowledge and background. The people might include people in the distant future, as when historians of mathematics interact with an historical text, or when a mathematician revisits an old text.

idealisations on proofs. In Chap. 9 we fill out the story by introducing a notion which I call 'fixtures' in mathematics. These are points of cross-communication and of checking one theory against another. They buttress the ideal notion of rigorous proof, and accommodate the variability.

Acknowledgements I should like to thank Batens and Primiero for each separately suggesting to me that I wanted to use adaptive, or another paraconsistent logic, metaphorically rather than directly.

References

Moore, G. (1988). The emergence of first-order logic. In W. Aspray & P. Kitcher (Eds.), *History and philosophy of modern mathematics* (Minnesota studies in the philosophy of science, Vol. XI, pp. 95–135). Minneapolis: University of Minnesota Press.

Sundholm, G. (2012). Error. *Topoi, An International Review of Philosophy, 31*(1), 87–92.

Chapter 8
Rigour in Proof

Abstract Rigour in proof is of utmost importance for the pluralist, since he has no solid ontology to ground his theory, and his conception of 'truth' is also relative (to a theory). In the first section we look at the pluralist's motivation for rigour. In the second section, we develop a characterisation of rigorous proof. There are several characterisations varying over the account of meaning we attach to mathematical claims and axioms. In the third section, we evaluate the characterisation with reference to our motivation. Lastly, we draw some general conclusions for the pluralist. With the analysis we discover that rigour is a regulative ideal, sensitive to philosophical inclinations.

8.1 Introduction

We shall begin by looking at the pluralist's motivation for increasing the rigour of a mathematical proof. Rigour is of utmost importance for the pluralist, since he has no solid ontology to ground his theory, and his conception of 'truth' is also relative (to a theory). Nevertheless, it would show a complete misunderstanding of mathematics, were the pluralist to say that 'anything goes' and 'it's all relative' in mathematics.

Mathematics distinguishes itself from other forms of enquiry by its precision and abstractness. The abstractness of the thinking, of the objects, relations, functions and functors is controlled, and validated, by proof. It is through proof that mathematical ideas are disseminated, accepted and understood by the community of mathematicians. But what is to be counted as a proof is itself a subject of study

This chapter is co-written with Pedeferri. A less pluralist oriented version of the chapter is in the form of a paper also co-written with him (Friend and Pedeferri 2012).

M. Friend, *Pluralism in Mathematics: A New Position in Philosophy of Mathematics*, Logic, Epistemology, and the Unity of Science 32, DOI 10.1007/978-94-007-7058-4_8,
© Springer Science+Business Media Dordrecht 2014

in mathematics.[1] Standards of rigour of proof vary with conceptions of rigour. The pluralist has to be able to say something quite firm about those standards, while accepting that there is some variation. This is what we do in this chapter.

In the first section we shall look at the pluralist's motivation for rigour.[2] Here 'pluralist' is meant in the sense of maximal pluralist, or pluralist of third level, as characterised in Chap. 6. In the second section, we develop a characterisation of rigorous proof. In the third section, we evaluate the characterisation with reference to our motivation. Lastly, we draw some general conclusions for the pluralist. With the analysis we discover that rigour is a regulative ideal, sensitive to philosophical inclinations.

However, we also observe the lack of rigorous proofs in the literature. Our diagnosis is that this is because there can be too much rigour in a proof! That is, what level of rigour is appropriate will depend on context and readership. This explains why, in practice the ideal standards are almost never met, and that this is accepted practice. It is a further question to what extent we should be complacent about the standards of rigour met in practice. A warning note was already sounded in Chap. 5, and we shall return to the matter of rigour in Chap. 12.

8.2 The Pluralist's Motivation for Rigour

The pluralist's motivation for increasing rigour in proof is *explicitness and honesty*. This motivation is not to be thought of as emotional but scientific, since it courts possible objections to given proofs. Explicitness and honesty are scientific virtues.[3]

Before we characterise 'rigour' let us explore how proofs are used and understood. We start with a working hypothesis.

A rigorous proof has no mysteries or shadows. There is a maximal, or optimal, display of the reasoning.

When we give a rigorous proof, and if there is doubt or a dispute, then it can be more easily checked than a non-rigorous proof. Consider Fermat's last theorem. On many occasions purported proofs of Fermat's last theorem were subjected to scrutiny. On every occasion, lacunae were found, and the purported proofs were deemed unsatisfactory. Wiles's proof uses results and techniques from several, quite disparate and new areas of mathematics.[4] His first proof was unsuccessful.

---

[1] There is an impredicativity in the account of the pluralist. This is also a subject for future study.

[2] For a discussion of motivations for rigour which are not necessarily pluralist see Friend and Pedeferri (2012).

[3] 'Science', as used here, is not confined to physics, chemistry and biology. Rather, it means any form of rigorous, earnest and open enquiry.

[4] For example, the proof uses Eilenberg-MacLane-Steenrod axiomatization of homology and cohomology in topology, yet it is a problem "in number theory". Nor is the proof restricted to

It was only once every move was made sufficiently explicit, that some members of the community of mathematicians were able to scrutinise the proof and judge it satisfactory. But the story did not end there.

Since few mathematicians were sufficiently familiar with all of the techniques employed by Wiles, few felt qualified to review the entire proof. So, there was a deferring to the expertise of the few; and this leaves, not so much doubt in the truth of the conclusion, but more a dissatisfaction because of lack of understanding. A simpler proof was sought. A simpler proof is a type of confirmation of judgment of 'success'. A proof might be simpler because it uses better understood techniques or because it is shorter. Many ideas in mathematics are represented as theorems, and enjoy several correct proofs. Each proof tells us something new about the idea. So, proof is not only there to tell us that a theorem is true in a theory, but also to explain something of the genesis or implications, or 'spread' of the proof.

Elaborating on genesis, implications and spread: genesis can be thought of as historical or conceptual. The first proof of a particular result is historical. We can learn from it the context within which the mathematician was thinking about the problem. The situation of the mathematician tells us something of how the mathematician found the truth of the theorem in question.[5] But we can have another type of genetic proof that shows 'conceptual origin'. These conceptual genetic proofs give a justification for the theorem in terms of some more primitive, or 'foundational' theory. That is, the proof fits the result into a different context, and that context is meant to hold some philosophical virtues, such as sparse ontology, logical primitiveness, intuitive appeal and so on. Arguably, Frege had this sort of idea in mind when he set up his logical theory, and proved *already accepted truths*. He was not confirming that they were true, rather he was giving an ultimate justification for our theory of arithmetic in terms of logical notions. His ultimate justification had the philosophical virtues of universality, analyticity and logical validity (Friend 1997). These were genetic proofs in the conceptual sense.

Since the pluralist is pluralist about foundations, he cannot remain silent in the face of such philosophical claims about 'ultimate justifications'. He will temper them by removing claims about absolute truth or essence, and replace these with implication and spread. The implications of a theorem concerns the spread 'internal to the theory', how the theorem affects our understanding of the ontology of the theory, or connects ideas in a theory together, demonstrating their relationship to each other.[6]

---

these areas of mathematics. In 2009, Mark Kisin simplified Wiles's proof "so it does not really use algebraic geometry, but is still all about the 'cohomology' that Grothendieck invented and which descends through Cartan and Serre from Eilenberg-MacLane-Steenrod." (McLarty, personal correspondence, 2010).

[5]There are exceptions. Under a formalist influence, a mathematician will try to hide the genesis of a proof or idea. Thus, he presents the proof as standing on its own.

[6]As a side note, it is interesting to think of *reductio* proofs as giving us the limitations of internal implications. Think about *reductio* proofs as telling us that if we introduce notion x (expressed as a

Spread, concerns the implications for other theories.[7] Spread to other theories becomes transparent if we make a proof in a theory which has wide scope, such as set theory or category theory. So, the rigour of a proof is not only conducive to our accepting the result *as true*, but also in understanding the result in its implications. Rigour is the first step towards such understanding.

How does this work? When we make the inferential steps in a proof explicit, we expose the reasoning to careful scrutiny. In mathematical proofs, it is not enough (except in some applications) that an inference is probably correct, it has to be definitely correct.[8] The notion of *correctness* used by the pluralist is not an absolute notion, transcending all of mathematics, since we have no absolute judge of correctness in the form of a foundation. Rather, we judge correctness, or evaluate a proof, by reference to background knowledge. This accords with reports on the practice of some mathematicians. The background knowledge might include a number of theories, some quite abstruse. For example, some proofs require the existence of some remote large cardinals, embedded in a big theory. Oddly, the big theory is "much shakier than the mathematics that we do" (Thurston 1994, 171). This is because of the presence of, for example, large cardinals, which many mathematicians do not feel they fully understand. Nevertheless, the appeal to the large cardinals in a proof of "less shaky mathematical ideas" helps us to better understand those large cardinals. In these cases, the confirmation of the theorem is only as strong as the most abstruse or 'shaky' notions in the proof. But seen the other way around, the proof acts as a re-confirmation, through a link with, the abstruse mathematics. It helps the abstruse part to come closer to the 'main stream', and be accepted, known and understood.

The importance of explicitness lies not in communication of true mathematical facts, but rather, in its acting as an invitation to judgment and further exploration of limitations. Under this motivation for rigour, there are three aspects to a proof and what it communicates.

(a) There is the *conjectural aspect* of the theorem, which is purportedly proved.
(b) There is the *exposure of the reasoning*, so that this can be scrutinised – leading to a judgment, not only about truth, but also about spread.
(c) There is the *invitation to explore the limitations* of the results: the limitations of the theorem and the justifications for the theorem.

Working backwards: exploring the limitations (c) is done by looking to the class of meta-theories. If we know something of the class of meta-theories within which the

---

wff), then the theory becomes absurd. Here, since we are sensitive to the paraconsistent logicians, 'absurdity' can mean non-triviality, rather than non-contradiction.

[7]Wright refers to this as the width of cognitive command, and the greater the width, the greater the objectivity of the notion (Wright 1992). Squaring the pluralist account with Wright's sensitive treatment of objectivity is the subject of future research.

[8]This is because the 'probability' being evoked here is subjective probability, and depends on knowledge and experience.

proof is acceptable, we know a lot about the scope of the proved theorem. Of course, the relation is not tight or complete in practice, since we might not know what the maximal meta-theory is, so what the maximal scope of the theorem is. Also, we might not know what the minimal meta-theory is, so the minimum needed to justify a theorem. Reverse mathematics only goes some way to helping us to be precise.[9] Worse still, we might not have a particular meta-theory in mind at all, and might just help ourselves to 'whatever it takes' or 'whatever seems appropriate' to verify a proof, and defer learning about some relevant meta-theories. We saw an example of this in Chap. 5, when we discussed the classification of finite simple groups. So the 'theory' in 'meta-theory' might not be an explicit theory at all, but more the background knowledge, and general context, which we assume can be made more explicit if necessary.

What we do know is that if the proof of a theorem is acceptable in a meta-theory (or a class of meta-theories), then we know it will be true along with every theorem of theories verifiable by that meta-theory or that class of meta-theories. The exploration of the limitations, constraints of a theorem or idea is where a lot of the interesting work is done in mathematics. Referring to (b), what is nice about giving a very transparent and meticulous proof is that the justification is exposed. It can then be scrutinised, questioned, corrected, developed *etcetera*. Sometimes we confirm, and sometimes we innovate, as a result of scrutinising our justification. In this sense, scrutiny is a means of understanding or making links between mathematical theories. The scrutiny exposes (a) the conjectural aspect of the idea, or theorem. We shall not say more about this here. Discussion is deferred to Chap. 12.

Before moving to the next section, it might be worth distinguishing justification from explicitness. 'Justification', as it is used here, is a term we take from the constructivists, such as Martin-Löf and Sundholm (Sundholm 2000, 6–7). They believe that judgments are made of theorems *known*. For a constructivist, knowledge does not come in degrees. Justification is a wholly epistemic notion; and while a particular justification, (or set of justifications) could be in error, if we are assured that no error has been made, then we have certain knowledge.

The notion of 'explicitness' is looser. The pluralist 'rigour as explicitness' view is not just comprised of the twins: justification and error in proof. For, the justification can itself be questioned and scrutinised. For the pluralist, knowledge rests on justification, and justification is not absolute. While justification and knowledge are mirrors of each other, they both vary together. There are degrees of knowledge, as there are degrees of justification, and these degrees depend on degrees of explicitness. Unlike for the constructivist, for the pluralist, there is no presumption

[9] To complicate matters, we should also recall that the *minima* of one person is not the *minima* of another: so, for example, we might show that we only need three axioms and one rule of inference to generate a proof of a certain conclusion. Someone else might say that this is not minimal in an interesting sense, since some of the axioms are very powerful. It would therefore be better, according to that person, to show that the same conclusion can be deduced, maybe using more axioms or rules, each of which is less powerful. Reverse mathematics, as it is practiced, makes particular assumptions about what counts as 'minimum'. The choice can be questioned.

that there is a 'bottom' or end to our investigations. There is no atomic, or most primitive unit of knowledge. There are only relatively stable units; ones where we cannot imagine calling them into question. Knowledge, justification and explicitness are revisable, not only in virtue of the possibility of error, but also in virtue of other concerns as well, such as goal or increased understanding in another field. We might know everything we need to know about a theorem, for now, but we might well discover later that we did not know everything, that the justification can be, or even needs to be, deepened, to be re-secured in light of new concepts, theories or ideas. For example, we might learn that a term, axiom or constant, is ambiguous. If said term, axiom *etcetera*, is used in a justification, then that justification is not so much *in error*, but rather, it needs to be re-examined with disambiguations in mind. For example, take the word 'set'. 'Set' might mean 'set' as implicitly defined by ZF, by ZFC, by Gödel Bernays set theory, by Cantor, when he used the term and so on. Moreover the set theory we use to 'define' 'set' makes its own assumptions, concerning the law of non-contradiction, for example. 'Set' is ambiguous, and a new meaning is added with every new 'set' theory.

Having said this, it might seem mysterious that rigour is a virtue of mathematical proofs, since it rests on shaky ground.[10] However, it is exactly rigour which gives us stability, albeit momentarily. Rigour is something we can, and do, come to agree upon. Moreover, under our motivation of explicitness, there will be (ordinal) degrees of rigour displayed by particular proofs. It is the general agreement about these degrees of rigour that lends stability to our notions of truth and objectivity in mathematics. Now that we have our motivation, let us characterise the virtue: rigour.

8.3 A Characterisation of Rigour

The notion of rigorous proof we shall develop is inspired by Frege's notion of a gapless proof. Frege's gapless proofs start with axioms of logic (Basic Laws), which are presented to us as indubitable. Each step in the proof is either a substitution of one term for another, or is an instance of modus ponens. Frege's notion of gapless proof was very strict, in fact, too strict as a standard for mathematical practice. Thus, the idea of 'gapless proof' has been loosened by logicians after Frege, and can be recognised in its modern cousin: 'logical deduction'. But since the notion of logical deduction is usually made in reference to a particular formal theory, pluralists broaden this notion further, to more optimally reflect mathematical practice. Our notion is optimal in the sense of setting an ideal standard of proof while accommodating logical pluralism. We shall introduce the term 'rigorous proof' as a new technical term. Here is the definition.

[10]It can also be a disadvantage (Goethe and Friend 2010, 285).

Definition An *rigorous proof* is a proof that proceeds from axioms or premises, and in which every line of proof is accounted for by reference to a rule of deduction or by appeal to what is presented as an axiom of a theory, a premise or a definition. Each of these has to be of the *right sort* to qualify.

The rules, axioms, premises and definitions have to meet certain criteria. The criteria are the same for axioms and rules since an axiom can be expressed as a rule, and a rule can be expressed as an axiom.[11] The difference is that rules have the connotation of an action, whereas axioms have the connotation of a fixed eternal truth. Since we can inter-translate axioms and rules, the connotations can be taken quite separately from the mode of expression, and we can ignore them here. Henceforth, most of the time, we shall simply use the term 'axiom' as standing for 'axiom and/or its corresponding rule'. The criterion for a formula to qualify as an axiom is that it be an immediate judgment, in the sense of self-justifying. Self-justification is justification in terms of meaning.

Slogan A self-justifying axiom is an axiom that is true in virtue of the meaning of the symbols used in the wff.

The 'meaning' can be located in one, of at least four places, giving at least four interpretations of the slogan. We elaborate on the four, in order to fend from misinterpretation. The pluralist will favour the last.

(i) The meaning might be quite independent of the formal representation.

The account of meaning then proceeds: we have a prior (to formal representation) understanding of a concept (say that of conjunction). We then represent the concept with a symbol, and write out an axiom, or several axioms, which reflect, or represent, the meaning of the symbol. The axiom captures and gives formal expression to the prior understanding. This account suits the realist conception of formal system. In this case, our slogan with this interpretation can be made more precise:

Slogan (i) An axiom is self-justified just in case it represents the real meaning of the symbols in the axiom, where 'real' is independent of the formal representation, and independent of us or our abilities to know it.

For an alternative account of meaning, we shall switch the language back to rules rather than axioms, since this account of meaning is better suited to a constructivist rule, in the style of Martin-Löf's constructive type theory, and is not far from, in this respect, Gentzen's system of sequent calculus or Dummett's intuitionist system or Heyting's intuitionist formal system.

[11] There are some delicate exceptions. For example, the rule of induction is not as strong as the second-order axiom of induction. There are some theorems we can prove using the axiom which we cannot prove using the corresponding rule. The difference has to do with the universal quantifiers in the axiom, since this is different from the implicit 'quantifier' of when we can apply a rule. Tait pointed this out to me in conversation at the AMS (American Mathematical Society) meeting March 2012. These subtleties need not concern us here.

(ii) The meaning of the connectives is constituted by the rules governing them. The rules are a type of logical action.

Notice that when we say this, we have more than a mere stipulation. The rule is meant to capture a type of *act* of reasoning which is *natural* to humans and is knowledge preserving and knowledge conducive. In its formal guise, the rule is developed in order to bring reasoning about all of the logical symbols of the language in harmony. Formally this is easier to verify if there are very few logical connectives, or symbols to define.[12] Each stipulated rule, governing one symbol as main connective in a wff, is made with the idea of what distinguishes that type of logical inference from inferences associated with the other symbols in the language. Part of the meaning is given by the introduction and elimination rules by themselves, but part of the meaning is also given by way of contrast to the other rules.[13] So part of the meaning (what the symbol could have meant, but does not mean) is given by the system of rules in which the rule is imbedded.[14] That is, the rule has to make sense in the context of the whole of the formal system, and it has to make sense with respect to the acts of reasoning of humans. In particular, the whole system should not be trivial, and should resemble our natural reasoning. We shall call deploying such a rule, 'making an immediate judgment'. The new version of our slogan is:

Slogan (ii) A rule is self-justified when it is the result of an immediate judgment.

Note that 'immediate' is not meant in the temporal sense, but in the sense of 'un-mediated'. It might take us quite some time to come to an immediate judgment! This happens just in case it takes some time for us to appreciate that it is true and to appreciate that there is nothing more we can say to justify the rule. Thus, for the constructivist, there is a *last* thing we can reasonably say in a justification, and the justification is entirely strong. There is no regret that we cannot say more. There is no search for more to say.

---

[12]There is a limit. Because of the 'naturalness to humans' clause, a logic with the Sheffer stroke as the only connective is technically nice, but is not very natural because of the awkward match with natural language. Moreover, different linguistic communities will have very different notions of what counts as natural, for example concerning negation.

[13]This second part is not explicitly acknowledged by Dummett et. al. because they wanted meaning to be strictly additive. That is, it should be possible to understand one rule in isolation of the others, in order to respond to the manifestation requirement, dear to Dummettian anti-realists. We think that put starkly, this is naïve. Context does a lot of work in helping us to understand a term or concept. Put another way: while it is quite right that we can take a formal set of rules, and modify them one at a time, it does not follow that the meaning of, say, conjunction is necessarily unaffected if we 'only' explicitly modified our rule for conditional introduction.

[14]Constructivist mathematicians and logicians try to give an elimination and introduction rule to one connective symbol at a time. This is in order to separate the meaning of, say, conjunction from that of, say, disjunction. One motivation for this is that if we find that we have problems with the system – it generates unwanted conclusions, we can make minimal re-adjustments. All of this is correct, however, I do not think we can escape the idea that part of the meaning is implicit in the other rules in the formal system. See Appendix 2, where we discuss Prior's rules for the Tonk connective.

For the pluralist, in contrast, this sort of immediate judgement can only indicate temporary stability. The pluralist regrets not being able to justify a claim further, and assumes that this is a temporary state of affaires, and that we shall later find further justification.

(iii) The meaning of an axiom is all and only stipulated by the axiom.

This is more of a formalist take on axioms. The notion of 'meaning' here is very light. It is unrecognisable to the realist or the constructivist. At most (formal) 'meaning' is 'use' here, and the 'use' is all and only governed by a rule or axiom. A rule or axiom can be arbitrarily stipulated provided, in principle, it *could* be used. It follows that the *collection* of axioms or rules is subject to very few restrictions. One traditional restriction, thanks to Hilbert, is that any combination of axioms should not allow contradictions to be derived (or be self-contradictory). As we saw in Chap. 6, 'avoiding triviality' might be a better limitation, since we can have perfectly effective and non trivial formal systems which do allow some contradictions. Effectiveness is also often used as virtue of a formal system, since effectiveness is one way of specifying what we mean by 'possible use'. 'Effectiveness' is conceptually related to Hilbert's notion of 'finiteness', and could be thought of as the modern counterpart. As we saw in Chap. 6, the idea, then, is that a person could be introduced to a formal system without any appeal to applications or intuitive underlying ideas. Better: we could set up a computer to deploy the formal system. We can manipulate the symbols blindly, or mindlessly. Our slogan then becomes:

Slogan (iii) An axiom can only be self-justified in a particular formal context. It is justified in virtue of being a wff in the language, and when added to a given effective formal system does not turn the formal system into a trivial system or ineffective one.

Lastly,

(iv) Meaning is not a fixed entity.[15]

It changes and evolves. It is dynamic. Under this thinking, we might draw up the following account: we start with a concept, which is not completely precise. We might find that we are dissatisfied with the imprecision, or that we are encountering conflicts or difficulties. We could detect this phenomenologically, in attempting to communicate or justify an idea. As a bid for clarity we try the strategy of giving the concept a formal representation. We give axioms that are supposed to capture the essential aspects of the concept. What the formal representation does is to give a very rigid and precise temporary (or formal) meaning to the concept. We then become familiar with the rigid concept. We can then re-check it: does the rigid version represent the informal version? The answer might not even be a 'yes' or

---

[15]I am aware that some readers will start to feel distinctly nauseous at this suggestion. If you feel this way, then please, rest a while, have a wee dram, and when you feel a little stronger, consult the section on nausea in Chap. 11.

'no' answer. It might be a discovery that there were subtle nuances, which we were not aware of before we introduced the formal concept.[16] It might be that we have discovered that the informal concept was hopeless in some sense: say leading to other problems, so we let the rigid version take over the meaning,[17] or we might say that our informal understanding has been modified in light of the formal concept.[18] Here the meaning emerges as a result of the interplay between the initial concept, the formal version of the concept, our language (formal or informal) and other concepts. The meaning grows and dwindles and shifts with use and spread of connections. The mathematical idea or concept is shared through communication by means of formal representation, and the meaning changes. This is a complex, nuanced and dialectical notion of meaning. It is both 'meaning' in the first person, and in the third person. Our slogan now becomes:

Slogan (iv) An axiom is self-justified iff after an honest search, we have found no further justification, and it makes better sense of the concepts in the axiom than the alternatives we have looked at. Self-justification of an axiom indicates a temporary and relative stability. The stability is relative to the whole theory and greater context.

In other words, the axiom is accepted when it has been carefully scrutinised for fit with our other conceptions, axioms *etcetera*. However, ultimately, it can be re-examined.

Under the first three notions of meaning canvassed above, the axiom or rule governs the symbols used in stating the axiom. There is nothing to say to someone who doubts the rule, since the rule cannot be justified by appeal to other rules, to facts or to other logical systems. One simply accepts the rule or one does not. Acceptance is a pre-condition for using the formal system (correctly). However, for case (iv) we have to say a little more. Since, under the last conception of meaning, meaning is dynamic, adjustments are sometimes made, and are at least thought to be possible. In the last case, we might 'justify' a self-justifying axiom negatively by looking at failed axioms, or axioms which look very similar and are found lacking. We do this by changing the symbols one by one, or several at a time. We ask the questions: do we really mean to use a biconditional here, and not just a single conditional? Do we want the universal or the existential quantifier? Do we need to introduce/develop more symbols, rules or axioms? In some sense, this is the sort of scrutiny we will have used to come up with the axiom in the first place. The only difference here is that coming up with the axiom does not close discussion, since the terms we used to introduce the symbols, which make

[16]The view is that the exercise of giving a formal representation of an idea, or group of ideas, is an exercise in deepening understanding, since it gives us something relatively precise and fixed to measure our original concepts against.

[17]In this respect the pluralist distinguishes himself from a Brouwerian intuitionist who locates all meaning and real mathematics in the mind and never on the written/typed, page. For the Brouwerian, formal representation fixes and therefore necessarily distorts real mathematics.

[18]For interesting conjectures of how metaphor and symbol influence mathematical development see Johansen (2010, 193–194).

up the axiom are also, potentially, changing meaning. For example, we can ask the further question: how are we understanding the quantifiers in this axiom? Are they substitutional, objectual, does the universal quantifier imply the existential (are domains non-empty)? So re-checking (albeit negatively) that an axiom is self-justifying; re-checking that there really is no more to say, is always an option. Following Cellucci (2008, 12), we can say that an ‘axiom’, as the term is used in mathematics, is, philosophically, a hypothesis.

We discussed these four notions of meaning as qualifiers to the idea that an axiom or rule in a proof has to be of the right sort. It has to be self-justifying, and this can be understood in many different ways. To finish filling out this notion of rigorous proof, the last comments we owe concern definitions and premises. We shall make life easy for ourselves, and say that definitions, in a rigorous proof, are simply shorthand expressions.[19] So nothing new is introduced when we appeal to a definition. Premises are simple too. They have to be truth-apt wffs.

We made all of these qualifications concerning criteria for being an axiom, a rule, a premise or a definition in order to understand the definition of a rigorous proof, which we repeat is:

A rigorous proof is a proof that proceeds from axioms or premises, and in which every line of proof is accounted for by reference to a rule of deduction or by appeal to a self-evident axiom, a premise or a definition.

What is to be counted as a rigorous proof varies with what counts as a self-evident axiom or rule, and this varies with one’s philosophical inclinations. Thus, what is recognised as a rigorous proof for one person does not so count for another. We can now appreciate that there will not be uniform consensus over whether a particular proof is rigorous or not. The pluralist favours the fourth notion of meaning, but recognises that the others are used. The fourth variation will also admit of degrees of rigour in a proof. Degree of rigour is measured by reference to degree of scrutiny.

8.4 Evaluating the Characterisation

Rigorous proofs, as they are defined above, seem to set a good and high standard of rigour. We are now in a position to evaluate the characterisation with respect to the motivation of explicitness and honesty. Rigorous proofs are certainly explicit. The justification for the conclusion of a rigorous proof can be traced without ambiguity. However, there will be practical limitations which interfere with our always writing out rigorous proofs. For example, some proofs are too long to follow. Consider the four colour map problem: that any map on a two dimensional surface

[19]We could make the notion of proof more complex by allowing impredicative definitions or contextual definitions, where a contextual definition is one where the biconditional of definition is within the context set by quantifiers, and where one side of the biconditional of definition is an equivalence relation.

can be coloured-in using four colours in such a way as to never have the same colour adjacent. A computer has worked through all of the possibilities, and has confirmed that, indeed all we need are four colours. The proof is too long for any one person to check. However, one person can check the program, which generated the proof. That is, we can check that there is a rigorous proof, where an effective algorithm is thought to count as a type of rigorous proof. Checking that there is an acceptable proof is often what we do in mathematics. As Bostock writes: "one does not actually construct such [axiomatic] proofs; rather one proves that there is a proof, as originally defined" (Bostock 1997, 239). If we have a proof that there is an underlying proof of a proof of a certain sort (our favoured version of a rigorous proof) what does this indicate?

If the meta-proof is a rigorous proof, then we have met the motivation of explicitness and honesty, even if we have accounted for this only at the meta-level, and not directly. The meta-proof is explicit and honest concerning the object-level proof because all there is to following the underlying rigorous proof is to use rules, which are approved of in advance, and these are revisable, if there is further difficulty. But, as we mentioned, rigorous proofs are seldom found in mathematics at the object level or at the meta-level. There are limitations to the concept. In particular, what is indicated when a meta-proof, itself, is not a rigorous proof?

8.5 The Limitations of Rigour for the Pluralist

We have characterised rigour, for the pluralist, and, in general, it is a virtue. But it has limitations. In particular, in practice, we do not write out rigorous proofs in mathematics. So the notion does not reflect practice. But should we not take rigour to be normative and therefore, bring practice in line with the notion? There are practical reasons of time, space, resources and so on which weigh against writing out rigorous proofs. We skip some steps if they have been accounted for in advance, in the form of lemmas. We also refer to proved theorems and lemmas from the work of others. There is no need to reproduce these every time. But all of this meta-accounting implies that we *could* make a full rigorous proof, if called upon to do so. The question then remains, what happens when a meta-proof is not itself a rigorous proof, and we do not have sufficient evidence that the proof really can be turned into a rigorous proof? Is this adequate?

A rigorous proof forms the core of, underlies, or is an idealised version of a proof. For this reason, the pluralist thinks of rigorous proofs as regulatory ideals. The motivations for rigour reach beyond the proof itself. They reach to the context: once we have as much as is practicable of the proof displayed in full rigour, we need to be explicit about the context of proof. This requires us to say something about the proof in a meta-language. Moreover, the meta-language has to include justification for the axioms and rules. We can set up a protocol, but again, as a regulatory ideal. That is, the ideal is there to regulate disputes or assuage doubt. If no doubt is present, then there is no need to follow the protocol.

Say we come across a dodgy proof, moreover, the theorem proved raises doubt. Here 'doubt' is a trained professional reaction, not just a subjective feeling. What sort of protocol might we follow in this instance?

1. Re-read the purported proof.
2. Think about the conclusion independently of the proof.
3. Fill in doubtful moves in the proof, as per the notion of a rigorous proof. Failing this, rigorously prove that there is a rigorous proof.
4. Look to the context, or class of meta-theories. Check the proof that proves that there is a rigorous proof. Go to (1), but this time you are looking at the proof of a rigorous proof. When you get back to (4), if you are still in doubt, move to (5).
5. Think more widely about the class of meta-theories or the general mathematical context of the proofs. Maybe the underlying logic should be re-examined.

In practice, mathematicians do not follow this protocol, although it is a description of what one could do, in order to be rigorous. As Thurston observes:

> Mathematicians apparently don't generally rely on the formal rules of deduction as they are thinking. Rather, they hold a fair bit of logical structure of a proof in their heads, breaking proofs into intermediate results so that they don't have to hold too much logic at once. (Thurston 1994, 164)

Worse, they might find very little value in a formal proof.

> Fabio Conforto described his colleague and co-author's [Enriques'] "powerful intuitive spirit" and unalterable belief in "an algebraic world that exists in, and of itself, independent and outside of us" – a world in which "seeing" was the most important implement in a mathematician's toolbox: Enriques *did not feel the need of a logical demonstration of some property*, because he 'saw'; and that provided *the assurance about the truth* of the proposition in question, and *satisfied him completely*. (Babbitt and Goodstein 2011, 242)[20] (Italics mine)

So, generating rigorous proofs is rarely done. 'Seeing' is one way of meeting steps 2 and 5 of the protocol. However, we want some idea of rigour in order to check on reliability of proofs and results, especially if there is a dispute.[21] Again, however, the protocol is not followed in practice. It is only a regulatory ideal. Thurston again: "reliability [of results] does not need to come from mathematicians formally checking arguments: it comes from mathematicians thinking carefully and critically about mathematical ideas". This is presumably justified because the mathematician's working system is trustworthy in general. We think this because

---

[20]Enriques' most important proof was that the "characteristic series of a "good" complete continuous system of curves on a smooth algebraic surface F is complete. (Here "good" in the old Italian terminology means not superabundant, or, in modern terms, the first cohomology group of a divisor class should be zero)." (Babbitt and Goodstein 2011, 244.) The proof had a gap. This was not 'filled' until the appropriate algebraic tools were developed much later. This is one of the many interesting cases of mathematicians feeling that they are right, and being proved right much later.

[21]There was a rather heated dispute between Enriques' and Severi who criticised his proof in print.

it is "quite good at producing reliable theorems that can be solidly backed up" (Thurston 1994, 170). Sometimes, the 'solid back up' is a promissory note, and the truth of the theorem is an article of faith.

> A sceptical attitude [because of gaps in the proof] towards these ideas [of infinitely close points and curves] is easy to have but it is not very productive. Instead, those who are more trusting of what these concepts can yield, will, I am sure, discover new results in other fields. Quoted in Babbitt and Goodstein (2011, 246), from Enriques (1938)

This is why we do not think we need to follow the protocol in most or any instances. But we have to be careful about what 'solidly backed up' means here. It either means with reference to a rigorous proof, or something like it, or it means that we have found application – so we have an 'independent' check. The application might, however, turn out to not be very independent. For example, we might find an implementation for a result in computer theory, but the computer theory shares some preconceptions, and at least an underlying logic, with the result, so the 'backing up' is not so independent after all. Similarly, the idea of 'solidness' is not very convincing. All we have is a spread of results over a number of independent theories, but they are independent for reasons of evolving in different traditions, or having other goals, not for reasons of mathematical independence. In fact, they could not be mathematically wholly independent, because if they were they would be incommensurable, and therefore incomparable, and not reassuring at all! To be more careful, we would say that they are epistemically independent, or "we are not aware of the connections at present".

Nevertheless, were we to make rigorous proofs, we would find that they are maximally rigorous within a given axiomatic theory. Within that theory, we cannot be more rigorous. Every step in the proof is accounted for. Where the notion loses its stability is when we step out of the theory and question the axioms or rules of inference, where we question the definitions or licence for substitution (Cellucci 2008, 165). In fact, we can think of a rigorous proof as a way of exposing every move to invite greater scrutiny. The scrutiny has to come from an external perspective – usually from a meta-theory, meta-perspective, or transcendental perspective, possibly from intuition, or the mathematical world we 'see'. The external perspective, the notion of crosschecking in mathematics, is the subject of the next chapter. There I say something quite precise about the nature of the external perspective. Nevertheless, what this discussion of rigour has taught us is that we can think of a gap in a proof in this way too: as an invitation to either fill in the gap, or to seek an independent outside perspective. One way of anchoring the outside perspective is through invariance; another is through parameters. Since invariance and parameters can be used to better anchor our judgement of rigour, we turn to this in the next chapter. The next chapter, and Chap. 12 are both close companions to this one.

References

Babbitt, D., & Goodstein, J. (2011). Federico Enriques's quest to prove the "Completeness Theorem". *Notices of the American Mathematical Society, 58*(2), 240–249.

Bostock, D. (1997). *Intermediate logic*. Oxford: Oxford University Press.

Cellucci, C. (2008). Why proof? What is proof? In G. Corsi & R. Lupacchini (Eds.), *Deduction, computation, experiment. Exploring the effectiveness of proof*. Berlin: Springer.

Enriques, F. (1938). Sulla proprietà caratteristica delle superficie algebriche irregolari. *Rend R Acc Lincei, 6*(27), 493–498.

Friend, M. (1997). *Second-order logic is logic*. Ph.D. thesis, University of St. Andrews.

Friend, M., & Pedeferri, A. (2012). An analysis of the notion of rigour in mathematical proof. *L&PS – Logic and Philosophy of Science, IX*(1). Section 2. http://www2.units.it/episteme/

Goethe, N. B., & Friend, M. (2010). Confronting ideals of proof with the ways of proving of the research mathematician. *Studia Logica, 96*(2), 273–288.

Johansen, M. W. (2010). Embodied strategies in mathematical cognition. In B. Löwe & T. Muller (Eds.), *PhiMSAMP philosophy of mathematics: Aspects and mathematical practice* (Texts in philosophy 11, pp. 179–196). London: Individual Author and College Publications.

Sundholm, G. (2000). A plea for logical atavism (Manuscript of lecture presented at LOGICA 2000).

Thurston, W. P. (1994). On proofs and progress in mathematics. *Bulletin of the American Mathematical Society, 30*(2), 161–177.

Wright, C. (1992). *Truth and objectivity*. Cambridge, MA: Harvard University Press.

Chapter 9
Mathematical Fixtures

Abstract The pluralist sheds the more traditional ideas of truth and ontology. This is dangerous, because it threatens instability of the theory. To lend stability to his philosophy, the pluralist trades truth and ontology for rigour and other 'fixtures'. Fixtures are the steady goal posts. They are the parts of a theory that stay fixed across a pair of theories, and allow us to make translations and comparisons. They can ultimately be moved, but we tend to keep them fixed temporarily. Apart from considering rigour of proof as a fixture, I discuss fixed models, invariant notions and fixed information about objects across theories. There are other fixtures, but it is enough to start with these.

9.1 Introduction

In the last chapter we ended with the need to buttress the notion of rigorous proof. I propose as a buttress, the notion of a fixture. A formal logic is an obvious 'fixture', since it fixes the reasoning allowed in a mathematical proof. It upholds the mathematical theory. We discussed this in Chap. 5 and in the last chapter, so I shall not discuss it here. Instead, we shall look at other sorts of fixture.

'Fixtures' are notions which stay fixed across mathematical theories.[1] They are preconditions for crosschecking in mathematics.

The ideas for this chapter came from discussions with Pedeferri and Mourad. I should like to thank Priest for checking the material on the collapsing lemma and on chunk and permeate.

[1] The word 'fixtures' is supposed to be suggestive of the notion of a fixed point in mathematics, but it is a little looser than that of fixed point.

M. Friend, *Pluralism in Mathematics: A New Position in Philosophy of Mathematics*, Logic, Epistemology, and the Unity of Science 32, DOI 10.1007/978-94-007-7058-4_9,
© Springer Science+Business Media Dordrecht 2014

Mathematics, more than any other discipline, involves crosschecking.[2] We check one theory against another.

In this chapter, I discuss three sorts of fixture other than that of rigorous proof. I discuss: fixed models, invariant notions and fixed information about objects across theories. There are many more. But even in looking at three we can form an impression of the breadth and variety of crosschecking in mathematics.

I advance the thesis that the crosschecking supplants the need for absolute truth, absolute and independent ontology, a foundation or for a single orientation.

This thesis overhauls many preconceptions in the philosophy of mathematics, and I think that it offers a better account of the practice of mathematics.

The first fixture comes from structuralism, and is given general expression in model theory. It follows on the heels of the chapters on structuralism and formalism in Part I of this book. The second fixture comes from the Erlangen programme. The work of Klein inspired Lindenbaum and Tarski to generalise on his methods and apply them to questions in logic. They developed an idea of invariant, or fixed notions in logic. This is different from developing a fixed formal logic, since the logical notions are common to several different formal systems. The third comes from Lobachevsky. In Chap. 5, we witnessed Lobachevsky importing hyperbolic geometry into a 'foreign' context (Euclidean geometry), for the purposes of a proof.

There are other ideas we could develop along these lines. In general, the idea of 'fixtures' testifies to mathematicians applying one part of mathematics to another. Contemporary mathematicians do this all the time; they analyse a mathematical theory with reference to, or with the tools of, another theory. Moreover, the theory giving the analysis, interpretation or insight is not always consistent with the theory being analysed, and this sets off alarm bells, especially for philosophers.

More traditional philosophers are inclined to think that the only explanation for the presence of together inconsistent theories (such as spherical and plane geometry) is in terms of embedding one theory into a foundational theory or into one orientation, which they then identify with a foundation for 'mathematics'. They aspire to show that we can reduce mathematics to a (unique) foundation, or capture all of mathematics by means of an orientation; so all of mathematics is 'consistent' (or coheres) in the end. What look like together inconsistent theories, are simply sub-theories of the bigger theory. The sub-theories might contradict each other in some respects, for example, commutativity might hold for 'addition' in one sub-theory but not in another. But the foundational theory compartmentalises

[2]This was less the case in the past, especially when geometry and arithmetic were kept quite separate, but not completely. The interaction between the two makes for fascinating history. Today, we do not so much use arithmetic to check geometry and geometry to check arithmetic, but rather, we use set theory, and the ultimate tool: model theory to do this. However, as we saw in Chap. 4, and as we shall see more precisely, model theory has its limitations.

the sub-theories so that there is no explosion through *ex contradictione quodlibet*. The compartmentalisation is what allows the definitions and axioms of the particular sub-theory to also conflict with those of the foundational theory, or orientation.

However, as we have seen in the first chapter, the foundational project, as it is philosophically conceived, should be reconsidered. The evidence is simply lacking for the sort of foundations philosophers have been looking for: one providing an essence, ontology or absolute truth for all of mathematics. Similarly, the idea of a unique orientation is limited. We saw this in Chaps. 4 and 5. If we are ambitious to overcome the limitations, then we need a new idea, which can account for the otherwise seemingly casual borrowing between theories, the lack of rigour in accepted proofs, the variation in what is to count as a rigorous proof or the applying of one mathematical theory to another. The pluralist replaces the idea of foundation with fixtures. There are many of these, and there will be more developed in the future. As long as there are some, mathematics, as a discipline, is cohesive.

Of all the orthodox philosophical positions, the formalist and the structuralist come closest to suggesting this idea, but both positions are more restrictive than necessary, and than is desirable, if we want to account for present day mathematical practice. Because the formalist and the structuralist come the closest, we shall start with the formalist and structuralist philosophies, and develop the idea of fixtures from these. We end up with a more fluid, accurate and tighter conception of mathematics. The conception is more fluid because it allows for different types of fixture. The view is more accurate because it highlights what mathematicians are in fact doing, and the conception is tight because it accounts for the different ways in which mathematics re-enforces other mathematics, and that this can be done in several quite different ways.

More specifically, the formalist project is limited, and it is in overcoming the limitations that the pluralist replaces the notion of finitistic, or effective, rigorous proof and consistency of theory, with the notion of fixture. The pluralist can then account for the practice of using together inconsistent theories in a much more direct way, although he also endorses the formalist strategy of fixing an underlying logic *when this will work*, and he endorses the structuralist strategy of finding a meta-structure, *when this will work*. The formalist and structuralist strategies are possible strategies amongst others, and we need more than those two strategies to develop an accurate account of mathematics as it is practiced.

In the third section of this chapter, the pluralist will look the structuralist's broadly formalist approach. But in the following ones, I shall take an increasingly piecemeal approach. I shall appeal to: invariance under transformations of the domain of objects onto itself, and finally, to some recent developments of paraconsistent formal logics, where I shall also appeal to the idea of using logic metaphorically, as discussed in Chap. 7. This chapter thus follows the chapters on formalism, structuralism, rigour and the metaphorical use of logic. The chapter also anticipates remarks concerning the discourse of mathematics and communication, such as we find in Chaps. 11, 12, and 13.

In the conclusion, I discuss the philosophical lesson we can learn from the discussion in this chapter. The pluralist turns the traditional philosophical analysis of mathematics on its head. Rather than think of the fixtures as *evidence for convergence* due to an underlying ontology, truth or one foundational mathematical theory, or one orientation; the pluralist takes the fixtures as evidence for fixtures, and nothing else. Fixtures tell of the *coherence, cohesion and soundness* of the discourse. More important, they *supplant* the urgency to look for a unified ontology, absolute truth or foundational theory.[3] They play the role that consistency plays for the formalist, by sanctifying a mathematical theory, proof or methodology; but even this sanctifying is not done once and for all. As the sainthood of a person is revisited on occasion by the Catholic Church, so the endorsed mathematical theories and proofs are always up for scrutiny. It is in the multiplicity of ways of crosschecking mathematics that we find stability and security in the discourse – security against nonsense.

This is not to say that the pluralist denies the *possibility* of finding such philosophical prizes as a unified foundation. Rather, the pluralist does not feel that the time is right to favour such a search, for lack of evidence. But for all that, the pluralist does not despair. He does not turn to nihilism, and he does not become a quietist. In fact, as we shall see, he has a lot of work ahead.

9.2 The Path from Formalism, Structuralism and Constructivism to Pluralism

Here is one route to the general idea of this chapter. Start with a formalist conception of mathematics. For the formalist, the guarantee that one is doing mathematics is twofold. (i) A formal mathematical theory must be consistent (which we now think of as interpretability in a model) and this gives us freedom (from genesis, history and metaphysical baggage). The freedom comes with (ii) a stringent conception of proof in mathematics (the 'idealised conception'). The idealised conception is discussed in more detail in Chap. 12. For now, it suffices to have in mind a rigorous proof, as defined in the last chapter, or the sort of natural deduction proof we ask our students to produce in a first logic course.

This idealised formal proof is almost never found in published papers in mathematics, and holding mathematicians to that standard is unrealistic. Moreover, the pluralist is ambitious to reflect mathematical practice in his philosophical position. As a result, as pluralists, we need to revise our ideas and go beyond the formalist philosophy.[4]

The pluralist suggests that rather than hold mathematics hostage to the conception of an idealised proof, we hold it hostage to an open disjunction of 'fixtures', *one* of which is the idealised notion of proof; another, is to use model theory as

[3]This thought is perfectly reflected in Wright's discussion of convergence and cognitive command (Wright 1992, 88–93).

[4]More detailed evidence for this is left to Chap. 12.

our vehicle for finding fixtures. So, here, we leave the formalist behind and turn to the structuralist strategy. Using model theory, we can individuate theories as structures, and then compare them in a common language (of second-order logic). But as we saw in Chap. 4, structuralism, so conceived, is limited, since there are aspects of mathematics (as practiced) which cannot be recognised in a classical model theory framework. Pluralists want to reach beyond structuralism too. They aspire to maximal, not optimal, pluralism. Model theory with the identification of meta-structures is another fixture. But it is one amongst many more.

There are many fixtures used now, and there will be more developed in the future. To give an extensional definition of fixture would be premature, and might fix ideas too rigidly for us to properly recognise future mathematical developments, so instead of giving an extensional definition, let me give a schematic one[5]:

A 'fixture' is some mathematical idea, expressed as a constant across theories. It is used as a basis for comparison between theories. In particular, it is a necessary condition for the comparison of theories – ensuring some commonality.

Further justification for the use of one theory, technique, methodology *etcetera* for analysing, or proving, a result in another theory, involves demonstrating that the borrowing is not dangerous. 'Danger', here, means a number of things, ranging from triviality at one extreme, to departing from some goals or criteria, which are independently determined, by philosophy or practical considerations. Examples of these less extreme goals are: to only use constructively acceptable proof techniques, to be useful in an application in computer science, to remain within what is conceivable or within what we can 'picture' or imagine. Infinite dimensional space counts as 'unimaginable'. The pluralist is pluralist about these dangers, but recognises them within the context of an outlook or philosophy, since he is pluralist about philosophies too. This makes the pluralist sensitive enough to investigate which fixtures are appropriate for what purposes and inappropriate for others.

For example, constructivists could require that we only use 'constructive methods', where this is philosophically determined, so the 'danger' lies in departing from methods which are constructively acceptable. For all that, they might still allow a range of methods, or constants, which disagree with each other, *prima facie*. More specifically, a constructivist might use a rule of inference which is only allowed in certain contexts. For example, the law of excluded middle, or the axiom of choice, are allowed in very restricted contexts, where we can tell for any particular instance whether it or its negation holds, or where we give a choice function, respectively. The formalist is not a constructivist. A formalistically acceptable proof might not be constructively acceptable. A structuralist can accommodate the formal work of the constructivist, but does not fully share, or recognise, the constructivist's

[5]The definition is not extensional in the sense of determining its extension. It is not intensional, in the sense of 'depending on the mode of presentation'. It is not intentional, in the sense of being sensitive to an attitude. It is more of a scheme. The extension will depend on what we count as an idea, a theory, a comparison, and what we can recognise as common to two theories.

motivation. The structuralist thinks of constructive mathematics as a subset of classical mathematics. Thus, he can identify constructively acceptable mathematics, but cannot recognise the philosophical motivation because his philosophy is not philosophically pluralist in this sense.

The constructivist's practice is not well accounted for by the formalist or the structuralist. Nevertheless, constructivists are doing 'good' mathematics, at least *prima facie*. Since formalism and structuralism are too restrictive in the fixtures they consider, the pluralist widens the field. We shall begin with a familiar fixture, since it is recognisable by both the formalist and the structuralist. However, we shall then explore less familiar territory in the section on Lobachevsky.

9.3 Model Theory and Structures

The invoking of model theory concerns the structuralist thread in this book. Thus, this section re-enforces the work of Chap. 4. For, here I show more technically, and more precisely how structuralism has inspired pluralism, and how it can be thought of as a complement to pluralism. Since model theory is not central to the thesis of pluralism, I shall not give a summary introductory course on model theory. I refer the reader to Hodges (1997, 1–69) for this. Suffice it to say that we use model theory in order to compare theories, models or structures. It should also become clear how pluralism goes beyond structuralism in techniques, sensitivity and scope.

For the modern mathematician, a popular way of comparing mathematical theories is to organise them in terms of structures, and then to see what mappings hold between the structures. This way of comparing theories comes with a warning. The whole model theory approach assumes the language of set theory and is wedded to classical logic. These are the limitations of the model theoretic approach. Nevertheless, it is useful, provided we bear these limitations in mind.

A structure is an n-tuple consisting in a set of object constants, a set of objects over which variables range, a set of predicate, relation and function constants and variable predicates, relations and functions varying over the powerset of the domain of objects. With Shapiro, we are assuming a second-order language. To our n-tuple, we add an assignment function which maps object constants to objects in a domain, first-order predicates to sets of objects in the domain, binary first-order relations to ordered pairs of objects in the domain, and so on. A theory is a set of formulas closed under some operations. The interpretation, or application of a theory is quite separate. The interpretation is the semantics of the theory. The semantics is thought of in terms of models. Models satisfy, or fail to satisfy sets of sentences. We attribute properties to theories by investigating the relations (mappings), which bear between models which satisfy the axioms and, therefore, theorems (assuming the syntax is sound with respect to the intended semantics). We compare theories, by comparing their models. The comparison is made formally, using functions defined in the language of set theory. Using model theory, we learn the limitations of theories: we prove limitative results about a theory or

Table 9.1 Truth table for paraconsistent disjunction

p	q	p ∨ q
T	T	T
T	Ŧ	T
T	F	T
Ŧ	T	T
Ŧ	Ŧ	Ŧ
Ŧ	F	Ŧ
F	T	T
F	Ŧ	Ŧ
F	F	F

language. For example, we use model theory to study: embeddings, consistency, equi-consistency, completeness, compactness, soundness, the Löwenheim-Skolem properties, categoricity, decidability and so on.

Let us start with a simple example. A model theorist will think of a group as a structure. The structure contains an identity element, 1, two binary function symbols, $+$, $\times$, naming a group product operation and one unary function symbol, $^{-1}$, naming the inverse operation (Hodges 1997, 3). The theory (the set of formulas that are true in the theory) is closed under some version of addition and multiplication. This structure can be embedded in a larger structure which includes everything in a group plus the operation of division.

Take a slightly more interesting example. Propositional LP, the propositional logic of paraconsistency, includes truth-value gluts in its semantics. That is, propositional variables vary over true, false and both true and false sentences, or propositions. An example of an English sentence, which is plausibly both true and false is: 'This sentence is false'. LP was conceived to be able to cope with reasoning over such sentences, without engendering triviality. The motivation for developing a logic which can help us reason over such sentences at all, is that logic should be able to reason over *any* sort of truth-apt sentence, not just ones with one truth-value.[6] The formal system leaves in place all of the reasoning of classical logic. That is, if we look at the truth-tables, which define, say, '∨', in LP then the classical part of it will have the same results as in regular classical logic. See Table 9.1.

Here, we have defined V as follows: p ∨ q is true iff either p is true or q is true, and, p ∨ q is false if p is false and q is false. Ŧ is used to indicate both true and false. Notice that if we blot out all rows with p or q Ŧ, we get the classical table. We do the same for the other connectives. The peculiarities of minimal LP only show up in the presence of inconsistencies, in the rows with Ŧ. This is what minimal means in

[6]This invites a second question: whether logic should be able to help us reason over sentences with no truth value (a truth-value gap), but this is a separate question, since, arguably, no truth-value is not *a truth*-value.

So the question is really should logic help us to reason over something other than truth-apt sentences.

Table 9.2 Truth table for LP negation

p	~p
T	F
Ŧ	Ŧ
F	T

Table 9.3 Truth table for a classical tautology in LP

p	p ∨ ~p
Ŧ	Ŧ
F	T

this context. It is a minimal disruption to classical reasoning. Nevertheless, LP is a bit peculiar beyond the immediate presence of contradictions. It has no tautologies in the sense of formulas that are always and only true. If we loosen the definition of tautology to be a formula that is always true, then tautologies, can contain falsehoods (which are also true). So we can have two sorts of LP, one with tautologies and one without. The one with will be minimal since it together with classical logic will share in the tautologies. For example, p ∨ ~p is a tautology in classical logic (it is all (and only) ever true); it is also a tautology in minimal LP, because it is always true. But it could also be false. The reason is that if we consider the case when p is a sentence with both truth-values, then p ∨ ~p is true, but it is *also* false, since negation changes a true sentence to a false one, and a false one to a true one, so one that is both true and false, remains both true and false. See Tables 9.2 and 9.3.

Notice that I have been a little tricky here, since I am discussing a paraconsistent logic, and I am supposed to be using *classical* model theory. This is acceptable since the meta-language, the one I am *using* to analyse and compare the two theories is classical. The odd issue about truth-values is considered, from this meta-perspective, to be an unusual artefact of LP, but it does not infect the meta-language. Notice, that there is still something odd going on, since meta-languages are usually thought of as being as strong as, or stronger than, the object language, and to show this they are thought of as containing all the elements of the object language, and might have some extra symbols added on. This cannot be the case here, since in a classical meta-language, we do not have the resources to say that a formula in the meta-language has two truth-values (without thereby committing our theory to triviality). So, we actually have to do a little fancy footwork to account for the added richness of the object language over the meta-language.

We have two notions of truth-value. One is the classical one of the meta-language, the other is partitioned off as the 'mentioned' truth-value, which belongs to a theory. To alleviate confusion, we might replace 'truth' (of the object language) with '1' in the used meta-language, and 'false' with '0'; or we could be more elaborate and make up a new symbol altogether, so it becomes a technical term, and we think of LP as only a formal theory; and ignore the genesis of LP (to allow us to reason formally over contradictory formulas). Having done this, we can use our model-theory to tell us that the set of tautologies in classical logic is the same as that in LP. This will be enough to show that the theorem of deduction, for example, holds in LP;[7] and we will have learned something. Our understanding of LP will have increased, and our understanding of classical propositional logic will have increased.

What we have learned from our foray into model theory is that it is a well-crafted tool for comparing theories. We use it often, and it is very adaptable. But it does have its limitations, for example, having to ignore (or fail to be convinced by) the genesis of LP, in order to work with it, and compare it to classical propositional logic. The ignoring of such issues is also an advantage of model theory, for, it is that which allows us to make our comparison. We can then return to the more philosophical and intuitive ideas, but better informed as to some of the costs of, say, adopting LP. Note again, that the 'costs' will only be those recognised by model theory, i.e., that we cannot understand truth in quite the same way as we do in the meta-language, or in a classical language. A structuralist, *qua* structuralist, will not be able to do justice to the genesis of LP, since she cannot properly recognise truth-value gluts, since they cannot belong to the classical meta-language.

9.4 'Logical Notions' and Invariance

Tarski gives a very practical and interesting answer to the question: what are logical notions? The answer is ambiguous; and I shall return to this. To answer the question, Tarski extends Klein's technique to logic, saying that the technique can be further generalised to other sciences, in particular, logic (Tarski 1986, 146). First, I give some of Tarski's discoveries. It is evident that this technique and conception can be applied more or less widely, and encounters some interesting limitations even within the application Tarski made. I then address Tarski's conclusions and the pluralist's conclusions.

Klein aimed to find a unified approach to geometry by means of the study of space invariances with respect to a group of transformations. This idea was pressing because of the new non-Euclidean geometries. Not only did these force us to revise our compliance and faith in our geometrical intuitions, but the non-Euclidean geometries introduced new considerations on groups (therefore, introducing new algebraic ideas). The algebraic ideas then fed back to the geometries by way of finding invariances across geometries. The 'finding of the invariances' was essentially an analysis of the logical structure of geometry. Following Klein's insight, Tarski used the notion of invariance under a permutation of a domain of objects (on to itself) to identify logical notions (within the foundational theories of Whitehead and Russell's type theory and von Neumann set theory and Gödel-Bernays set theory).

---

[7]The theorem of deduction just states the relationship between the conditional and provability in a theory. If Σ is a set of wffs and α and β are particular wffs, the theorem of deduction is: $\Sigma, \alpha \vdash \beta \iff \Sigma \vdash \alpha \rightarrow \beta$.

We treat the domain over which we reason as a set, and we may transform it on to itself with any function, which we can define in the language. We discover that with some transformations the objects of the domain remain the same, such as with the identity transformation, some notions remain stable under some transformations and not others, such as the 'lesser than' relation. The notions (properties, relations between objects) which remain the same under *any* (recognised) transformation of the domain on to itself are invariant. These are the logical notions, for Tarski. In this very precise sense, we say that the 'meaning' of these notions is invariant. It turns out that these are the logical connectives, negation, identity and the quantifiers. Or rather, subsequently, we have worked out that the invariant notions can be reduced to these. Tarski himself did not make this discovery directly. Instead, he found that there are only two invariant notions of class: the notion of a universal class, the notion of the empty class (Tarski 1986, 150). There are four invariant binary relations: "the universal relation which always holds between two objects, the empty relation which never holds, the identity relation which holds only between 'two' objects when they are identical, and its opposite the diversity relation." (Tarski 1986, 150). Ternary, quaternary relations, and so on, also have a small, finite number of invariant notions under transformations, similar to the invariant notions over binary relations. The last notion which shows invariance is that of the cardinality of the domain.

Tarski had the Whitehead and Russell project very much in mind when he presented this material, so he went on to speculate whether this shows us that mathematics is really logic, and logical notions (so defined) are what tie mathematics together. For Tarski, such a conclusion is too hasty. These results should be taken with a pinch of salt. I shall discuss two reasons for this. One is that the methodology will not determine for every notion whether it is invariant, or 'logical'. Tarski is quite frank about this, for he asks the question: is the (set theoretic) membership relation a logical notion or not. It turns out that it *is* if we consider the membership relation in Whitehead and Russell's type theory. But it is *not* if we consider the membership relation in von Neumann set theory (Tarski 1986, 152–153). As Shapiro puts it: "... on this [model theoretic] account, the logical-non-logical distinction would be an artefact of the logic [of the meta-theory]." (Shapiro 1991, 7). Tarski concludes that the technique cannot answer the question whether membership is a logical notion. He then, a little quickly, concludes that "The answer is 'As you wish'!" (Tarski 1986, 152). The conclusion *would* be justified, if we had both an accompanying proof that there is no *alternative* formal theory (such as a super set theory) by which to decide the matter, and if we are convinced that 'logical notion' means precisely 'invariant in this super theory'. Not everyone would agree to these assumptions. Tarski discusses this problem, and the reader is referred to Tarski (1986) for details.

Instead, let us turn to the second problem with Tarski's approach, since it is more serious; it affects the notions Tarski identified as invariant, and the limitations of his technique. The technique is dependent on several assumptions to which Tarski is not entitled, since these are assumptions about logic (or meta-logic), and presumably this is what is at issue when we ask the question: "What are the logical notions?"

Here is one assumption which, if modified, changes what is counted as an invariant notion. Tarski assumes that every wff in the language is either true, false, never both and never neither. If we allow wffs with two truth values, (wffs which are both true and false) as we find in some paraconsistent logics, and we consider only theories without functions,[8] then the cardinality property will no longer be invariant! Consider the collapsing lemma as a permutation of a domain of objects on to itself.

> Let $\mathcal{U}$ be any interpretation with domain D, and let $\sim$ be any equivalence relation on D. If $d \in D$, let [d] be the equivalence class of d under $\sim$. Define a new interpretation $\mathcal{U}^{\sim}$, whose domain is $\{[d]; d \in D\}$. If c is a constant that denotes d in $\mathcal{U}$, it denotes [d] in $\mathcal{U}^{\sim}$. If P is an n-place predicate, then $<X_1 \ldots X_n>$ is in its positive [negative] extension in $\mathcal{U}^{\sim}$ iff $\exists x_1 \in X_1 \ldots \exists x_n \in X_n$ such that $<x_1 \ldots x_n>$ is in the positive [negative] extension of P in $\mathcal{U}$. What $\mathcal{U}^{\sim}$ does, in effect, is simply identify all the members of D in any one equivalence class, forming a composite individual with all the properties of its components. I [Priest] can now state the:
>
> *Collapsing Lemma*:
>
> Let φ be any formula; let v be 1 or 0 [T or F]. Then if v is in the value of φ in $\mathcal{U}$, it is in its value in $\mathcal{U}^{\sim}$.
>
> In other words, when $\mathcal{U}$ is collapsed into $\mathcal{U}^{\sim}$, formulas never loose truth values they can only gain them. The Collapsing Lemma is the ultimate downward Löwenheim-Skolem Theorem. (Priest 2002, 172)

The Löwenheim-Skolem theorem is:

If ***T*** is a countable theory[9] having a model, then ***T*** has a countable model (Shoenfield 2000, 79).

The downward Löwenhem-Skolem theorem is:

If ***T*** is a theory of cardinality ***k*** and having a model, then it has a model of cardinality less than ***k***.

If we are operating in a logic where some wffs enjoy two truth values, such as liar sentences, then, with the collapsing lemma, these can be collapsed into the equivalence class with the true sentences. The transformation exercised by $\sim$ will not always change the cardinality of the domain, but sometimes it will, and quite dramatically. In fact, it turns out that, under some circumstances, there will be domains of *every* cardinality that is lower than the cardinality of the original domain and is greater than zero!

---

[8] This is stipulated for technical reasons we need not explore here. It turns out that the collapsing lemma no longer works in the presence of some functions (Priest 2002, 172).

[9] A 'countable theory' is a theory expressed in a countable language: with only countably many constants, variables and predicates. If a theory is expressed in a countable language, then there are only countably many wffs. If we say that a theory just is the set of wffs which follows from the axioms by means of countably many rules of deduction (usually a small finite number), then we can see that the theory will only contain a countable number of wffs (Read and Wright 1991, 231–232).

Identity and universality will remain invariant in these logics, but emptiness will not. Under some transformations the empty relation will soon pick out a plethora of inconsistent objects. So, while in a *classical* setting the "empty relation never holds", it will hold in a Meinongian setting which is more indulgent towards the notion of an impossible object.[10]

What the pluralist learns from this is that invariance is subject to certain assumptions, and is therefore a useful concept in some contexts. It is not universal, and the idea of calling all of the invariant notions, under certain meta-logical assumptions 'logical', begs the question about what to count as 'logical notions'. Invariance, is not, for all that, useless as a concept. It can, as Klein discovered, be used to classify notions across theories. The method is a technique for identifying which notions stay fixed; but we have to be careful about the assumptions under which they stay fixed. Change the assumptions, and the classification changes.

9.5 Lobachevsky, Constrained Contexts, Chunk and Permeate

Recall the example we looked at in Chap. 5, Sect. 5.5, sub-section three. There we discussed Lobachevsky's importing a model of hyperbolic geometry to solve the problem of finding exact solutions for indefinite integrals. Hyperbolic geometry is non-Euclidean, but the problem of finding solutions for indefinite integrals arose in a Euclidean context. Therefore, Lobachevsky was using a piece of mathematics, which is inconsistent with the context of the problem; and, therefore, his proposed solution warrants scrutiny.

Indulging in an anachronistic exercise, where we forget all of the confirmation of Lobachevsky's results, we might suppose that the purported 'solution' to the problem is inconsistent, or arises *directly* from an inconsistency. We are working in a classical context, so from a contradiction anything follows, so it is easy to derive an 'exact solution' in a trivial context! The problem is that any course of values counts as a solution to any indefinite integral. In a trivial theory, we suffer the embarrassment of riches.[11] But recall that Lobachevsky claims to be careful about this. He gives two assurances of consistency. One is

[10] I should like to thank Priest for looking over both the material on the collapsing lemma and on chunk and permeate, later on in the chapter.

[11] Trivialism is the position that every grammatical, categorically correct, sentence is true. A sentence is categorically correct if it makes no 'category mistakes': where we confuse what type of object we are talking about. For example, it makes no sense to talk of water dreaming, angry chairs, kilograms travelling *etcetera*, unless, of course, we are in a fantastical/super-natural setting or using a metaphor. Trivialism does not treat of this sort of incoherent discourse. Rather it is about a discourse. Rather it treats of a truth-apt discourse. It is the dual of scepticism. Under global scepticism every grammatical, categorically correct sentence is subject to doubt. While trivialism is the dual of scepticism, it is logically much worse. Unlike the sceptic, the trivialist position does

that the problem involves triangles, and the relevant properties being exploited (the relations between angles and sides) behave 'in the same way' in the two geometrical theories. The other assurance is that he turns the geometry into what he calls "analytics". The problems are then straightforward calculations, and with this, the "calculations are necessarily coherent and one cannot discover anything what (sic!) is not already present in the basic equations. It is then impossible to arrive at contradiction," (Lobachevsky 1914, 34), unless it is to be found in the basic equations. "Analytics" is an oblique reference to the (pre)-formalist-type of presentation of a theory, where enough ground work is laid out, so that all that remains are mechanical calculations (Rodin 2008, 12). If, to repeat, we ignore subsequent developments by, for example, Beltrami (1868), which confirm Lobachevsky's solutions,[12] we could be forgiven for remaining doubtful of Lobachevsky's assurances, in light of the philosophically received view that we ought always to work consistently *within* a consistent theory to avoid triviality. So, what do Lobachevsky's assurances amount to, especially at the time of his writing?

First we should be clear about how Lobachevsky thought about his own assurances and compare this to the received view. The *received* view, today, is that Lobachevsky's hyperbolic geometry was developed as a rival to Euclidean geometry (Rodin 2008, 1). But as Rodin argues, this is not at all how Lobachevsky himself conceived of the situation. Lobachevsky thought that hyperbolic geometry is the more fundamental geometry, and that Euclidean geometry is a special case (Rodin 2008, 10).[13] With this sort of thinking it becomes clearer why Lobachevsky felt quite justified in 'importing' the 'foreign elements' of hyperbolic geometry 'into' the context of Euclidean geometry. He thought he was still working *within* his own more general theory. The problem Lobachevsky was trying to solve, concerning indefinite integrals, was easy to solve in hyperbolic geometry.

The received view sees things quite differently. Readers of Lobachevsky missed, or did not accept, the point that hyperbolic geometry is more fundamental, or more general. Therefore, under the received view, we have two rival theories, which, if put together, produce an inconsistency, and therefore, Lobachevsky's solution was highly suspicious. This is also why it was important to receive independent

---

not 'implode' since its own very trivialism is true, by its own lights. It is an entirely robust and stable position. However, it is highly uninteresting to maintain it.

[12]There were several confirmations of Lobachevsky's results. We could turn the tables, and ask why several? Well, the pluralist answers, because there was still some doubt remaining, the doubt that accompanies not full understanding, or 'unsatisfying' explanation. This theme will be developed.

[13]Kagan (1957) is more ambivalent about this, tracing the doubt and fluctuations in Lobachevsky's remarks. The doubt was quite normal, if one considers the intellectual setting for Lobachevsky's new geometry; it was challenging the *doxa* of more than 2,000 years. Nevertheless, Kagan's description of Lobachevsky's mature view accords perfectly with Rodin. For the context here, it will make sense to work with the mature view.

confirmation of the solutions.[14] Why did the 'received view' prevail? This is probably because, in the wider community of mathematicians studying geometry, Euclidean geometry was still considered to be fundamental. When 'other' types of geometry were developed, they had to show their connection to Euclidean geometry. Geometry on a sphere could be justified, because the sphere was conceived of *as* occupying Euclidean 3-dimensional space. That is, the sphere was placed in the context of Euclidean geometry. Hyperbolic geometry was different. It was not thought of, by Lobachevsky, as contained *in* Euclidean space; but the other way around.[15] Taking both the received view and Lobachevsky's own view into account what we learn is that the two geometries are not rivals, what we have are rival perspective on geometry: one where hyperbolic geometry is more primitive, the other where Euclidean geometry is more primitive. These perspectives were then later replaced with the more Hilbertian formalist view, where the interpretation and genesis of a formally presented theory is considered to be quite independent of the formal theory. Once we couch geometries in arithmetic, we see hyperbolic geometry as an equal rival to Euclidean geometry. Neither is the more fundamental.[16] Whose view prevailed, and why, is historically interesting, however, there is also a philosophical point.

It is not immediately clear how to compare two theories, (especially if we are not agreed as to how to present them: axiomatically, genetically, *etcetera*). Today, we are accustomed to axiomatic presentations, or rule-based presentations, and this uniformity of protocol of presentation invites comparison. This is one of the main reasons the Bourbaki approach was so important and interesting, it allows for easy comparison. The *form* of presentation of pairs of theories *determines*, but also *prejudices*, our judgment as to which is the more general. This was one of the goals of the uniformity of presentation: to determine such judgments! If we present hyperbolic geometry and Euclidean geometry, both axiomatically, then, the two geometrical theories differ over the parallel postulate. With this presentation, the two theories are simply rivals, and we can pin-point exactly where they differ. However, we should be aware that the determination is not absolute and independent.

For, if we have a more genetic, synthetic or conceptual presentation of the two geometrical theories, then we might well come to the judgment that hyperbolic geometry is more fundamental than Euclidean geometry, or the other way around, since Euclidean geometry is quite intuitive. Rodin remarks, for example, that Lobachevsky did not *at all* give an axiomatic presentation of hyperbolic geometry, but mixed up (what we would now distinguish as) definitions, axioms and theorems

[14]The independent confirmation came from Riemann who gave a model for both the Euclidean and the hyperbolic geometries (Katz 1998, 781).

[15]Beltrami's modelling of hyperbolic geometry in Euclidean geometry, does give us a sense of hyperbolic curved space (couched in Euclidean 3-dimentional space), but this was a later development.

[16]For an interesting comment comparing Lie's presentation of geometry in terms of groups, to Hilbert's approach see Hilbert (1971, 150–152).

(Rodin 2008, 3, 4). They were all presented on a par, to give a feel for the content of geometry. As a result of this different presentation style, due to a different conception of mathematics, we would determine that Euclidean geometry works only under some special circumstances; 'special' relative to hyperbolic geometry. More precisely, the postulates of Euclidean geometry hold in the horosphere of hyperbolic geometry, where the horosphere is a limit area of the geometry.

Killing two birds with one stone, we return to the axiomatic or analytic presentation, and show that, regardless of this, Lobachevsky was entitled to bring in the 'foreign' hyperbolic geometry to solve a problem in Euclidean geometry. I shall show this, not in the traditional way of modelling one in another, because this presupposes the received analytic view. Rather, I shall model the reasoning of Lobachevsky using a method of analysis called: 'chunk and permeate' – a model of reasoning which is useful for rational reconstructions of this kind. If the rational reconstruction works, then we do not need to decide whether the genetic presentation is 'better' than the axiomatic presentation, because, either way, by 'or-elimination',[17] Lobachevsky's solution stands, and there is nothing suspect about it.[18] In fact, the question is not whether Lobachevsky did find a solution or not, since this has not been in doubt since Beltrami. Rather, the strength of the independent analysis lies in its generality. We can use 'chunk and permeate' to scrutinise any manner of proof independently of the overall theory in which they are couched, and independent of whether we have fixed an overall theory at all! We saw in Chap. 5, the very extensive mathematical projects, where problems are divided into cells, which each prove a theorem, by independent means of one another. Chunk and permeate might be a good tool for reconciling and scrutinising the proofs in these projects.

Again, let us be quite clear about the structure and strength of this part of the argument. What I shall do in the following paragraphs is give a rational reconstruction of what Lobachevsky was up to in solving his problem about finding exact solutions for indefinite integrals. The rational reconstruction gives us justification for Lobachevsky's results, independent of Lobachevsky's justification, or of a more modern 'formalist' justification. In Chap. 14 we shall look at Rodin's own justification of Lobachevsky's method, and discuss what each contributes to our understanding.

---

[17]'Or-elimination' is an inference rule in some systems of natural deduction. It is for reasoning from a disjunction (in our case: either the genetic presentation is better or the axiomatic presentation is better) to a conclusion (in our case: Lobachevsky's solution stands up to scrutiny, and was perfectly good reasoning). The rule is that you should arrive at the conclusion separately from each disjunct, you can then claim on the strength of the disjunction alone that the conclusion follows regardless of which one is true or whether both disjuncts are true.

[18]There was independent confirmation of Lobachevky's solutions developed after Lobachevsky, by Riemann, Klein, Helmholtz and finally Beltrami (Katz 1998, 767, 779, 783). However, because of the mode of presentation of the solutions, these results would not help for the argument here, since their presentation of geometry sits between the Hilbert-style presentation and the more synthetic style of Lobachevsky (Rodin 2008, 23). Therefore, the confirmation is not completely 'independent'. To what extent, or rather, when, this is a problem is a deep and interesting question concerning types and degrees of objectivity. I shall address this in a future paper.

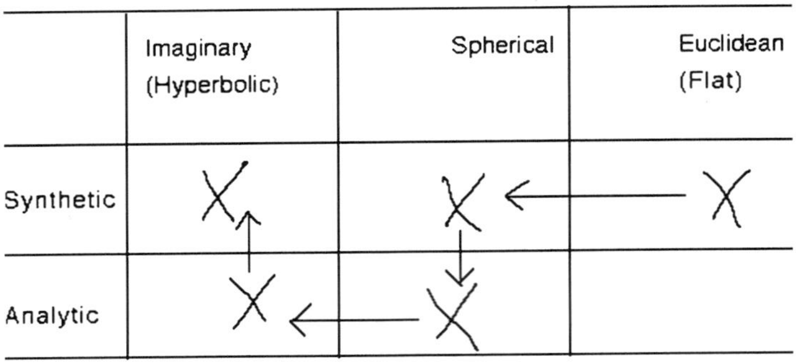

Fig. 9.1 Rodin's diagram of the structure of Lobachevsky's argument

9.5.1 Chunk and Permeate

The rational reconstruction concerns splitting up a mathematical problem into chunks, but letting some information flow between pairs of chunks, and this flowing is called permeating. Thus the chunks are not hermetically sealed, rather it is as though there is a screen, or filter, on some information. Moreover, what information flows between one pair of chunks might be quite different from the information which flows between another pair of chunks (otherwise we would have something like an *underlying* logic, theory, or set of assumptions). The trick then is to reassure ourselves that the particular mix of chunking and permeating, does not lead to triviality. Lobachevsky assures us of consistency, and consistency implies non-triviality, so we shall work with consistency rather than non-triviality.

I follow Rodin's analysis of Lobachevsky's argument (Rodin 2008, 20), where he is careful to distinguish the synthetic (intuitive, content based) presentation from the analytic presentation, of three geometrical theories: Euclidean, spherical and hyperbolic. See Fig. 9.1.

There are five chunks used in the reasoning to find the solution: (1) the chunk in synthetic Euclidean geometry, where we recognise and formulate the problem. We move from that to (2) where we see the problems translated into a synthetic presentation in spherical geometry. We then go to (3), where we have an analytic presentation of spherical geometry, followed by (4), where the results are translated into the analytic presentation of hyperbolic geometry. In the last two, all we have are straightforward calculations, which is why Lobachevsky calls this "analytics". Indeed, the chunk and permeate technique will apply rigorously to the analytic chunks only. For the synthetic chunks, we have to rely on our geometrical intuitions; I shall have to use the chunk and permeate method metaphorically, as in Chap. 7. We end with (5) the synthetic presentation of hyperbolic geometry, since this is the more fundamental for Lobachevsky. The solution, can then be transferred directly back to Euclidean geometry, since, as we noted, at least seen synthetically, Euclidean geometry is a special case of hyperbolic geometry, not a rival.

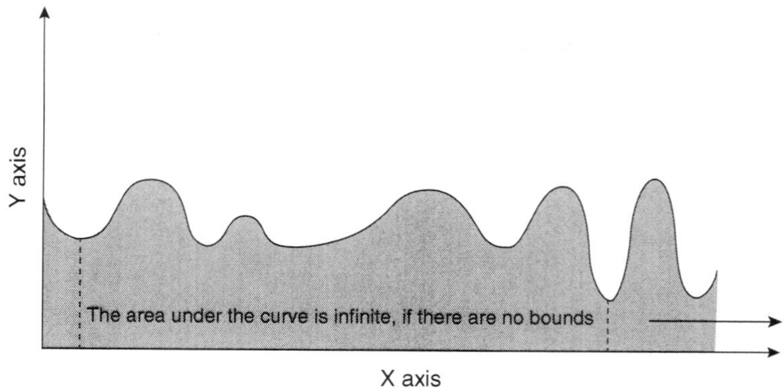

Fig. 9.2 An indeterminate integral

We begin with the permeating from the first to the second chunk. In Euclidean geometry, the calculation of indefinite integrals made no sense (Rodin 2008, 19). We have a continuous function, but the area underneath could not be calculated in general. It would be infinite regardless of the shape of the function line. That is, without a lower and upper bound on the arguments, the space under the curve is infinite. See Fig. 9.2.

Note that ideas of 'sense', or 'making sense of a calculation' belong to the synthetic presentation of a theory, not to its calculation. To make 'sense' of the calculations, Lobachevsky invites us to think about spherical geometry.[19] We transform the sides of triangles, a, b and c into their analogues in spherical geometry: ai, bi, ci, where 'i' is the square root of -1. In respect of the *relations between* angles and sides of triangles, the analogy holds. So the information that permeated from the first to the second chunk concerns only the relations between sides and angles of triangles. This is perfectly all right, since we think of spherical geometry as a proper part of Euclidean geometry; spheres are shapes in Euclidean space (Rodin 2008, 21).

The analytic presentation of spherical geometry was developed with the synthetic part in mind. Stronger than that, the analytic, calculating, part is directly responsible to the synthetic presentation. That is, if the analytic part were to give us a result outwith[20] the synthetic presentation, then the analytic presentation would be altered, not the other way around. In this sense, the analytic presentation is a perfect image of the synthetic presentation of the geometry. So we are now in analytic spherical geometry, and want to go to analytic hyperbolic geometry.

---

[19] Strictly speaking this is incorrect. Lobachevsky's understanding of the analogy between spherical and hyperbolic geometry is not in terms of the curvature of space. He took the analogy to be formal. It was Lambert who helped us to understand the analogy more clearly (Rodin 2008, 19).

[20] 'Outwith' is a word preserved by the Scots, but forgotten by the rest of the Anglo-Saxon speakers.

For the analytic chunks, I follow Brown and Priest (2004) closely. The following quotations are all from Brown and Priest (2004, 380). Chunk (3): the language $\boldsymbol{L}$ is that of spherical geometry. $\vdash$ is the consequence relation. Σ is the set of axioms, which govern angles and the relations between sides of triangles in spherical geometry. $\Sigma^{\vdash}$ is the closure of Σ under $\vdash$. "A *covering* of Σ is a set $\{\Sigma_i \, ; \, i \in I\}$, such that $\Sigma = \bigcup_{i \in I} \Sigma_i$, and for all $i \in I$, Σ_i is classically consistent." Each I is a Gödel code for a formula, so we presuppose some sort of standard normal form, and Gödel coding for sentences in the language. We can then make inductive arguments over the set of sentences which follow from Σ_i. The covering is all such formulas. We have well established and independent confirmation of the classical consistency of trigonometry on a sphere, and therefore, any subset of sentences in spherical geometry will also be classically consistent, by straightforward induction on the set of sentences. "If $C = \{\Sigma_i \, ; \, i \in I\}$ is a covering on Σ, call ρ a *permeability relation* on C if ρ is a map I X I to subsets of the formulas of $\boldsymbol{L}$." In other words we allow all of the formulas, from spherical geometry, which concern the relationship between angles of triangles and their sides to permeate into the next chunk: that of hyperbolic geometry. What does not permeate through is information about parallel lines. The other axioms are common to Euclidean, spherical and hyperbolic geometry. This is a clear and easy case for the chunk and permeate analysis. We work out the solution to integrals in hyperbolic geometry, again only using information common to both spherical and hyperbolic geometry.

Moving from chunk 4 to 5, we make a move from the analytic presentation of a geometry to the synthetic presentation. Again, we remind ourselves of the genesis of the analytic presentation of hyperbolic geometry (chunk 4). This was developed by Lobachevsky, and he was no friend of the formalist approach to mathematics. He did, however, recognise the benefits of developing a mechanical calculus within a theory. Assuming he did this responsibly, i.e., the calculations are sound, then the move from the outcome of the calculations to the sense of what they mean, i.e., the move from the analytic to the synthetic part of hyperbolic geometry, is straightforward, since they mirror each other (Rodin 2008, 19).

What have we learned from this? We have not learned that Lobachevsky's practice was legitimate. We have known that since 1868, thanks to Beltrami, since it was Beltrami's demonstration which convinced the wider community of mathematicians. What we have with the chunk and permeate analysis (at least in the analytic case of moving from chunks 3 to 4) is a new type of confirmation, and one with wide scope.[21] It has wider scope than model theory because we can suppose different chunks to have different underlying logics, including paraconsistent ones (Brown and Priest 2004, 386). This is also why the technique is more amenable than formalism to analyse a lot of modern proof techniques. After all, if we are

[21] In respect of scope, it is similar to model theory methods of choosing a meta-structure which shows the relationship between to object-level structures (up to isomorphism); more of this in the next section.

interested in proofs, and organising these into chunks, then we should not be restricted to thinking in terms of structures, with the structuralist; and we should not be restricted to thinking in terms of an idealised conception of proof, with the formalist.

If we accept the more metaphorical version of chunk and permeate, then it becomes an easy tool for verifying chunking-type reasoning. Chunk and permeate gives us an orientation from which to check reasoning which uses together inconsistent theories. If we are careful with our individuation of a chunk, and with the permeability relation, then we can verify that the reasoning is in fact legitimate, and we can accept the results. Or, more carefully in the metaphorical case, chunk and permeate will tell us where to look for dodgy chunk-type reasoning. This is immediately useful, since, modern mathematics more often engages in 'big' mathematical projects, where problems are too long and difficult for one person to solve. Thus, the large problem is sectioned off into cells (which might constitute a chunk, or might be divided into further chunks) where independent mathematicians 'solve' particular problems, using whatever means they are trained in, and are familiar with. Often the means used in one cell, or chunk, is inconsistent with the means used in another. To ensure against triviality, chunk and permeate then becomes a useful, and natural, tool for analysing flow of information (permeability) between cells, or chunks.

9.6 Analysis by Way of a Conclusion

For philosophers who are looking for a 'unified account' of mathematics, there are many mysteries left unexplained in existing attempts to provide such an account. The mysteries concern mathematical practice.

For example, we might ask why there are so many proofs for the some theorems. There are well over 100 proofs for Pythagoras' theorem which are non-equivalent, except in the conclusion. They are not all 'suspect' proofs, made in some obscure part of mathematics equi-consistent with some other obscure part. In other words, new proofs are not developed to assuage *doubt as to the truth*, or robustness, of the result: the square of the hypotenuse is equal to the sum of the squares on the other two sides. Put another way, they are not meant to be part of some *inductive* argument for Pythagoras' theorem. Each is a deductive proof. Even in the case of theorems we judge less certain, *some* proofs are useful for assuaging doubt, but some are not; and yet, the body mathematical accepts them as valuable contributions to the field of mathematics. So what is the surplus information we gain from a proof, over the truth of the theorem proved?

The pluralist thinks of this question as indicative of a wider phenomenon. There is something unique and interesting about mathematics. Mathematics 'hangs together'; it seems to be objective and non-circular. It is not like a Popperian pseudo-science, but neither is it checked against physical phenomena, and our observations. Moreover, we do not have the evidence to attribute the hanging together to: absolute truth, consistency, embedding in a unique foundational theory or ontology. Ultimately, these are the wrong places to look, although they might be useful locally, at the first or second levels of pluralism. Instead, what is salient at the third level is that mathematical theories can so often be applied to other mathematical theories. In fact, the claim is stronger than this. It is not that they *can* be so applied, it is that modern mathematics largely *consists in* such applications. Moreover the wealth of crosschecking is exactly what leads to our producing several proofs of the same theorem; but more important, crosschecking, applications and fixtures is sufficient to warrant our confidence in mathematics in the *absence* of a unique foundation, and all that that entails philosophically.

In this chapter we merely *introduced* the notion of 'fixtures'. Once we see some of these, others will suggest themselves to the reader when he, or she, revisits mathematical texts or articles. I propose the exercise of looking at the table of contents of a recent journal in logic or mathematics, and count how may articles are about limitative results, connections between theorems in different theories, in applications of one methodology to a 'foreign' area of mathematics and so on. There will be a significant percentage of such articles. It turns out that a lot of cutting edge contemporary mathematics is of this nature. Moreover, the crosschecking is not all part of a unified outlook. The previous section of this chapter testifies to this. There, I used the paraconsistent method of analysis: chunk and permeate, to make a rational reconstruction of Lobachevky's thinking.

Even by examining only three sorts of crosscheck (which pre-suppose 'fixtures'), and imagining that there are others, we learn two lessons.

One is that mathematics does 'hang-together' and forms a distinctive discourse (at least distinctive in character, and not necessarily because of a realist ontology).

The second lesson is that we do not need to rely on an ontology, a notion of absolute truth or a unique theory to play the role of foundation, or place undue emphasis on an idealised conception of proof in order to justify pluralist mathematical practice, and recognise its importance.

Fixtures show that we do not need to appeal to the philosophical triumvirate: ontology, knowledge or truth. Instead we explain the triumvirate in other terms. The pluralist offers a rich account of the 'hanging together' of mathematics. No other discipline has developed a web of crosschecks as keenly as mathematics, and it is these crosschecks which allow us to loosen the stringent constraints of formalism or of foundationalism.

The pluralist diagnoses that traditional philosophical approaches, inclinations and tools sometimes misfire, when brought to bear on the subject of mathematics because they take the triumvirate as primitive, or already understood, when this is not at all the case. Why have traditional approaches in philosophy misfired? It is partly a matter of (philosophical) temperament. Philosophers tend towards a certain temperament (towards monism). Such philosophers think of crosschecking as *evidence for* the fact that there is a deep underlying truth or that there is an underlying mathematical ontology, or logic, or major umbrella theory, or something to *explain* the miracle of application. As Tarski writes:

> The conclusion [that there are different answers to the question whether $\in$ is a logical notion] is interesting, it seems to me, because the two possible answers correspond to two different types of mind. A monistic conception of logic, set theory and mathematics, where the whole of mathematics would be a part of logic [or some umbrella theory, in our case], appeals, I think, to a fundamental tendency of modern philosophers. [Tarski was giving this talk in 1936]. Mathematicians, on the other hand, would be disappointed to hear that mathematics, which they consider the highest discipline in the world, is a part of something so trivial as logic; and they therefore prefer a development of set theory in which set-theoretical notions are not logical notions. (Tarski 1986, 153)

Mathematicians would be disappointed, not so much because logic is thought of as trivial, since it is not so judged today, but by the constraints on their creativity. Nevertheless, present day mathematicians do share Tarski's mathematician's concern about being held to a foundational standard, and that is because so much of the development of mathematics has no explicit roots in set theory. That is, mathematicians have, for the most part, quite disregarded the foundationalist aspirations. Instead, they rely on the crosschecks as confirmation of their results.

Moreover, the crosschecking is robust since it is rigorous. There are plenty of contexts where attempts at cross application do not work. It is not the case that everything in mathematics fits together in any way we choose, and it is the *failure* of cross-application which is evidence for the objectivity and non-triviality of mathematics. This sort of objectivity is not grounded in an ontology. Rather, some successful instances of fit, or convergence, are evidence for some successful instances of fit and convergence, nothing more.

I can hear the honourable opposition saying that there is much more successful fit and convergence than we might have thought *prima facie*. But, this is just to admit that we are not good predictors. The pluralist replies that the fit occurs where the fit occurs, and often it is not perfect, as in the case of renormalisation, or in the case of reducing calculus to set theory.

What we *do* have evidence for is that there is a crosschecking, and a conversation in mathematics; that the crosschecks are as rigorous and thorough as we choose. Sometimes we are slack, for example when we try to apply mathematics to physics, and find that we have to gerrymander the mathematics to fit the physical theory – see renormalisation. In fact, mathematics is the discipline where the crosschecks are the most rigorous of any area of research. This accounts for the *phenomenology* of objectivity, absolute truth and independence of mathematics. But phenomenology is not evidence for objectivity, truth or independence! Our phenomenology sometimes misleads us.

Moreover, returning to the quotation from Tarski, the mathematician might well (depending on temperament, again) feel either unconcerned by metaphysical notions underlying her subject matter, or she might feel that raising such questions is alien to her, or, as we saw in the chapter on formalism, she might adopt a schizophrenic attitude. All of these possible attitudes point to the variety of attitudes held by practicing mathematicians, and that there is such a variety testifies to the confusion, or lack of good and coherent *traditional* answer to the philosophical questions. The pluralist philosopher has a different temperament, better aligned with Tarski's modern mathematician.

References

Beltrami, E. (1868). Saggio di interpretazione della geometria non-euclidea. *Giornale di Matematiche, 6*, 284–312.

Brown, B., & Priest, G. (2004). Chunk and permeate, a paraconsistent inference strategy. Part I: The infinitesimal calculus. *Journal of Philosophical Logic, 33*(4), 379–388.

Hilbert, D. (1971). *Foundations of geometry*. LaSalle: Open Court.

Hodges, W. (1997). *A shorter model theory*. Cambridge: Cambridge University Press.

Kagan, V. F. (1957). *N. Lobachevsky and his contribution to science*. Moscow: Foreign Languages Publishing House.

Katz, V. J. (1998). *A history of mathematics; An introduction* (2nd ed.). Reading: Addison-Wesley.

Lobachevsky, N. I. (1914). *Geometrical researches on the theory of parallels*. Chicago: Open Court Publishing Company.

Priest, G. (2002). *Beyond the limits of thought*. Oxford: Clarendon Press.

Read, S., & Wright, C. (1991). *Read and Wright formal logic; An introduction to first-order logic* (4th ed.). St. Andrews: University of St. Andrews, Department of Logic and Metaphysics.

Rodin, A. (2008). *Did Lobachevsky have a model of his imaginary Geometry?*. http://philsci-archive.pitt.edu/4099/1/loba8.pdf

Shapiro, S. (1991). *Foundations without foundationalism; A case for second-order logic* (Oxford logic guides 17). Oxford: Oxford University Press.

Shoenfield, J. R. (1967, re-edited 2000). *Mathematical logic*. Natick: Association for Symbolic Logic, A.K. Peters.

Tarski, A. (1986). What are logical notions? J. Corcoran (Ed.), *History and Philosophy of Logic, 7*, 143–154.

Wright, C. (1992). *Truth and objectivity*. Cambridge, MA: Harvard University Press.

Part III
Transcendental Presentation of Pluralism

Chapter 10
The Paradoxes of Tolerance and Transcendental Pluralist Paradoxes

Abstract In this chapter, I look at three types of paradox which attend pluralism: the external paradox, the internal paradox and the transcendental paradox. The external paradox is generated from the following question: if we are tolerant towards other positions than our own, then what if the other position is intolerant of our own? In the name of tolerance, if we accept that the other position is, in some sense correct, then the intolerance towards us is correct, and we should give up our original position. If we decide that the position is incorrect, then we have failed to show tolerance.

The second, internal paradox, is generated within pluralism. The pluralist wants to show tolerance and interest in other positions, to learn from other positions, to entertain them seriously. But he is also insistent on certain issues of protocol or attitude. Is this not itself, dogmatic, and therefore intolerant? The external and the internal paradoxes are solved in a relatively benign way.

The transcendental paradox is generated by deploying the 'inclosure schema'. (Priest, Beyond the limits of thought. Clarendon Press, Oxford, 129, 134, 2002) This is a paradox at the limit of thought, and it is by deploying this paradox that the pluralist transcends his own position. The solution to this paradox is non-standard.

10.1 Introduction

Definition *Paradoxes* are thoughts or ideas, represented by sentences, or wffs, which, upon initial analysis, appear to be both true and false.

In this chapter, I shall work on the notions of 'analysis', and 'appear to be', as they are used in the above definition of paradox. As I have defined them, paradoxes are not necessarily insoluble. Sometimes we use the word 'antinomy' for a solvable paradox, but I shall use the word 'paradox' as ambiguous between having a simple, classical solution and not having such a solution.

M. Friend, *Pluralism in Mathematics: A New Position in Philosophy of Mathematics*, Logic, Epistemology, and the Unity of Science 32, DOI 10.1007/978-94-007-7058-4_10,

© Springer Science+Business Media Dordrecht 2014

The reason for leaving the ambiguity is to reflect sensitivity to several ideas as to what can be counted as a solution. If we leave the field open, then 'solutions' might not be quite as we expect. Sometimes, we can discover, on *further* analysis, that a paradox was only apparent and it is true and not false, or it is false and not true.[1] At other times, such as in the set theoretic paradoxes or some of the semantic paradoxes, we end up in a constant tension between two pulls or attitudes. Our finest logical analysis and reasoning bring us to the conclusion that such paradoxes are both true and false. In these cases, if we also assume the validity of *ex contradictione quodlibet*, we think of the paradoxes as bringing disaster, which is hardly a 'solution'. Therefore, in these cases we are motivated to shift to another context or theory.

For example, we re-adjust the naïve comprehension axiom of set theory, or introduce a notion of proper class (for the set theoretic paradoxes), or we adjust our theory of truth and language (for semantic paradoxes).[2] In the semantic case, we restructure our language(s) so that certain interpretations of paradoxical sentences are forbidden – such as when, following Tarski, we organise our languages into a hierarchy, and stipulate that we may only attribute truth (or other semantic properties) to a sentence at a 'lower' level in the hierarchy.

However, there is a third possibility. Instead of adjusting or modifying the theory, there is a more radical 'solution'. We can reject *ex contradictione quodlibet* and accept the paradox, since (without *ex contradictione quodlibet*) contradiction does not bring disaster. In this case, we do not treat paradox as a disaster, but as a feature, as a dialetheia: an idea that is both true and false, a true contradiction. In the last section of this chapter I shall discuss this more radical solution to paradoxes.

In Sects. 10.2 and 10.3, I discuss the two paradoxes of tolerance. One paradox of tolerance, I shall call 'external' and the other, I shall call 'internal'. The motivation for the name 'external paradox' is that the paradox is generated by looking at a non-pluralist theory. In contrast, the internal paradox is generated by looking at pluralism itself.

The external paradox is generated from the following question: if we are tolerant towards other positions than our own, then what if the other position is intolerant of our own? In the name of tolerance, if we accept that the other position is, in some sense correct, then the intolerance towards us is correct, and we should give up our original position. If we decide that the position is incorrect, then we have failed to show tolerance. I address this external paradox in Sect. 10.2.

The second, internal paradox, is generated within pluralism. The pluralist wants to show tolerance and interest in other positions, to learn from other positions, to entertain them seriously. But he is also insistent on certain issues of protocol or

[1] The implication, in these cases, is that the analysis which led to a paradox was not very thorough. Of course, from a certain epistemic situation, it might be almost impossible to give a more thorough analysis.

[2] Priest (2002, 142–155) argues that there is no substantial difference between the set theoretic and the semantic paradoxes. I agree with this, but the distinction is a well accepted one. The distinction plays no role here, so my drawing it is *sans importance*.

attitude. Is this not itself, dogmatic, and therefore intolerant? This paradox will be addressed in the third section. Both paradoxes are solved in a relatively benign way.

However, there is a third paradox, and this is of a different type. I call it a 'transcendental paradox'. This one, we generate by deploying the 'inclosure schema'. A version of it (Priest 2002, 129) was devised by Russell as a diagnosis of what is common to (at least) the set theoretic paradoxes. Priest (2002, 134) generalises Russell's schema, and uses it to demonstrate many more paradoxes: paradoxes at the limits of thought. The general idea is that in any attempt to give a theory of all and every: description, number or conception, we approach a limit threshold, which we at once respect, and also transcend.

> Limits of this kind [of expressibility, conception, iteration, knowledge . . .] provide boundaries beyond which certain conceptual processes (describing, knowing, iterating etc.) cannot go; a sort of conceptual *ne plus ultra*. . . . [S]uch limits are dialetheic; that is they are the subject or locus of true contradictions. The contradiction, in each case, is simply to the effect that the conceptual processes in question *do* cross these boundaries. Thus, the limits of thought are boundaries which cannot be crossed, but yet which are crossed. (Priest 2002, 3)

It is in the fourth section that I treat of this more radical dialetheic solution to paradox in pluralism. I *generate* a transcendental paradox of pluralism, using the inclosure schema, thus demonstrating the existence of at least one (not so easily 'solved') paradox of pluralism. I then treat the paradox as a dialetheia, thereby giving a type of solution, or reconciliation. Thus, in the fourth section, we learn to treat some paradoxes as a 'transcendental' feature of pluralism. Moreover, there is not just one such paradox of pluralism. There are many. But since similar paradoxes can be generated for any large 'all encompassing' theory or philosophy, it is not only pluralists who are motivated to look closely at the dialetheic solution. For those who feel queasy, I refer you to the section on nausea at the end of the next chapter.

10.2 The External Paradox and the Argument from Modesty

In the introduction I wrote: "in the name of tolerance, if we accept that the other position [which is intolerant towards pluralism] is, in some sense, correct, then the intolerance towards us is correct, and we should give up our original position." Pluralism is therefore unstable in the presence of *any* position that is intolerant towards pluralism. From the standpoint of pluralism we end up having to shift position, and completely betray our pluralism and tolerance, and adopt the other position, *just in virtue* of its being intolerant towards ours.

This cannot be right. Intolerance is no guarantee of correctness! The evidence is that there are several intolerant positions that all make conflicting claims about the same subject matter. In fact, very often, intolerance, which leads to dogmatism, hides a lack of further argument or resources to defend a position. If one cannot defend a position through convincing argument one either shifts to a more defensible position, or one becomes dogmatic; one starts posturing and insisting, maybe resorting to force (although this is rare in mathematics or philosophy). This is not to say that dogmatism is always wrong. It can be justified for practical reasons. One might resort to dogmatism because one's interlocutor is simply not bright enough, within some time constraints, to appreciate the point one is trying to communicate. If we set aside concerns about: time constraints, lack of resources and level of mathematical and philosophical sophistication of an interlocutor, then dogmatism indicates, not correctness, but lack of rational persuasive resources.

Let us move to the specific case of dogmatic philosophies of mathematics: the monist and dualist philosophies. Echoing Chap. 4: by adopting a philosophical position, or a foundational mathematical theory, we have the resources to make claims at different levels. Rehearsing the levels: at the lowest level we have claims strictly internal to the theory. For example, we have theorems of a mathematical theory or principles that characterise the philosophy. "$2+2=4$", or "addition is commutative" are examples on the mathematical side. "Since numbers are objects of (our foundational theory of) mathematics, their existence is independent of our conceiving of them" is an example of claim that could be made by a realist in ontology. The paradox of tolerance will not arise at this level.

It will arise at the next level where we meet claims to the effect that this mathematical theory or philosophical position is the only correct, or true, one. That is, the 'dogmatism' is internal to the theory, and so directly depends on our first accepting the theory, at least temporarily, for the sake of argument. At the second level, or at a meta-level, we have claims of the form: "first-order arithmetic has non-standard models" on the mathematical side, or "realism is the correct philosophy" or "logicism is an unstable position" on the philosophical side. It is at this level that we find the intolerant claims, such as "ZF set theory is the (only possible/admissible) foundation for mathematics". Such statements preclude other positions, in particular they preclude pluralism, and the pluralist faces the external paradox. Move up to the third level.

At the third level, we either re-trench our dogmatism and continue to maintain that our own theory is the only acceptable one. We are dogmatic 'all the way up'. Or we are pluralist and refuse the dogmatism of the lower levels. The refusal blocks the dogmatism 'all the way up', but only *from* the pluralist perspective. That is, it will not convince the dogmatist, only the pluralist. Nevertheless, the threat of paradox is mitigated. Under this third-level pluralism, each mathematical and philosophical theory at lower level is tolerated *up to* the point of dogmatism, that is, up to the point where the dogmatic claim is made. It is a third-level pluralist claim that the pluralist (occupying this third level) is intolerant towards intolerant claims of others at second level. By refusing to recognise as legitimate, particular dogmatic claims, the pluralist solves the external paradox of tolerance for the pluralist.

The solution has an added subtlety, not only is it a solution only for the pluralist, but the pluralist does not even have to insist on the dogmatism (of rival positions) being *incorrect*. It is enough to remain agnostic, and insist on scientific honesty: that unless we have further evidence for the truth of the dogmatic position, we remain pluralist. Should such evidence present itself, then it is, of course, correct to give up pluralism. Remember that pluralism includes a principled agnosticism, not fanatical agnosticism.

One argument for pluralism is the 'argument from modesty'. It runs: there are several different, conflicting mathematical theories and philosophies of mathematics. None has succeeded in persuading everyone, or even every 'rational' person (unless one defines 'rationality' very narrowly and cleverly).[3] There does not seem to be an immanent convergence in positions either. Depending on how finely we want to distinguish positions, we might even think that there are an increasing number of plausible positions.[4]

They are not equally plausible; even if we consider 'plausibility' to be a subjective term where what is plausible to one person is not to another. Why not? Plausibility also depends on background knowledge, so this is one way of resolving disputes about plausibility: we can supply the necessary information. The knowledge will include knowledge of arguments put forward in defence of various positions. Plausibility is an educated judgment, not a mere judgment of taste. So, we can partially rank theories or philosophies, one as more plausible than another through argumentation and by increasing our knowledge and considerations. But even amongst the most educated, we do not have convergence concerning the ranking. Future arguments and future information might lead to convergence, divergence, or convergence followed by divergence, divergence followed by convergence; we simply do not know. Therefore, we have no rational basis, nor authoritative basis (based on amount of education)[5] upon which to make a final choice for one dogmatic philosophy. The pluralist is aware that he located in time. Right now, while he is a pluralist, he is not convinced by one dogmatic theory. Otherwise he ceases to be a pluralist. Thus, one can change from one position to pluralism and back again. Taking into account this possibly temporary aspect of pluralism, the pluralist is someone who, right now, demurs from making a choice, and accepts all of the plausible theories – mathematical and philosophical, *as* plausible theories.[6] The pluralist may then rank them according to comparative degrees of plausibility, but better, we also work on the measure of plausibility. We work to be explicit about the respects and measures *by which* one theory is

---

[3]The definition would have to be quite clever to avoid the charge of begging the question.

[4]We have to be careful about judgments about "recently increasing numbers of positions". Population increase accompanied by a little mathematical education, together would suggest an ever increasing number. However, such claims should be moderated, by looking at numbers of people in a position to develop and publish on a position, publishing ethos *etcetera*. In other words, it might seem as though there are an increasing number of positions just in virtue of our ignoring the past positions, which were not published, or have become less available, and seeing only the published and publicized positions in an age where the university ethos is to 'publish or perish' which acts as strong incentive to publish. But the point remains that one should be alert when faced with naïve claims about quantities of positions.

[5]'Amount of education' is, of course, not to be confused with number of degrees or prestige of award granting institutions. Here 'education' is meant in the basic sense of pursued, sustained and critical enquiry. University degrees are a rough *indicator* of education.

[6]This is reminiscent of Hellman's notion of 'possible mathematical theory', but the pluralist is less constrained about the parameters on possibility (Hellman 1989). The pluralist is also more sensitive about measures for ranking, and rating those.

more plausible than another, and then work to come to determinations about those respects and measures through critical argument. The upshot, once one has adopted the pluralist position at third level is to accept a position at second or first level in the philosophy of mathematics, but excuse the dogmatic second-level claim about correctness or absolute truth of the non-pluralist theory. For, such a claim is thought to be premature. The pluralist counters dogmatic claims of a theory by pointing out the existence of alternative theories, and running the argument from modesty.

Note that pluralism is not dogmatic; instead it is 'principled'. That is, upon examination of a position, and in light of serious and open critique, one might well decide that one position is more plausible than all the rest, and that the measure of 'plausibility' is a perfectly good one! In this case, the pluralist should adopt the position, and give up pluralism. However, he will do so in an open minded way, that is: he will always be willing to change position, change his mind, in light of countervailing evidence. Even the judgment that X is *the most plausible* philosophy of mathematics is revisable. The judgment is indexed to (or ultimately depends upon) knowledge and a set of arguments.

10.3 The Internal Paradox

The internal paradox of tolerance comes from the idea that the pluralist wants to show tolerance and interest in other positions, to learn from other positions, to entertain them seriously. But he also insists on certain issues of protocol or attitude, he *insists* on a *fair*, *honest* and *rational* open-mindedness. Is this not, itself, dogmatic, and therefore intolerant? It seems that pluralism is dogmatic. Oh, oh!

In Chap. 6 I made the following claim, which I said I would address in this chapter. The claim is:

> Third-level pluralism includes a set of attitudes, amongst which, we find an avoidance of dogmatism, in favour of qualification and clarification. One by one, dogmatic claims are replaced by careful explanation that justifies (and shows the limitations of) what was stated as a dogmatic claim. (Friend 2014, 113)

If we insist on this, then, surely, we are dogmatic. We are dogmatic about attitudes of enquiry and protocol. We witnessed this in the preceding section too when we insisted that if, from a pluralist position, one comes across an overwhelmingly plausible non-pluralist position, then one *should* adopt that position. Moreover, one *should* adopt it in an open-minded way, that is, in the spirit of revisability. Almost any philosopher of mathematics will agree to such a claim. However, usually they are also convinced of a particular position. Their behaviour is not contradictory, they are simply convinced of their position, but are willing, at least in principle, to change their minds. The very fact that they go to conferences where other positions are represented is testimony to the open-minded attitude of philosophers. Put so mildly, one might suspect that I count almost everyone as a pluralist! Maybe I do – provided they leave off the dogmatic claims!

How can a pluralist defend a theory or philosophy? A pluralist can do this because he thinks of the protocol and 'should' claims as normative, not as dogmatic or absolute. The difference between normativity and dogmatism is this. A norm sets a standard of behaviour or of attitude. In contrast, a *doxa* is not merely a standard but is an absolute set of rules. They have to be obeyed blindly, whereas normative claims are ideals of behaviour to which we aspire, or which we seek to follow. Both adhering to them and following them requires judgment. The particular norms advanced here are common to philosophical practice. This is why so many philosophers will agree to them (and are quite pluralist unbeknownst to them!) They will agree to the protocol *as norms* and not as *doxa*, and it is in this spirit that the pluralist urges the philosopher to consider these protocols and attitudes. These norms fit well within philosophical practice, and should be familiar. Looked at this way, our threatening internal paradoxical tolerance dissolves into a banality. When debating a theory or a philosophy, we have to keep some goal posts steady. We may revise these one by one, or even several at a time. In principle, we can question anything, even axioms of a mathematical theory, or protocol, but not everything can be questioned *all at once and all of the time*, such would count as unruly behaviour, and would be impracticable.

Thus, by appeal to the distinction between normativity and dogmatism, the pluralist avoids the accusation of inviting an unsolvable internal paradox. Nevertheless, there are more serious paradoxes, which cannot be so easily resolved. For their solution, we turn to the dialetheist. A dialetheist pluralist is a pluralist at third level with a paraconsistent logic underlying his pluralism. He does not consider himself to be in a trivial setting, and he has a dialetheist attitude towards some paradoxes. The paradoxes in question are ones that we do not seem to be able to resolve into one truth-value.[7] Examples are the liar paradox or some of the set theoretic paradoxes. These are then considered to be true contradictions, and are called 'dialetheias'. We can generate these liberally by using the inclosure schema. Thus, we turn our attention to the generation of dialetheias, and the dialetheist's attitude towards them.

10.4 Dialetheism: Paradox as a Feature, Not a Disaster

We might have found some resolutions to the paradoxes of tolerance, but we suspect that there might be other paradoxes facing the pluralist, not least because he entertains, discusses and takes seriously, not only pairs of theories which contradict

---

[7]It follows that, faced with an arbitrary paradox, the dialetheist may try to resolve it classically, i.e., into one, and only one, truth value. Whether she takes this option or not will depend on her attitude towards classical solutions and dialetheias. It is a question of weighting the options. It is when the price is 'too high' (too much distortion of the original theory), or when no devices are present or even on the horizon, that the dialetheist will consider the paradox to be a dialetheia. Thus, ultimately, identifying a dialetheia is an inductive process. We give up trying to give a classical solution at some point, and decide that it will be more advantageous to accept the paradox as a dialetheia.

each other but also quite disastrous theories, such as trivial theories. We might be afraid that the trivialism and contradictions infect the rest of the discourse, and so there is logical risk in 'being tolerant'. The risk is that our very own discourse collapses into trivialism. We have already seen, in Chap. 9 that there is a perfectly robust distinction between a theory containing contradictions and a trivial theory. The distinction is preserved simply by displaying a formula or sentence which is only false. We shall call this 'Post-non-triviality'. The reason I call it this is a little involved, but not uninteresting. Here is the reasoning.

Post gave us the following definition of completeness of a theory:

> A theory is Post complete iff "every time we add to it [the theory] a sentence unprovable in it, we obtain an inconsistent system." (Mancosu et al. 2009, 426)

In other words, there is something maximal about the theory, we cannot squeeze anything more into it without generating inconsistency. Of course, Post was assuming that *ex contradictione quodlibet* holds in the theory. Post then defines a formal system to be inconsistent "if it yields the assertion of the variable p" where p allows us to derive any sentence, as a sentence in the theory (Mancosu et al. 2009, 426). Again the theory has as many sentences in it as possible without collapsing into triviality. We can modify the definition a little, to make it serve our purposes.

> A theory is Post non-trivial iff there is at least one formula in the language of the theory which we can display and is only false.[8]

Of course, it will not be a very useful theory if there is only one falsehood, which is not also true. What we want for workable theories is that there is a substantial set of only true sentences, a substantial set of only false sentences, and 'very few' sentences which have two truth values, at least on present thinking.[9] Even the notion of 'very few' is only metaphorical. What we really want is to be able to distinguish the class of only true, only false, and dialetheic sentences. The formal system LP presupposes that there is such a distinction. It is a separate issue whether or not we can *detect* the difference in an application of LP. We can go some way towards the detection, by specifying that the dialetheias are constructed using the inclosure schema. But we have to be careful even here, since some paradoxes generated by the inclosure schema are less interesting and less important than others.[10]

---

[8] We have to say something about 'displaying' because the trivialist will admit that there is a formula, or sentence in its language which is only false, but it will not be able to generate or display the sentence since it thinks every sentence in the language is true.

[9] This could change. However, even in a book such as Mortensen's Inconsistent Geometry, where he is interested in exploring the structure of inconsistent geometrical shapes, lines, spaces, *etcetera*, he is still assuming that what he says (in the meta-language) is largely only true (Mortensen 2010).

[10] What I am really concerned about is that we cannot distinguish between different sorts of paradox: the dialetheic ones and the ones we should continue to work to 'solve' in a more conventional way. We shall return to this issue in Chap. 13. To foreshadow: there is no easy answer, but there are partial answers.

Modifying our definition:

a Post consistent theory is consistent, if it is decidable and counts *ex contradictione quodlibet* as valid, and non-trivial otherwise.

In the latter case, we use the term 'Post non-trivial'. Not every theory containing contradictions is trivial, especially in the light of a relevant logic. A dialetheist adds to relevance the idea that some contradictions are true (as well as false). Also, as we have said in Chap. 6, depending on which underlying logic a pluralist adopts, he will bring a different philosophical stamp. The dialetheist stamp includes the inclosure schema. This is a formal schema for generating paradoxes. The inclosure schema can be used in the presence of any large or 'all encompassing' theory. All we need is a little ingenuity in interpreting the schematic letters.

I shall give the full definition of the schema in a moment. Let me first make a few general remarks. The inclosure schema is a way of generating paradoxes, some of which are also dialetheias.[11] The schema was originally developed by Russell as a diagnosis of what is common to the set theoretic paradoxes. Priest (2002) generalises Russell's schema, and uses the inclosure schema to expose paradoxes implicit in many philosophical positions. Some of these paradoxes are then thought of as dialetheias: conceptual limits to the philosophy. Here, I shall use the inclosure schema to produce a dialetheia about pluralism.

There are three conditions that have to be met to produce a contradiction (which might then be susceptible to dialetheic treatment). The first (1) is existence. (2) (i) is transcendence, and (2) (ii) is closure. For those unfamiliar with the inclosure schema, I'll quote it verbatim, and then discuss it.

A contradiction fits the inclosure schema iff it has two characteristics. [Existence is not a characteristic, so it is a separate pre-condition.]

1. $\Omega = \{x; \varphi(x)\}$ exists and $\psi(\Omega)$.
2. For all $x \subseteq \Omega$ such that $\psi(x)$:

    (i) $\delta(x) \notin x$,
    (ii) $\delta(x) \in \Omega$. (Priest 2002, 276)

Deciphering (1): this is the existence clause and sets out the terms for the next clauses. Ω is the set of all sets x which have the property φ, and some other property, ψ, can be attributed, in turn, to Ω. In our case, the property φ will be 'is a characteristic or combination of characteristics found in philosophies of mathematics'. The set Ω exists. It is the set of *all* collections of characteristics found in philosophies of mathematics to date. At any one time, there are only finitely many of these. We can think of this global set as giving us the materials

[11] I am not certain about the quantifier. Some dialetheists might even say that all inclosure schema generated paradoxes are dialetheias, or even all and only inclosure schema generated paradoxes are dialetheias. The more conservative 'some' is used here. It is enough for our points about pluralism that we generate one dialetheic paradox using the inclosure schema.

for a global pluralism.[12] In our case, the property ψ will be 'could be made into a philosophical position'. So, if we took all of the philosophies of mathematics, we could have a new philosophy of mathematics, i.e., a global pluralism. ψ is a modal property, since 'could' is modal. There is no guarantee in advance that whatever could be made into a philosophy will be an interesting, strong or stable position. Some possible philosophies will be weak, unstable, boring, trivial or nonsensical. So the claim $\psi(\Omega)$ is that it is possible to make a global pluralism out of all of the philosophies of mathematics; success or interest is a separate matter. Maybe 'candidate philosophical position' is better, but we omit 'candidate' as a distracting complication.

Deciphering (2): we start with the subsets, x of Ω, i.e., all of the (second-level) philosophies of mathematics, and all their combinations plus third level pluralism and so on up to global. The subsets x are potential philosophies. These can be scrutinised in the light of pluralism, since many of these are made up of combinations of other philosophies, monist, dualist and 'sub-optimal', 'optimal' and 'maximal' philosophies. Some combinations of these will lead to nonsense, such as a fictionalist theory which is 'combined' with a philosophy which is realist in truth value. Such a combination is nonsense because both fictionalism and realist in truth values philosophies are classical, and therefore endorse *ex contradictione quodlibet*, and therefore the combination results in a trivial theory.

'Optimal', here, refers back to Chap. 4, Sect. 4.8, and the distinction between optimal and maximal pluralism. An optimal pluralist is a pluralist towards successful mathematical theories, and 'success' is measured in a certain way – as being accepted by the mathematical community, or being describable as a structure of model theory, for example. A sub-optimal pluralist will develop a pluralist philosophy (not monist or dualist), in the sense of including several mathematical theories but is neither interested in 'success' nor in maximal pluralism. For example, we might think of a geometrical pluralist: someone who accepts all of the received geometrical theories on a par. There are many sub-optimal pluralist philosophies, and I suspect that some mathematicians who call themselves 'pluralist' are pluralist in this sub-optimal sense.

Thus, the first part of (2) says that for all subsets of a global theory which are possible philosophies, we can transcend the subset theory (i) while staying within the global theory, clause (ii). More specifically, we start with the transcendental part, (i). δ is our diagonaliser.[13] In our case it will read: 're-organise the philosophy by adopting a different underlying logic' at the third (meta-level) of discourse, for

[12]Note that we have said nothing on how to individuate philosophies or theories, thus some might be unsuccessful, or even disastrous. Global pluralism was found uninteresting philosophically in Chap. 6.

[13]The term 'diagonaliser' is meant to be suggestive of Cantor's diagonal number: the one that proves that the list of reals, which we originally tried to make, is incomplete. It is constructed from the existing listed numbers, but is not a member of what was supposed to be (*per impossible*) a complete list of members. So, 'diagonalisers' use what falls under a concept in order to take us beyond that concept. This is the transcendence part of the inclosure schema.

example, we might use a relevant logic to underpin our fictionalism, rather than use the more common classical logic. By diagonalising we step out of x. We transcend our sub-theory.[14] Moreover (ii) the diagonalised x is a member of the notion of global pluralism, Ω. This makes sense since Ω is a global theory.

We can now use the inclosure schema to create a paradox. Diagonalise Ω, by switching underlying logic. The diagonalised Ω transcends Ω itself, but it could not have since $\delta(\Omega) \in \Omega$ according to (ii); contradiction. If we consider global pluralism, then we can generate many pluralist philosophies, and we can generate a paradox. The idea of a 'position' suggests a finished entity, the global pluralist position. But it is not fixed, because by its very nature, because of its pluralism, it grows. Call this the transcendental paradox. What are we to make of it? First, is it a dialetheia? I.e., is it a true contradiction? We found global pluralism to be unsavoury in Chap. 6, but here we know why. This is why we opted, instead, for maximal pluralism as being much more reasonable and workable as a position. The paradox we have generated using the inclosure schema gives us a reason to find global pluralism unsavoury because dialetheic. Nevertheless, unsavouriness is not a principled reason for ignoring the theory – by our own pluralist precepts. Therefore, we *should* (normative 'should') consider it if we have the time and energy. So it is a true contradiction. The next problem is how to cope with true contradictions of this type.

The reaction of finding global pluralism 'unsavoury' is a conditioned reaction.[15] It is conditioned by our upbringing in classical logic. In the literature, there are enough arguments supporting the thesis that this particular aspect of our upbringing is not defensible, and I refer the reader to these. See especially (Priest 2002, 2006a, b; Mortensen 2010). Overcoming our upbringing, or turning to other traditions, we can think of global pluralism as an extensionally growing philosophy. It is not, for all that, intensionally growing. The intension is fixed. That is, global pluralism is an intention to entertain, and evaluate any proposed philosophy of mathematics, including ones made from a combination of others. The task cannot be completed, and we can never reach a final decision. The 'best' we could hope for is relative stability. For this reason, global pluralism is not as chaotic as appears at first glance, since we take time to develop and evaluate philosophies, and there are

[14]Note that the subset is not a proper subset, so we can diagonalise on Ω. In fact this is what we shall do to generate the contradiction.

[15]Note that it is not conditioned in every upbringing of every person who ever lived. There are people who never receive the necessary level of education to be conditioned. They are only conditioned, insofar as they are, by 'luck' because Aristotle's law of excluded middle is followed by enough influential people in society, again, this reaches only as far as it reaches, not every 'uneducated' or 'socially removed' person is affected, partly because the 'influential people' are so for socio-political reasons, and therefore might not themselves fully adhere to the law of non-contradiction. But there are also sophisticated examples of alternative upbringings. See, for example, the writings of Nāgārjuna, (Priest 2002, 249–270). Moreover, the non-standard upbringing is not restricted to the far and exotic East. When Aristotle argues for the Law of non-Contradiction, he is making an *argument*, and this suggests that he has a worthy opponent. In particular, Aristotle was addressing Heraclitus and Protagoras (Priest 2002, 11).

only a small number of people who do this. So at any one time, we have relative stability. It is not the case that 'anything goes'.

We build and consider positions which have some pedigree, positions which have been received by the community of philosophers of mathematics. We build on each other's shoulders, not in the void. Construction and evaluation take place within a community, and therefore, to develop the metaphor of chaos, we do not have chaos, but anarchism, that is, a self-regulating community. We, therefore, have the option, nay constraint, to *confine* our research. Our historical situation, our finiteness, and the small number of people in the community, together bridle global pluralism, so it should not be thought of as threatening.

10.5 Conclusion

To sum up: if we take a paraconsistent logic to underpin our pluralism, and we think that pluralist philosophies of mathematics (taken globally) contain dialetheias, then we think that pluralism commits us to recognising some contradictions as true. The dialetheic response is to think of pluralism as a transcendental position. It is a family of positions, and as soon as it is taken as a whole it is then transcended. The global theory grows. The situation is not as bad as it might seem. For one, we are in good company: the company of almost every leading world philosopher: Plato, Aristotle, Aquinas, Leibniz, Sextus Empiricus, Berkeley, Kant, Hegel, Russell, Frege, Quine, Davidson, Wittgenstein and Heidegger to name a few (Priest 2002). Apart from the good (or bad) company argument, we can think of the dialetheias this way: moving up a level from third-level dialetheic pluralism, we take seriously the idea that there might be rival logical underpinnings, each meriting its own take on the lower levels of analysis. After all, we could just choose a different paraconsistent logic, since the term 'paraconsistent' is adopted in quite different philosophical and logical traditions. Each has its own motivations, and will bring its own philosophical stamp to bear on its version of pluralism. For this reason we can, and cannot, really discuss global pluralism as a fixed entity, and this might not be so difficult to concede.

However, we should be careful. The paradoxical situation is worse than I have let on. There are not just three paradoxes. There are more. We can use the inclosure schema to generate other paradoxes of pluralism. It is simply a matter of choosing good combinations of properties and a diagonaliser, instances of the schematic letters: Ω, φ, ψ and δ! There might, for example, be a way of using the inclosure schema to generate a paradox about maximal pluralism too, and maybe the same applies for some of the optimal pluralisms. For, each professes to set some sort of limit to what is being considered, and that limit invites transcendence. Under the desperate prospect of generating ever more paradoxes, dialetheism becomes *more* attractive (and maybe also less)!

Why more attractive? When we say that 'a pluralist is pluralist about pluralism' we are expressing the closure of pluralism. But pluralism transcends pluralism too. How do we interpret this? Pluralists are well aware that philosophy and mathematics are historically (conceptually) situated. That is, they are developing with reference to each other, what looks like a good philosophy of present day mathematics, might not look so good of future mathematics, *mutatis mutandis* for applying present philosophies of mathematics to past mathematics. Our interpretation, or rational reconstruction of past mathematics in terms of present mathematics is a delicate issue.[16] More important dialethically, as pluralists at the third level, we take seriously the possibility that there could be significant improvement to the logic or class theory underpinning our philosophy, in which case we would revise the underlying logic. In saying this we show how we can transcend our maximal pluralism. We shall not transcend everything all at once, but rather stage by stage. It is piecemeal development situated in history. Paradoxically, change and transcendence are what save us from particular paradoxes being completely unsolvable. There is a dialetheic 'solution', or if one prefers, 'resolution'.

References

Friend, M. (2014). *Pluralism in mathematics: A new position in philosophy of mathematics* (Logic, epistemology and the unity of science 32). Dordrecht: Springer.
Hellman, G. (1989). *Mathematics without numbers*. Oxford: Oxford University Press.
Mancosu, P., Zach, R., & Badesa, C. (2009). The development of mathematical logic from Russell to Tarski 1900–1935. In L. Haaparanta (Ed.), *The development of modern logic* (pp. 318–470). Oxford: Oxford University Press.
Mortensen, C. (2010). *Inconsistent geometry* (Studies in logic 27, Series Ed. D. Gabbay). London: Individual author and College Publications.
Priest, G. (2002). *Beyond the limits of thought*. Oxford: Clarendon Press.
Priest, G. (2006a). *Doubt truth to be a liar*. Oxford: Oxford University Press.
Priest, G. (2006b). *In contradiction* (2nd ed.). Oxford: Clarendon Press.

[16]We address an example in Chap. 14.

Chapter 11
Pluralism Towards Pluralism

Abstract In this chapter we visit the subtle question of whether the pluralist is pluralist towards himself. We answer this question in two ways. The first is technical, and we develop this answer in Sects. 11.2 and 11.3. The second answer is general, and we develop this answer in Sect. 11.4. A reader could skip Sects. 11.2 and 11.3 without loss of coherence, especially if said reader is not wedded to a particular formal paraconsistent logic. The final two sections are for those who feel queasy from following the conceptual gymnastics of pluralism.

11.1 Introduction

We mount to the fourth level. In this chapter we visit the subtle question of whether the pluralist is pluralist towards himself. We answer this question in two ways. The first is technical, and we develop this answer in Sects. 11.2 and 11.3. The second answer is general, and we develop this answer in Sect. 11.4. A reader could skip Sects. 11.2 and 11.3 without loss of coherence, especially if said reader is not wedded to a particular formal paraconsistent logic.

To give a technical answer to our question. I assume LP, as a well-established, tried and tested paraconsistent logic. LP bears witness to the following claims: (1) contradiction does not have to entail triviality, and (2) when well managed, paradox does not lead to *incomprehension* in the long run, although it might lead to a period of puzzlement. Both claims rest on the fact that we can work in a paraconsistent setting. For example, there are more and less interesting non-standard, paraconsistent, models of arithmetic, geometry and set theory. See Mortensen (1995) for arithmetic, (Priest 2002, 174) for set theory and Mortensen (2010) for geometry.[1] Nevertheless, those working in this area admit that it's hard

[1]One complaint I have heard against this sort of work is that the 'interesting models' are not so interesting since they have not been used to tell us anything about other parts of mathematics. To

M. Friend, *Pluralism in Mathematics: A New Position in Philosophy of Mathematics*, Logic, Epistemology, and the Unity of Science 32, DOI 10.1007/978-94-007-7058-4_11,
© Springer Science+Business Media Dordrecht 2014

to think in contradictory situations. But, this should not dissuade us. That 'it is hard', is simply a subjective, psychological report, and certainly does not mean that it is impossible. I doubt it is impossible, since, to all appearances, some people seem to manage. The people in question are, at least, all paraconsistent logicians.[2] To accustom oneself to such thinking, it is not a bad exercise to work through some of the technical parts of this chapter, as a way of introducing work in inconsistent settings, but strictly speaking, these sections can be skipped. To introduce work in paraconsistent settings, Mortensen and Priest use the collapsing lemma to show that what we think are 'consistent formal systems' have inconsistent and non-trivial interpretations. Priest calls these *coherent* interpretations. I shall do something else with the collapsing lemma, since we were already introduced to it in Chap. 9.

I shall use the collapsing lemma on LP. This will tell us *not* that LP is coherent – we suppose this.[3] Rather, it tells us that there are *several* coherent interpretations, or models, of LP. Therefore, if someone is wedded to LP as the logic to underpin pluralism and provided the reasoning goes through, then one has to acknowledge the existence of several LP interpretations. This is enough to show that a pluralist (at third level) who uses LP will have to be pluralist towards interpretations. That is, he will be semantically pluralist towards third-level pluralism. The models generated by the collapsing lemma will tell us different things about the pluralist deployment of LP. Here, we only look at one non-standard interpretation generated using the collapsing lemma and the inclosure schema. The result of the technical work is modest. Nevertheless, it bears witness to the possibility that similar results can be had with different paraconsistent logics underpinning pluralism.

In the fifth section, we step back. That there should be several versions of pluralism should not be surprising since there are other candidate formal systems. Using others will give a different flavour to pluralism. The pluralist has to be

---

answer this criticism, first: when Priest et al. claim that the models are interesting, they mean this in a specific sense. The *uninteresting* models are those that are trivial everywhere except for a small consistent part. The *interesting* models are ones that recover consistency after a certain inconsistent limit. The inconsistent limit might re-surface again later. The inconsistency is a fixed point (Priest 2002, 173). Therefore it is correct that this is not enough to interest other mathematicians. It is a completely legitimate demand, since acceptance and inclusion of new mathematical areas is only achieved through crosschecking. As things stand at present, paraconsistent logics and mathematical theories are generally treated as a mathematical curiosity. Regardless, it is *philosophically* important for distinguishing trivial from inconsistent mathematics. But there is a second answer: working out links with other areas of mathematics takes time and resources. There happen to be very few people working in this area. I am certain that as soon as one of them finds something interesting for others in the mathematical community, he, or she, will let us know. The time it takes is a feature of communication in the mathematical community, and how that works.

[2]On explaining something about paraconsistent logic to a Romanian ecological economist, he assured me that all Rumanians think this way. So the group might well include much more than only the paraconsistent logicians. In fact, this is one of the motivations for studying paraconsistent logic: paraconsistent reasoning is empirically observed, especially in philosophy classrooms, but also on the street in Romania.

[3]LP is Post non-trivial, and this is enough for coherence.

pluralist towards himself just in virtue of admitting alternative logical formal systems to underpin a version of pluralism. Again, *qua* programme, here we see that we can make different versions of pluralism by adopting different underlying formal logical systems. Even the indulgent reader might by this stage have renewed her doubts. So we re-discuss trivialism and the difference between paraconsistency and trivialism. We include a final section called 'nausea', for those who feel queasy.

11.2 The Collapsing Lemma

Priest and others call it the 'collapsing lemma'. Meyers preferred the name the 'bubbling lemma'. Both are suggestive. The lemma collapses a domain into a smaller domain to give a new model or interpretation. But once we do this, it bubbles up, to produce all sorts of weird and wonderful paraconsistent models.[4]

Re-stating the collapsing lemma (familiar from Chap. 9):

> *Collapsing Lemma*
>
> Let φ be any formula; let v be 1 or 0. Then if v is in the value of φ in $\mathcal{U}$, it is in its value in $\mathcal{U}^{\sim}$.
>
> In other words, when $\mathcal{U}$ is collapsed into $\mathcal{U}^{\sim}$, formulas never loose truth values they can only gain them [they can become both true and false]. (Priest 2002, 172)

I'll explain each symbol by way of re-expressing the lemma for those who would like to refresh their memory. If you have understood it either here or in Chap. 9, then skip the rest of this section.

An interpretation $\mathcal{U}$ is a model <D, I>. Interpretations are usually used to 'satisfy' sets of formulas, i.e., theorems. Classically, they make all the theorems only true, here we have them make all the theorems true, but they could also be true and false. The interpretation makes a proposition or wff, true, or false, or both. An interpretation satisfies a set of formulas iff it makes them all (at least) true. If the formulas are first-order formulas, then the interpretation will include a domain of objects, over which variables range. Object constants pick out particular objects in the domain. There might also be predicate or relation constants in the interpretation. These pick out subsets, or ordered subsets of the domain. These subsets of D are the extensions of the constants. We leave out functions from the language being interpreted because the collapsing lemma cannot be proved if both: there are functions in the language and $^{\sim}$ is not a congruence relation on the functions in the language (Priest 2002, 173 n. 6). We can think of an interpretation $\mathcal{U}$ as a pair: <D, I>, where D is the domain and I is a (meta-level) *function* which maps individual constants in the language into D and maps predicates into their positive and negative extensions in D. Elaborating on the notion of 'negative extension': a positive extension is the set of objects in D

[4] I was torn about what to call it myself. On the one hand, the 'collapsing lemma' is more current, so it would be less confusing to use that name. On the other hand, the 'bubbling' lemma is more fun.

which have the property in question; the negative extension is the set of objects in D that do not have the property. We have to specify both separately, since there could be (inconsistent) objects, which are in both extensions, and this is how we can tell that they are inconsistent. Now, $\sim$ is an equivalence relation on D. $\sim$ will partition D, lumping some objects together because they are interchangeable (equivalent) under $\sim$. $\mathcal{U}^{\sim}$ will look exactly the same as $\mathcal{U}$ if $\sim$ partitions the domain by identifying every member with itself. $\mathcal{U}^{\sim}$ will make a different domain, $D^{\sim}$, if it has some properties which have more than one object in their extension. It will be smaller, if D is finite. If D is infinite, then $D^{\sim}$ will be a proper subset of D. For example, let D be comprised of the natural numbers. If $\sim$ says: "take every number greater than 10000, and stick them together (because they are too big, and I do not care about them) leave all of the other numbers alone" then $\mathcal{U}^{\sim}$ will be quite small and finite, with 10000 members, or if 0 is in D, then 10001 members. Our new $\mathcal{U}^{\sim}$ domain $D^{\sim}$ will be the set: $\{[1], [2], [3], [4], \ldots [\{10000 \ldots \}]\}$. Priest uses square brackets around d: [d], to indicate that each member of $D^{\sim}$ is an equivalence class made from D.

The mapping $\sim$, has to preserve a lot of the original structure of D. All constants are preserved as such under $\sim$. They are just repeated in, and are each separate members of $D^{\sim}$; similarly for predicates which preserve both their positive and negative extensions. We add a valuation function 'v'. This assigns 0, 1, 0 and 1, to formulas in the language based on the interpretation.

The lemma says:

all formulas which are true, given $\mathcal{U}$, are still true under $\mathcal{U}^{\sim}$.

So, for example, if $2+2=4$ is in $\mathcal{U}$, it will still be true (given value 1) in $\mathcal{U}^{\sim}$. It might, of course, be contradictory in the new $\mathcal{U}^{\sim}$, in which case it will also have the truth-value 0. The collapsing lemma without an accompanying result about Post consistency is not interesting, since we then have no guarantee that we have not collapsed into triviality, placing all objects in the domain into both the positive and negative extension of individual constants, predicates and relations. The trivial model is made by collapsing the domain into one object, by stipulating that every object is identical to every object: $\forall x\, \forall y(x=y)$.

11.3 The Collapsing Lemma and LP

Consider LP as a formal theory of mathematics, with: formulas that are true, formulas which are false, and formulas which are both true and false under given interpretations. See Appendix 1 for more details. We shall show that while LP has a standard interpretation, we can use the collapsing lemma to make a new (collapsed) interpretation of LP.

We take the first-order version of LP, since we want domains. The language of LP has the logical connectives, a quantifier, variables, a finite number of predicate letters, identity and no functions. The finite number of predicate letters, and the absence of functions is needed for technical reasons, the details of which are not important here.[5] First, LP has its interpretations $\mathcal{U} = \langle D, I \rangle$. The interpretation is made in a language. This is not something normally mentioned, but it will be important for us. The language of $\mathcal{U}$ includes all of the connectives, the universal quantifier, the constant '=' and so on, just as in LP. The semantics of LP is the standard $\mathcal{U}$ of LP. We shall change two things. In the domain of $\mathcal{U}$ members are also names, and each name will be given a Gödel number. The names can thus be ordered: d_1, d_2, d_3, and so on. This should be harmless, since we think of a language as something deployed and therefore essentially finite, so it is enough if there are fewer than ω names in a given formal language. The second change is that in $\mathcal{U}$ the valuation *function* is a *relation* between formulas and members of the set of possible truth values: $\{\{0\}, \{1\}, \{0, 1\}\}$. The valuation relation relates a formula to 0 or to 1, the 'or' is inclusive.

Moreover, in concert with the usual interpretation, $\mathcal{U}$ has it that every formula which is valid in classical logic (i.e., true under all interpretations where formulas are only related to 1) is 1 (only) in LP. There are, however, formulas with both truth values, and these are the contradictory formulas; ones whose valuation relation relates the formula to 0 and 1. The standard or intended domain of interpretation of LP is all objects which can be characterised.

The characterisation principle tells us that an object just is the bearer of characteristics (expressible in a language).

Of course, there will be contradictory objects, namely those characterised by contradictory properties, pairs of properties which preclude each other.

Collapse the standard domain in the following way. Put together, into one class all of the contradictory objects and all of the objects never experienced. We do not need to know what these are (in fact we cannot) they are just the bearers of the characteristic 'has not been encountered'. Provided we are not strict constructivists, we can make sense of the idea that we know that this new interpretation $\mathcal{U}^{\sim}$ is a proper subset of $\mathcal{U}$.[6] We also know that the resulting interpretation is not trivial. This is so for two reasons. If you block *ex contradictione quodlibet* proofs in your system, then you have non-triviality because there will be some formulas that are not provable. That is, we have Post non-triviality. We can show this by induction on proofs. Secondly, what we have done is to give a type of interpretation to LP where we have put all of the impossible and unencountered objects together. There are some quite standard encountered objects, and there are predicates under which they fall. Moreover, since they are standard encountered objects, there are predicates

[5]The finite number of predicates is important for ordering minimally inconsistent LPs (Priest 2006b, 227). The lack of functions is needed for the proof of the collapsing lemma to go through. Of course, we can re-express functions as relations (Priest 2002, 173). 'Predicates', here, are predicates or relations. That is, a relation is a two-or-more place relation.

[6]There are more careful ways of saying this that are acceptable to a constructivist, but they are elaborate, and therefore, using them (since there are different versions of constructivism) would risk increasing confusion.

under which they do not fall. So, again, with a little philosophical indulgence, we have a non-empty class of formulas which are only true, a non-empty class which are only false, and a non-empty class which are both true and false.

By this re-interpretation, we have shown that we can give a second interpretation of LP, a non-standard one, and it has some philosophical interest. This is because what it amounts to is a way that the realist can make some sense of some versions of constructivism. It is not a sense the constructivist will recognise. Our exercise is also a demonstration of pluralism at work. The cheeky conclusion is that the LP monist has to acknowledge that it is coherent to give an alternative interpretation of LP. Maybe this is no surprise, since there is at least some sense in which dialetheists can 'understand' constructivists, by, for example, discussing the fact that intuitionist logic allows *ex contradictione quodlibet* proofs. The understanding is partial and concerns the common territory, the objects of which are both possible and encountered. Meyer however, see Chap. 13 for a further twist to the story.

I should add a last note on what has transpired in this section. Meyer developed the collapsing lemma. Mortensen uses it to produce paraconsistent models of arithmetic. Priest uses it, in a similar fashion to Mortensen to produce a paraconsistent model of first-order set theory. They do this for the following reason: "one way to show a notion to be coherent is to produce models of it," (Priest 2002, 170). Of course, the models for a paraconsistent arithmetic or set theory will be paraconsistent, and not trivial. They will not be classical, consistent models.

My purpose is quite different. I am *not* starting with a supposed consistent formal system and then showing that the paraconsistent version of it has a paraconsistent model, and therefore the paraconsistent version is coherent. I am *starting* with a paraconsistent logic, so the standard model is already paraconsistent. Nevertheless, the logic too, has several interpretations. Therefore, what I show is that the notion of 'several interpretations of paraconsistent logics' is coherent. It is therefore *coherent* to be pluralist about a pluralism (of third level) underpinned by LP. The reason this matters is the subject of the next section.

11.4 Who Cares? Widening the Picture, Other Paraconsistent Logics

We showed in the last section that LP has more than one non-trivial interpretation. In some ways this should not be too surprising. The world could be many ways and still LP would be a logic we could deploy in that world. But the point is that this thought is *coherent*, unless one shows that the reasoning in Sect. 11.3 is irredeemably wrong or that the technical modifications made in order set up the demonstration of Sect. 11.3, are too distorting of LP. Assuming that the demonstration is accepted, then a fan of LP would have to be pluralist about the different interpretations, since LP does not determine one interpretation. There is no way to decide between them without stepping outside the reasoning, and bringing some *extra-logical* considerations to bear on our choice of interpretation. I do not deny that this can be done. Given the example above, it would be obvious to argue against collecting together the unencountered and impossible objects since this is metaphysically unsatisfactory. Nevertheless, it might serve some purpose. It makes some sense for the realist (as an attempt to understand the constructivist), and makes no sense at all for the constructivist *qua* constructivist. It can only make sense to the constructivist as an attempt to try to meet the realist part-way to explain constructivism to the realist. So, it is not a philosophically satisfying, or stable interpretation of LP.

Nevertheless, it is enough for me here to acknowledge that there are two *coherent* interpretations.[7] We could demure from making a choice of one interpretation over the other, on strength of coherence. That is, it is coherent to be agnostic about (at least some of) the different interpretations, i.e., be pluralist towards the different interpretations.

If one is not a convinced fan of LP, but of another formal system, it would be interesting to see if we could run the reasoning using the collapsing lemma, or some other means of generating different interpretations for that formal system. This, again, hails future projects. Note that, these would not show that we can reason coherently (non-trivially). We know that already. Or, we assume it when we accept or deploy a paraconsistent logic, even if we do so metaphorically. What comparable results would show is that once we accept a given paraconsistent logic, it is coherent, at least in some cases, to be pluralist about more than one interpretation of that logic. Indeed, the fourth-level pluralist thinks we do this implicitly whenever we are engaged in comparing discourses of together contradictory theories.

If we do not naturally fasten on one paraconsistent logic, to the exclusion of other paraconsistent logics, then we do not even need to go through all of the reasoning of Sects. 11.2 and 11.3, to reach this conclusion. We are already pluralist about pluralism. For we accept, implicitly, that we could underpin our third-level pluralism with any acceptable paraconsistent logic, together with all of its coherent interpretations.

Regardless of whether we arrived at this conclusion starting from a logically monist position, or from an already logically pluralist position, we are still faced with a choice about our *degree* of pluralism towards pluralism. If we default to the more radical position, of not favouring any paraconsistent logic over others, then it is worthwhile working through what sorts of differences we might see, and what will be common to all of the pluralist philosophies. These differences will determine our degree of pluralism, this time, regardless of our starting point. Let us begin with the similarities.

---

[7] It is plausible to ignore these if we realise a few facts. First, there are several interpretations of LP which we can make using the collapsing lemma. Second, philosophical considerations might not decide between all pairs of interpretations to favour one over the other. Moreover, third, there might not (yet) be *any*, non-*ad hoc*, philosophical or otherwise, means of making a determinate choice.

Similarities

1. All paraconsistent logics are motivated by the idea that we *do* encounter contradictions, or at least seeming contradictions. Moreover,
2. we seem to be able to behave quite rationally in such situations. We do not lapse into trivialism.
3. All successful paraconsistent logics avoid trivialism.

The differences between the systems concerns their diagnosis of:

Differences

1. what an inconsistency is, before it is given formal representation,
2. which formal representation of contradictions is best and
3. whether all apparent inconsistencies are only that, so they can all be dissolved.

An example of an apparent contradiction might be an inconsistent data set, where there was an error in entering some of the data. The error can be easily corrected, making the data set consistent again. The differences in diagnosis are usually manifested in the interpretation of the conditional in a natural language, the notion of implication and the notion of entailment. Some formal systems also target other connectives or logical notions such as conjunction or disjunction. Many include truth-value gluts or gaps in their semantics. Attending these differences, the reasoning will differ when we are confronted by a contradiction.

By now, some readers who have followed this far will have re-kindled their doubts about pluralism. In order to remind them of one of the crucial points, we revisit trivialism, and how the pluralist avoids it.

11.5 Trivialism, Relativism and Inconsistency

What is wrong with trivialism? And why might we think that pluralism is not trivial? After all the following criticism is quite general, and reminds us about the norms in the pluralist philosophy and how a lack of norms can degenerate into trivialism, which is a close cousin to rampant relativism.

Criticism 1 If the maximal pluralist is so loath to set norms or arbitrate between exiting norms, then everything goes, and the position is actually trivialist. For any theory, we can find a meta-theory or an attitude, which endorses the theory, so there is no real philosophical judgment, there are just relative judgments or descriptions. No one wants a trivial philosophy because it is the same as a rampant relativism.

There is a technical and direct threat to pluralism when the pluralist even just entertains or discusses trivialism.

Criticism 2 The pluralist might end up with a trivial philosophy because the pluralist takes seriously some trivial mathematical theories. Call this 'the argument from infection/ explosion'. After all, the language of these theories is a proper sub-part of the philosophy, so the triviality spreads through the philosophy.

This has not in fact occurred.[8] In practice we observe a clean break between trivial mathematical theories and philosophical positions which entertain them. The infection does not spread to the philosophy. But we need to say more. We need to explain why this is, especially in this case.[9]

Before we answer the criticisms, we should appreciate the danger. A trivial *philosophy of mathematics* holds that every well-formed mathematical formula, in any language of mathematics is true (and its negation is true), and any philosophical sentence about mathematics is also true.[10] Anything goes, and all judgments are as good or correct as the next.[11] Trivialism is pretty hopeless *as a philosophy*, although it is very easy to defend or maintain verbally or in writing (although a trivial piece might never be accepted for publication)! The strongest criticism against it, which is pertinent to this project is well expounded in Priest, and it is an argument from meaning (Priest 2006a, 68–69).[12] It is not clear that a trivialist can mean anything by his utterance or written statement, since there is no recognisable *judgment* attending sentences. They are all true. So there is no meaningful intentionality, since there is no distinction between a belief, a known fact, a subject of fear, desire or what have you. Since there is neither judgment nor intentionality attending the use of language, the philosophy of mathematics being presented is degenerate. According to someone who is not a trivialist, the trivialist theory renders[13] mathematics

---

[8]In know of no discussion *of* trivialism which has degenerated *into* trivialism, except in moments of jest.

[9]One might think that I am being somewhat unfair, and ignoring a lot of philosophical activity. For example one might point out that Russell was much aggrieved by the paradoxes, and theorised a lot about them; and I should not ignore this since Russell's investigation into the paradoxes shaped his philosophy and formal system. Moreover, some very important philosophical work has been done in looking very closely at Frege's trivial theory – such as the work of Dummett, Wright, Boolos and Heck. I appropriate such activity, and call it pluralist! What is anti-pluralist is any accompanying revisionism. So, we should be careful about our interpretation of the intention behind the excellent work cited above, we might say that these philosophers engage in pluralist work despite themselves.

[10]We might come to this position by supposing, say, that ZF contains a contradiction. More precisely, we need a theory which is considered to be foundational to mathematics, we need for it to be a classical theory: allowing *ex contradictione quodlibet* inferences, and we need to be able to derive a contradiction from the axioms using the rules of inference.

[11]For a good discussion of trivialism see (Priest 2006a, 56–71).

[12]In (Priest 2006a), Priest writes that he is not completely satisfied with the argument from meaning, and thinks that his argument from physical survival is stronger. Note that in (Priest 2006a), Priest is arguing about trivialism in general, not about trivialism as a philosophy of mathematics. For the purposes here, the argument from meaning is both satisfactory, and the stronger argument.

[13]The trivialist will 'hold', in the sense of 'assert', any position. This is not the point. Trivialism in mathematics arises from the idea that mathematics is classical and there is a contradiction in mathematics, and therefore (under our old classical reasoning) all of mathematics is true, we *then* get to the meaninglessness of any particular mathematical statement, and wallow in our degenerate theory. There is a sequence to the reasoning, which gets us to the degenerate position. Once there, reasoning, as such, is impossible.

meaningless, and the philosophy itself is meaningless. Provided that we hold that some wffs are false (and not true), we do not have a trivial theory; this is Post-non-triviality. An example of a wff, which the maximal pluralist holds false (and not true) is: $\vdash_{PA} 2+9=34$.[14] We read this: "in Peano Arithmetic, two plus nine equals thirty-four". These are the Ts (sentences, s, of a theory, T). This is enough to distinguish the maximal pluralist from trivialist philosophy of mathematics. Note that we have not *defeated* the trivialist with our argument. Rather, we have simply distanced maximal pluralism from trivialism, and this is enough to fend from the first criticism that maximal pluralism *is* a trivial philosophy.

However, the second criticism has not been answered. How can we entertain, and discuss trivial theories seriously without our own talk degenerating into trivialism? Trivial mathematical theories are the most controversial of the 'bad' theories the pluralist will discuss. A trivial *mathematical theory* is one where every well formed formula in the language of the theory is true. Different trivial mathematical theories are distinguished from each other by their language.[15] To distinguish between different trivial theories, we look to the differences in vocabulary, which are part of the characterisation of the theory. This is enough to clearly distinguish between, say, Frege's theory and Cantorian set theory. For a trivial mathematical theory two factors have to be in place. The underlying logic of the theory has to be classical (has to allow *ex contradictione quodlibet* inferences) and there has to be a contradiction derivable from the axioms using the rules of inference of the theory. Not all trivial theories are the same. They differ in language, and in structure. Paraphrasing Mortensen (2010, 4): "the inconsistent has structure".[16] This is literally the case for geometrical inconsistencies and arithmetic inconsistencies. We use inconsistent non-trivial models to give sense to that structure. The inconsistent has structure, but what of the trivial?

Historically, there are three (to my knowledge) mathematical theories that were trivial and had a profound impact on mathematics or logic. These are: Cantor's naïve set theory, Frege's formal theory of logic and the first version of Church's formal theory of mathematical logic. All three had repercussions on subsequent mathematics. None led to the collapse of 'all of mathematics'. None led even to

[14]The trivialist will, of course, agree that '$\vdash_{PA} 2+9=34$ is false', since the trivialist will agree to everything. The maximal pluralist will disagree that '$\vdash_{PA} 2+9=34$' is true. The quotation marks are important. The trivialist has to agree, and cannot disagree, except in quotation marks. This is all we need to distinguish the positions.

[15]If (what we suppose to be) two trivial theories have different languages, then they can be distinguished from each other, not otherwise. Some sentences will be true in one, but not recognizable in the other. I thank Priest for pressing me on this point at the Logica conference 2005.

[16]This is a paraphrase. What Mortensen (2010, 4) actually writes is: "The importance of inconsistent images is enormous, I think. Even sceptics who disbelieve in paraconsistency have difficulty in insisting that the inconsistent has no structure, when confronted with these examples [of images of inconsistent objects]." I do not think I have misrepresented him in my paraphrase.

the collapse of 'that part of mathematics infected by the theory'.[17] The important proofs contained in the exposition of the above trivial theories do not proceed as *ex contradictione quodlibet* inferences, which is *why* the theories are considered to be important despite their being (technically) trivial.[18] Thus, further paraphrasing Mortensen, for the pluralist who is indulgent towards trivial theories, 'trivialism has structure' too!

Moreover, this is recognised implicitly. The good trivial theories are studied and trawled for good ideas and insights. Witness the work of Dummett, Wright and Heck on Frege's trivial theory. In general, after spotting an inconsistency, mathematicians try to fix the theory with *minimal* changes. This is what Church successfully did, and what Frege tried to do. Why do mathematicians do this? It is partly due to the fact that there is a sense of 'correctness' of the theory independent of the inconsistency. That is, there is a difference between, say, a proof in Fregean notation where there is a mistake in the application of *modus ponens*, and the mistake in the whole theory: that we can derive a contradiction in it. With the first sort of error we correct it by changing the proof. With the second sort of error, we correct it by changing the axioms, or some other fundamental part of the theory.

To ignore the mathematical and philosophical influence of such theories, again, would be to provide a philosophy of mathematics that lacks in scope. Byers remarks: "No description of mathematics would be complete without a discussion of its *subtle* relationship to the contradictory (my emphasis)." (Byers 2007, 81). Our only hope of engaging in a subtle discussion is through reference to a paraconsistent logic, since other logics are anything but subtle in this respect! Byers remarks later:

> Moreover, paradox has great value. Thus (*sic!*) paradox should be seen as a generating force within the domain of mathematical practice. ... Where do that power and dynamism come from? Well, they come from ambiguity, contradiction and paradox. These things are therefore of great value. They need to be unravelled, explored, developed, and not excised.[19] (Byers 2007, 112)

Ambiguity, paradox and contradiction need to be unravelled if one wants to give an account of the practice and development of mathematics. This is partly a socio-psychological task, but it is also philosophical, since it raises epistemological questions largely ignored by traditional philosophies of mathematics. For, if Byers

---

[17]For example, we did not stop doing arithmetic when Russell discovered paradox in Frege's reduction of arithmetic to logic. This is also evidence against trivialism.

[18]Azzouni (2007, 599) says something similar about the triviality of natural language. Accepting that the semantic paradoxes make natural language trivial, it is then clear that "no one actually makes any inferences on their basis [the basis of inconsistencies arising from the papradoxes], and so the body of purported knowledge that speakers (collectively) are building up, is not ... tainted by such."

[19]Note that Byers makes no mention of paraconsistent or relevant logics. I therefore assume that he is not advocating a paraconsistent point of view or anything of the sort. Nevertheless, in the quotations I cite here, and in many other places in the book, I found support for the position advocated in this book. I do not know what Byers reaction would be to the mention of paraconsistent logics.

is correct, then it is *through* awareness of, and *in confrontation with*: ambiguity, paradox and contradiction that we *develop* mathematics. Of course, when we *use* established mathematical theories we can comfortably ignore ambiguity, paradox and contradiction. The latter are epistemological tools for developing new mathematical ideas. They are not strict limitations or parameters on reasoning or on the corpus of mathematics.

We can draw further distinctions. We have discussed some good trivial theories. There are also interesting but bad trivial theories. Consider Prior's "Tonk theory" (Prior 1960–1). It is interesting because it teaches us something about the limitations on choices of pairs of rules for connectives. So, even this trivial theory has some merit. In contrast, a *hopeless* and bad trivial theory is one where *all* of the proofs have to use *ex contradictione quodlibet*.

11.6 Nausea

> Things are bad! Things are very bad: I have it, the filth, the Nausea. And this time it is new: it caught me in the café. Until now, cafés were my only refuge because they were full of people and were well lit: now there won't even be that anymore . . . (Sartre 1964, 18)

Some philosophers, when reading about, or hearing about, pluralism experience a sense of nausea. This is professional nausea, not a subjective symptom. Here is the pluralist diagnosis. The diagnosis is partly psychological, and the psychology is almost ingrained in the profession. The psychology is that we hanker for certainty and definiteness. This is just part of the human condition. We particularly like definite answers to questions in the form of 'yes' or 'no'. We tend not to like 'it depends', and 'it might seem this way, but consider again', especially when these further considerations end with no resolution in the form 'yes' or 'no', but are just open-ended conditionals. But this only describes our feelings. Such feelings are allowed to determine our ambitions and to dictate what we accept as a solution in philosophy. However, the contention of the pluralist is that the feeling is a prejudice and should not be allowed to determine anything but our direction. It should not be allowed to determine what is to count as an acceptable outcome.

It does not follow from these open-ended situations that no resolution will, or can be had, or that no resolution *has been* reached! Provided an open-ended enquiry includes parameters, a protocol, a sense of correcting an enquirer who strays off the path, then there is a sense in which a resolution has been reached. Here, 'resolution' is to be contrasted to a 'yes'/'no' type of solution. If we have a resolution, then we have a direction, and a way of marshalling our thoughts. That is, 'resolution' is not to be considered to be a lesser (with a negative connotation) type of 'solution'. Instead, 'resolution' is just one form of solution, amongst others. In fact, it is quite common. We might even want to go further and distinguish between different sorts of resolution. As pluralists, we can note that resolution is a psychologically more demanding type of solution, but it is an exciting one too.

The pluralist urges a distinction between the philosopher's private feelings, reactions and motivations and the public philosopher. As a public philosopher, one's feelings are irrelevant. Moreover, assuming some form of free will, they can be brought under some control (of the will). Ultimately, we can choose whether to feel nauseous or excited. Either way, professionally, philosophy just is the sort of subject where there are many more resolutions than solutions. Therefore, professionally and publicly, philosophers had better just get used to it. In other words, the nausea or excitement is a professional hazard, and cannot be used as a refutation or endorsement of a position. Pluralist philosophers are committed to going where the argument takes us. This is our meta-dogmatism. Since we have a choice in the matter, it is professionally healthier to foster excitement than feelings of nausea.

References

Azzouni, J. (2007). The inconsistency of natural languages: How we live with it. *Inquiry, 50*(6), 590–605.

Byers, W. (2007). *How mathematicians think; using ambiguity, contradiction, and paradox to create mathematics*. Princeton/Oxford: Princeton University Press.

Mortensen, C. (1995). *Inconsistent mathematics* (Mathematics and its applications 312). Dordrecht: Kluwer Academic Publishers.

Mortensen, C. (2010). *Inconsistent geometry* (Studies in logic 27, Series Ed. D. Gabbay). London: Individual author and College Publications.

Priest, G. (2002). *Beyond the limits of thought*. Oxford: Clarendon.

Priest, G. (2006a). *Doubt truth to be a liar*. Oxford: Oxford University Press.

Priest, G. (2006b). *In contradiction* (2nd ed.). Oxford: Clarendon.

Prior, A. N. (1960). The runabout inference ticket. *Analysis, 21*(2), 38–39.

Sartre, J-P. (1964). *Nausea* (L. Alexander, Trans.). New York: A New Directions Paperbook.

Part IV
Putting Pluralism to Work: Applications

Chapter 12
A Pluralist Approach to Proof in Mathematics

Abstract We further explore the pluralist's conception of proof. In particular we contrast it to the so-called axiomatic conception of proof. The pluralist adopts an analytic conception of proof. Two claims are defended. One is that all proofs can be viewed as analytic. The second is that it is preferable to do so. The reason it is preferable is that proofs open our eyes to exploration not only towards further proofs in the same formal system, but analytic proofs also invite us to question the axioms and the contexts of proof. We exercise our sceptical caution, to lead us to much more fundamental types of exploration than we would have engaged in had we viewed proofs as axiomatic.

12.1 Introduction

In this part of the book, we set the pluralist to work. In this chapter, I revisit the notion of proof in mathematics. I develop further what happens conceptually when we accept that there is not one standard of rigour. I start by discussing two notions of proof due to Cellucci. They might be thought to capture competing conceptions of proof, but we shall see that this is not the case. The same proof could be seen as both. The two conceptions are the 'axiomatic' and the 'analytic' conceptions of proof. The first is more foundationalist, the second is pluralist, and it is so in an interesting way, which is connected to several ideas in this book.

It is valuable for any philosophy of mathematics to form a conception of proof because proof is one of the characteristics which sometimes distinguishes mathematics from other disciplines. Proof permeates modern mathematics. However, we should be aware that it has not always been so. See the web site for the International Study Group on Ethnomathematics. The history and geography of proof in mathematics is an interesting subject, but not one I am plumbing here.[1] Instead, I restrict my

[1]The notion of what counts as mathematics, can be geographically and historically extended if we ignore our modern, rigid, conception of proof. A suggestion, made by Bishop is to think

M. Friend, *Pluralism in Mathematics: A New Position in Philosophy of Mathematics*, Logic, Epistemology, and the Unity of Science 32, DOI 10.1007/978-94-007-7058-4_12,
© Springer Science+Business Media Dordrecht 2014

attention to that part of mathematics that distinguishes itself from other areas of research *because* of the emphasis on proof (François and Van Bendegem 2010, 116). That is, I am not so concerned with the notion of proof for all of mathematics, but rather, for that part of mathematics where proof is considered to be inseparable from mathematics. This is, at least, all of modern academic mathematics, but reaches far into the past since proof has its roots at least in Euclid.

In this chapter, the dialogue will run as follows. In the second section, I elicit the distinction, between axiomatic and analytic proofs. In the third section I argue that all proofs *can* be seen as analytic. In the fourth section I argue that they *should* be seen as analytic, and that some mathematicians agree with this. Finally, in the fifth section, I shall confirm the view by citing Rav's remarks concerning proofs, and discuss how these add richness to the pluralist notion of proof.

12.2 The Axiomatic View and the Analytic View

Cellucci introduces some useful terminology. Proofs can be *thought of* as analytic or as axiomatic.

Definition *An axiomatic proof* is one that begins with some axioms, or in its sequent calculus guise, rules of inference,[2] and proceeds using only sanctioned rules of inference to lead to a conclusion. Under the axiomatic proof *view* the axioms (alternatively: rules) are to be taken as absolutely true and completely basic and obvious.

For this reason, they are not to be questioned. (The formalist version of this is that the axioms are just stipulated, and it is for this reason that they are not to be questioned). Since the axioms should not be questioned, and the reasoning proves the theorem, the theorem is accepted as *final*.

The notion of axiomatic proof is reminiscent of Frege's gapless proofs and Gentzen's sequent calculus proofs. The former were based on basic laws, and these were presented as indubitable. The latter were presented as the natural way in which mathematicians reason. Hilbert's view is also, in some sense, axiomatic. We do not question the axioms because they are self-justified by being finitist and together non-contradictory. There is no further purely mathematical justification owed, although, of course, we *could* add further justifications in terms of applications. As we can see, the axiomatic proof *view* is both prevalent in, and common to, several otherwise quite different philosophical approaches.

---

of mathematics as an essential tool for *coping* with the environment. Coping includes: counting measuring, locating, designing, playing and explaining. If explaining is short of proving, then proving is not central to mathematics as practiced in many cultures, and most of the time by most of us. This point comes from François and van Kerkhove (2010, 129).

[2]It is understood that rules of inference and axioms are usually inter-definable, and therefore a natural deduction proof or sequent calculation are also axiomatic proofs.

The pluralist joins Cellucci, Manin and Poincaré in taking exception to the axiomatic view. Manin comments on Poincaré: "when Poincaré said that there are no solved problems, there are only problems which are more or less solved, he was implying that any question formulated in a *yes/no* fashion is an expression of narrow-mindedness." (Manin 2007, 13). That is, if we think that by giving a rigorous proof we have established the truth of a theorem for once and for all, then we are mistaken. Instead, we should broaden our gaze after a proof. What we *have* done in giving a proof is expose the justification internal to a theory for the theorem with maximal clarity, and we invite scrutiny and further intellectual probing. Manin quotes Samuelson with approval: "I will conclude with a penetrating comment of Paul Samuelson... One of the advantages of ... the canons of exposition of proof... is that we are forced to lay our cards on the table so that all can see our premises." (Manin 2007, 5), quoted from (Calzi and Basile 2004, 95–107).[3] The sentiment the quotation conveys is exactly the pluralist one concerning proof; that proof is about exposing reasoning and justification in order to communicate an idea and *invite further* scrutiny, not to close the original question. The further scrutiny, if successful, will lead to a deeper understanding. The pluralist sees proofs, not as axiomatic but as *analytic*.

Definition An *analytic proof* solves a problem by making hypotheses and usually using a mixture of deductive moves and induction (loosely construed to include diagrams *etcetera*) to present a solution to a problem (Cellucci 2008, 3).

It follows that analytic proofs usually contain an informal or gap-like element. Let me distinguish between internal and external gaps.

Internal gaps in analytic proofs are indicated by the presence of a diagram, an inductive argument, a genetic argument or some use of metaphor.

Mathematical writing at all levels is rife with such proofs – proofs with internal gaps.[4] But, even in a very rigorous looking proof, there are gaps. In this case they are 'external'. The 'gaps' are in the context, and the acceptance of axioms or rules.

External gaps in proofs concern context, they are gaps in the justification for an axiom or rule of inference.

Hypotheses (which are the counterpart of 'axioms' in axiomatic proofs) are always subject to revision, by definition. The existence of external gaps amounts to nothing more than a denial that the concept of self-evidence of an axiom or rule can be maintained indefinitely.

[3]The page number for the quotation is not given, nor is it very important.

[4]As (François and van Bendegem 2010, 117) remark, "We all know... that "real" mathematical proofs hardly reach this high quality level [of an axiomatic proof]. (The usual claim (sic) [observation?] is that the first chapters of any introductory book in whatever area of mathematics satisfy this standard, but from the third chapter onward the standard is left behind)."

12.3 All Proofs *Can* Be Viewed as Analytic

I challenge the impression that rigorous proofs are axiomatic and claim that:

Claim all proofs can be thought of as analytic, including rigorous proofs.

There are three cases to consider:

(i) Informal and non-rigorous proofs with no meta-proof to assure us that there is an axiomatic proof.

This is just a proof with gaps which stands on its own, although it is in a context, and has a genesis. We shall look at the history of Enriques' proof as an example.

(ii) The second case is when there is a meta-proof assuring us that there is an axiomatic proof.
(iii) The third case is where we are presented with an axiomatic proof.

12.3.1 *Informal and Non-rigorous Proofs Can Be Seen as Analytic*

Addressing (i): not all informal and non-rigorous proofs have an underlying rigorous proof. We are sometimes proved wrong, but then we did not have a proof in the first place, we simply had a purported proof, and subsequent investigation demonstrates the mistake. However, we might well suspect that if the purported proof is of some mathematical truth, then there must be an underlying rigorous proof. Sometimes it cannot be given immediately.

Consider the following example. Enriques' completeness proof for the theory of algebraic surfaces could not be made completely rigorous at the time that he wrote up the proof. Nevertheless, according to Mumford (2011, 250) Enriques "certainly had the correct ideas about infinitesimal geometry, though he had no idea at all how to make [the required] precise definitions." It was not until Mumford became aware of "Grothendiek's theory of schemes and his strong existence theorems for the Picard scheme [that he was able to see that] a purely algebro-geometric[5] proof was indeed possible." (Mumford 2011, 250). The 'filling in of gaps' in Enriques' original proof, done by Mumford, is a nice illustration of giving a much more rigorous and satisfying proof much later. We were fairly convinced it could be done, since we used Enriques' results in other proofs. Mumford's proof is of type (iii). So, all we did when confronted with a proof of type (i) was to look for the underlying proof, a proof of type (iii). We do this because we are not fully assured of the correctness of the result until we have an underlying proof or the proof that there is an underlying proof (a proof of type (ii)).

[5]An algebro-geometric proof uses algebraic tools which were 'alien to the intuitions' of the Italian geometers at the time. The algebraic tools were developed by Zariski, Weil and later by Serre and Grothendieck, after the death of Enriques (Mumford 2011, 250).

With such examples, we feel confident that there is an underlying *axiomatic* proof to any correct (in some sense of 'correct') result in mathematics. A meticulous proof is what allows us to check a result thoroughly and completely. So the notion of axiomatic proof stands solidly in the case of (i), since, if the result is 'correct', then we can find a corresponding proof of type (ii) or (iii). It might just be a long time in coming, but ultimately, for all correct mathematics there exists some sort of axiomatic proof. So far, modulo 'correctness', the axiomatic view holds.

Nevertheless, an inductive argument based on examples is no guarantor of underlying axiomatic proofs for an arbitrary analytic proof, nor is it a guarantor that there is a rigorous underlying proof that the result is incorrect. We can but search for either sort of proof, and have faith that we can find one. But until we actually find such a proof, we should view such proofs as analytic, and certainly not as axiomatic. So proofs of type (i) are *prima facie* analytic proofs, and this is *why* we look for a corresponding proof of type (ii) or (iii). The harder task for defending my claim is that proofs of type (ii) and (iii) can also be viewed as analytic.

12.3.2 When There Is a Meta-proof Assuring Us That There Is a Rigorous Axiomatic Underlying Proof, We Can Still See These as Analytic

The claim of the pluralist is that it is not certain that the first sort of proof *can* be given an axiomatic proof; that is why we feel it is necessary to give a meta-proof of this fact! So, until such an axiomatic proof, or meta-axiomatic proof is forthcoming, the pluralist remains agnostic as to whether or not it can be turned into a rigorous proof. Until an axiomatic proof is given, the proof can obviously be seen as analytic. But we might decide that it is enough to give a proof that there is an underlying proof.

Here is the argument in its favour. The idea is that we give a proof that we *could* fill in the gaps if asked. Just because, in print, or in a lecture, many proofs are not axiomatic proofs, it does not follow that there is no *underlying* axiomatic proof. In fact many proofs are proofs that there is an axiomatic proof. Moreover, we are well motivated to not make explicit axiomatic proofs at the object language level. The practical considerations are not made from a sense of laziness; although this is part of the motivation (it is no small matter to change a proof into an axiomatic proof). Repeating what was said in Chap. 6, even logicians (whose truck and trade is in logical, or axiomatic proofs) tell us that axiomatic proofs are hard to construct, and are often so lengthy that "one does not actually construct such proofs; rather one proves that there is a proof, as originally defined." (Bostock 1997, 239). That is, instead of giving an analytic proof of an informal proof, we construct a proof that our original proof with gaps can be turned into an axiomatic proof if we so choose.[6]

[6]Miller defended such a view in unofficial conversation at the Logic Colloquium meeting in Sophia in 2009.

Meta-proofs might be formal and rigorous, and so good candidates for being thought of as axiomatic themselves or they might be informal. In the first case, we have a proof of type (iii). In the second we have two proofs of type (i), and therefore, we are entitled to view the pair of proofs as analytic. They are still based on hypotheses, and can be revised. Proofs of type (ii), with a supporting meta-proof are only as strong as the meta-proof, and usually this is not an axiomatic proof. If it is not, then we are still owed an axiomatic proof. If it is already axiomatic, then we turn to the final case.

12.3.3 Rigorous Proofs Presented as Axiomatic

It is instructive to distinguish axiomatic, analytic and rigorous proofs. Rigorous proofs could be either axiomatic or analytic; although, *prima facie*, we might well think that all rigorous proofs are axiomatic.

A **rigorous proof** is a proof that proceeds from axioms or premises, and in which every line of proof is accounted for by reference to a rule of deduction or by appeal to an axiom, premise or definition. Each of these has to be of the right sort to qualify.

The criteria for 'right sort' are listed in Chap. 8. Summarising the discussion there, 'of the right sort' means self-justifying or self-evident; and what counts as self-justifying or self-evident is that it is in virtue of the meaning of the axiom or rule that we suppose it to be true, and there is no *further* justification. There is something self-explanatory, so our explanation comes to an end. This is the respect in which an axiomatic proof is final. However, as we discovered in Chap. 8, what counts as self-evident varies with our account of meaning. Therefore, what is self-evident to one person is not so to everyone. Thus, let us consider two cases. The first is when there is an internal gap in a rigorous proof. The internal gap will be there only according to someone who holds a different account of meaning than someone who sees the proof as axiomatic. The second case is when we are all agreed that we have a perfectly rigorous proof. The second is the harder case against the pluralist claim, so let us start with the first.

Consider a proof with a line which is justified by appeal to a diagram. A Euclidean proof might appeal to a diagram which indicates the idea of extending a line indefinitely in one direction. It corresponds to a rule of inference 'of the right sort' to some: provided the diagram is understood, and carries enough meaning that it is considered to be 'self-evident', then it is fine to include a diagram in a rigorous proof! But, since 'self-evidence' can be thought of as suspicious, as *per* Chap. 8, such a rigorous proof is analytic since we can question the self-evident steps. Of course, every step of the proof can be questioned. The appeal to a diagram is an easy case. But if we survey a sufficiently wide sample of formal 'logical' systems, then we will be struck by the fact that there is no axiom or rule of inference that is common to all of them. Even *modus ponens* is not universally accepted in its classical guise.

We could then retort that some of these formal systems are 'unreasonable', they do not conform to our deep intuitions about logic. The counter-retort is that calling an intuition 'deep' is hardly a good rational move in this argument, since it is exactly these intuitions which are at issue. Thus, all proofs, including the 'gapless', rigorous ones can be thought of as analytic, since self-evidence is not final. However, there will be harder cases for the pluralist, where intuitions and conceptions of self-evidence are shared, so it is quite obscure to see where we might question the proof.

Ultimately, all proofs have external gaps. Consider a rigorous proof that is presented as an axiomatic proof, and we are loath to question the internal steps because they accord with our deep intuitions about reasoning.

There exist some proofs which we can see *as* axiomatic and *as* analytic, and *all* proofs can be seen as analytic. Returning to our three cases, the axiomatic proof might be the promised 'underlying proof'. It might be a meta-proof, of the existence of the underlying axiomatic meta-proof or it might be presented as a nice axiomatic proof in its own right. If such proofs can be seen as analytic, then this covers all three cases.

To make the point, consider a particular rigorous proof scheme. Take X and Y to be wffs, ':' means 'you may infer', commas between wffs, indicates that they are found on separate lines of proof.

Axiomatic proof:
Rule 1: X & Y : Y.
Rule 2: X, Y : X & Y
Rule 3: X, X → Y : Y

From the premises: P → (Q & R), P, S, prove: R & S.

Proof:

1. P → (Q & R)	Premise
2. P	Premise
3. S	Premise
4. Q & R	Rule 3 (lines 1, 2)
5. R	Rule 1 (line 4)
6. R & S	Rule 2 (lines 3, 5).
	QED

The above six-line proof is a perfect (rule-based version of an) axiomatic proof scheme. Particular proofs are had by specifying wffs for the proposition letters. With such a proof scheme, we seem to be left in no doubt as to its validity. The rules are all perfectly good classical rules. Every line of proof is accounted for in an appropriate way. The constant symbols are all logical constants, so we do not even need to appeal to mathematical induction for this proof. The proof scheme is perfect. It, or its instances, could be entered in The Book of Proofs. Therefore, we could see it as an axiomatic proof. Is it? Or, rather, if it is, does this preclude us from viewing it as an analytic proof?

Consider the proof again. I re-write the proof to make this explicit, and then explain why it made sense to do this.

Analytic proof:
We want to solve the following *problem*: if in a given *context*, where *hypotheses* 4, 5 and 6, are *plausible*, does R & S hold? The hypotheses in bold are generally plausible (because logical). The hypotheses in 'light', are plausible in a particular context.

Hypothesis 1: X & Y: Y
Hypothesis 2: X, Y: X & Y
Hypothesis 3: X, X → Y: Y
Hypothesis 4: P → (Q & R),
Hypothesis 5: P,
Hypothesis 6: S.

Proof:

1.	P	Hypothesis 5
2.	P → (Q & R)	Hypothesis 4
3.	Q & R	**Hypothesis 3**
4.	R	**Hypothesis 1**
5.	S	Hypothesis 6
6.	R & S	**Hypothesis 2**.

Note that the new version makes it plain that the account proof is highly contextualised, relying on a number of hypotheses! The hypotheses have different status, some belong to a specific context, others to a general context. When we have such a proof, the *informal element* is not directly present in the lines of proof, but belongs to the context, or setting, of the proof. The informal element belongs to the hypotheses. If we do not believe R & S, then we can question the particular context, hypotheses 4, 5 and 6. That is, we might allow that the argument is valid, but unsound because, for example, hypothesis 6, i.e., S is clearly false.

Or, more interesting still, we might question the more general hypotheses. In particular, it might be explained to us that **Hypothesis 1** is plausible because '&' means 'intersection', where this is familiar from set theory. It is usual (set by logical context) to interpret → in **hypothesis 3** as material conditional. Hypotheses 1 and 2 interpret '&' as classical conjunction. In all classical formal systems of logic, these are the rules that govern these symbols. This is why we call these symbols 'logical *constants*' (maybe this is no more than an act of insistence). But we can question the rules of inference without inviting nonsense.

Here is an example of how we do this.[7] If, on an unusual day we were tempted, for whatever reason, to interpret '&' as *relevant conjunction* then the original

[7]It is probably more customary to question the material conditional, since it receives a lot of attention in the literature. Instead, here I choose to question classical conjunction, in order to stretch the point, and show that nothing I sacred.

argument is not valid and there is something wrong with the proof. In particular, a certain sort of relevant logician, such as (Mares 2004, 48) will reject **hypothesis 2**. He will insist that in order to conjoin two wffs, there has to be something in common between them (in the proof system we would show this using dependency numbers, or indices marking a situation). The rejection only makes sense on the presupposition that we are using hypotheses in proofs, not rules or unquestionable axioms. *Ad esse, ad posse*, actual therefore possible, as the Mediaevals taught us; so we *can* view an account proof as analytic, if we so choose. We simply have to suspend our deep intuitions, and we have plenty of tools for doing this in the form of formal systems of logic. These can be used formally or metaphorically to view any proof as analytic. In the next section we support the claim that it is *preferable* to think of all proofs as analytic.

12.4 All Proofs Are *Better* Viewed as Analytic

Why is the analytic view preferable? If we are faced with an axiomatic proof, and we are not convinced by it, then we are stuck. We have to modify our understanding of '&' and fall in line with the reasoning as it is presented. For, the rules define the use of the symbols, and we are not to question the rules or axioms. Under the axiomatic view, the first proof we have above is final and complete.[8,9] This makes the axiomatic view the poorer view. We would miss out on the development of relevant logic, and be poorer for it.

However, there is something deeper going on than just developing arbitrary logical systems. On our relevant days, we can use the proof to diagnose that our discomfort with the proof lies with **hypothesis 2**, rather than, say, the implausibility of hypothesis 6. **Hypothesis 2** is not a rule for *relevant* conjunction. To even advance this relevant criticism, we have to appeal to the meaning of the proof, and of the symbols. The meaning has to *precede* the axioms. Meaning is not (*pace* Hilbert, and the Dummetian intuitionist) wholly determined by the axioms and rules of inference. One of the problems with the axiomatic view is that it does not reflect the *modus*

[8]The conventionalist version of this runs: if we insist that the symbol '&' together with the rules implicitly defining it are a convention, then we are free to change said convention. Think of Hilbert: the symbols are essentially arbitrary, and are implicitly defined completely, and only, by the axioms and rules which mention them. Hilbert had an axiomatic view of proof. Regardless of what fuels one's axiomatic view, under it we would be stuck with the proof and could make no further moves.

[9]For a long time, in books on the history of mathematics, there was the view that mathematics was to be identified with what was developed in Europe. Other remarkable developments made "outside" were recognised only if they fed in to European mathematics. Moreover, mathematics is cumulative, and once something is proved it is forever true – in fact, it always was true. This view is being questioned by present ethnomathematics (François and Van Kerkhove 2010) and by revisions in, and new views towards, mathematics (François and Van Bendegem 2010). The new views are pluralist.

operandi of the working mathematician in search of proofs. The axiomatic view puts things backwards. Instead, mathematicians look for plausible hypotheses which will support, 'prove', an *already* plausible theorem.

> If the Pythagorean theorem were found to not follow from the postulates, we would again search for a way to alter the postulates until it was true. Euclid's postulates came from the Pythagorean theorem, not the other way" (Hamming 1980). In mathematics you "start with some of the things you want and try to find postulates to support them" (Hamming, 1998, 645). The idea that you simply lay down some arbitrary postulates [or even ones you are convinced are true] and then make deductions from them "does not correspond to simple observation" (Hamming 1980). (Cellucci 2008, 5)

Or as Byers (2007, p. 337) says: "One couldn't get started on a proof if one had no idea if or why the theorem in question is true." This concurs with Rav, who also stresses the prior understanding necessary for executing a proof:

> ... the structure of a proof does not depend even implicitly on a deductive calculus; ... [rather, the reverse,] it depends on an understanding of the terminology, i.e., of the *meaning* of the terms used in that claim and on the background knowledge. (Rav 2007, 313)

It is the background knowledge that will justify the axioms/ hypotheses. Moreover, the background knowledge can change. It is not absolute, unique and fixed for once and for all. It is the background knowledge that allows us to attribute meaning and purpose to the theorem. The quotation from Manin in the last section continues: "Moreover, since formal deduction strives to be freed of any remnant of meaning (otherwise it is not formal enough) it ends by losing meaning itself." (Manin 2007, 39). So, in some cases, we do not even *want* to realise the ideal of making axiomatic proofs. Why not?

The point of analytic proofs is to *communicate* something, to convey an idea, and to open a discussion. This makes proofs intensional and intentional. Intensionality: too much information is sometimes obscuring. Proofs written in textbooks for human readers, are there to communicate, not so much to get to the absolute truth or write up a proof for The Book of Proofs. Proofs with clearly inductive moves or gaps or diagrams are intensional. They are especially designed to present material to an audience. They are crafted. Some are better, some are worse, and whether they are better or worse, depends on the background knowledge of the reader. Intentionality: we choose which theorems to prove and to present. We do not indiscriminately prove theorems. Computers can do that. The proof is there to make a point, to supply background information and convey an idea. This makes analytic proofs intensional and intentional. It is better to view proofs as analytic because under this view we look for much more in a proof than truth and rigour.

12.5 Proof as an Invitation to Interpolative Enquiry

Most of the time, proofs establish a theorem, and we move on from the theorem to further enquiry in the form of further theorems. Together the theorems give us a sense of the theory, what we can assert in it, what its structure is. Going on to prove further theorems is what I shall call 'extrapolative enquiry'. What is less appreciated, is that we can also engage in interpolative enquiry. The term is borrowed from Rav. We shall see the quotation shortly. Because interpolative enquiry is less appreciated, I concentrate on it in this section.

Continuing in the company of Cellucci we observe that the questioning of axioms is made on the basis of background information. This information is imperfect, and therefore there is no absolute standpoint from which to accept axioms and *a fortiori* the conclusions which follow from them. Cellucci concludes from this: "the fact that generally there is no rational way of knowing whether primitive premises [or axioms] are true … entails that primitive premises of axiomatic proofs are simply accepted opinions, … or rather, that they are plausible propositions … Thus they have the same status as hypotheses in analytical proofs." (Cellucci 2008, 12).

It gets worse. When we question axioms, or other parts of a proof, when we are not convinced of a proof (however formal or informal), we dig deeper into the reasons for accepting a given theorem. For example, we might not be familiar with enough of a background theory, so we need to be given further explanations. Rav describes the process:

> If some reader wants or needs more details, as for instance concerning modular arithmetic [in a proof to show that $1+1=0 \pmod 2$] it can be provided by giving further explanations, as is done in teaching unprepared students. In principle, though, one could go through the whole development of Peano Arithmetic, develop modular arithmetic and what not. How far one has to go back in one's justification of an inference is a pragmatic question; *there is no theoretical upper bound* on the number of *interpolations* necessary for an absolute justification (whatever that would mean)." (Rav 2007, 313–314)

The explanations can run quite deep. Contrary to what we often tell our students, it does not stop with logical proofs, rigorous proofs or axiomatic proofs either. This applies not only to abstruse axioms of set theory, say postulating the existence of an inaccessible cardinal, but also to very fundamental axioms or rules. If a student claims to be unable to grasp *modus ponens*, for example (and I have had the pleasure of this experience), we might 'prove' it by appeal to truth-tables. But professional logicians' take on truth tables is not uniform! Joining the company of well-respected logicians, the student might legitimately ask why the truth tables are as they are. For example, we might ask why a wff can only have one truth value (as opposed to both: a truth-value glut) or why we can only choose between two (as opposed to having a third, or having none at all: a truth-value gap), or why we cannot have degrees of truth pertaining to a wff. So appeal to truth-tables as a definition of a connective is not the final word.[10] We might, instead 'prove' *modus ponens* by a definition and proof through disjunctive syllogism. But not everyone accepts disjunctive syllogism: classical logicians and intuitionists do, but not relevant logicians. So both depth and

[10] If we take the truth-table definition to be a stipulation, then that delimits the context when the rule may be deployed. This is fine temporarily, but sooner or later, it will be possible to go beyond the stipulation and ask what alternatives there might be and to what extent they make sense in particular contexts.

breadth of explanation is a pragmatic matter. It depends on what the student, or professional logician, will find convincing. But we can turn this around, and rather than find the student annoying and cast about for a reason, 'proof' or explanation which will satisfy, we can delight in the further exploration demanded by the student. Such delight is inconceivable under the axiomatic view of proof. But the further exploration is a net gain for the mathematician and is encouraged by the pluralist.

12.6 Conclusion

To sum up, the pluralist takes the analytic view of proof. A proof is not the final word, but a communicative act directed towards the community of mathematicians, what François and Van Bendegem (2010, 117) call "a proof community". The communicative acts form a discourse. The discourse is grounded in application (to areas independent of mathematics – independent in their ontology), a meaning context (which has to be thought of holistically, and made explicit, but includes the rest of mathematics) and rigour of argument: a distinguishing mark of mathematics. This is how the pluralist sees mathematical proof. In the pluralist view there no sense of absolute truth got at through proof (although this might be thought to be one (mistaken) aims of a proof).

Acknowledgement I should like to thank Aberdein for useful comments on an early version of this chapter.

References

Bostock, D. (1997). *Intermediate logic*. Oxford: Oxford University Press.

Byers, W. (2007). *How mathematicians think; using ambiguity, contradiction, and paradox to create mathematics*. Princeton/Oxford: Princeton University Press.

Calzi, M. L., & Basile, A. (2004). Economists and mathematics from 1494 to 1969. In M. Emmer (Ed.), *Beyond the art of accounting. Mathematics and culture I* (pp. 95–107). Berlin: Springer.

Cellucci, C. (2008). Why proof? What is proof? In G. Corsi & R. Lupacchini (Eds.), *Deduction, computation, experiment. Exploring the effectiveness of proof*. Berlin: Springer.

François, K., & Van Bendegem, J. P. (2010). Revolutions in mathematics. More than thirty years after Crowe's "Ten Laws". A new interpretation. In B. Löwe & T. Müller (Eds.), *PhiMSAMP philosophy of mathematics: Sociological aspects and mathematical practice* (Texts in philosophy, Vol. 11, pp. 107–120). London: College Publications.

François, K., & Van Kerkhove, B. (2010). Ethnomathematics and the philosophy of mathematics (education). In B. Löwe & T. Müller (Eds.), *PhiMSAMP philosophy of mathematics: Sociological aspects and mathematical practice* (Texts in philosophy, Vol. 11, pp. 121–154). London: College Publications.

Hamming, R. (1980). The unreasonable effectiveness of mathematics. *The American Mathematical Monthly*, *87*(2), 81–90.

Hamming, R.W. (1998). Mathematics on a Distant Planet. *Amer. Math. Monthly 105*(7), 640–650.

Manin, Y. I. (2007). *Mathematics as metaphor*. Providence Rhode Island: The American Mathematical Society.

Mares, E. D. (2004). *Relevant logic: A philosophical interpretation*. Cambridge: Cambridge University Press.

Mumford, D. (2011). Intuition and rigour and Enriques's quest. *Notices of the American Mathematical Society, 58*(2), 250–260.

Rav, Y. (2007). A critique of a formalist-mechanist version of the justification of arguments in mathematicians' proof practices. *Philosophia Mathematica*, Series III, *15*(2), 291–320.

Chapter 13
Pluralism and Together Incompatible Philosophies of Mathematics

Abstract In this chapter, the pluralist arbitrates between two philosophical positions: the extensionalist and the constructivist. Both are anti-realists of a sort. The extensionalist position is that of Quine, and is represented by Bar-Am. The constructivist position is that of Sunholm and Martin-Löf. The two merit comparison because they both give a sensitive account of the history of logic, moreover, they give much the same account. The two positions differ on their final judgment of the modern trend. The extensionalist sees progress where the constructivist sees emptiness. To draw out the differences, we shall also meet the formalist and the realist. We shall see the accord and disaccord between these positions, especially in respect of their attitude towards logic.

13.1 Introduction

So far in this book, not much has been said about anti-realists as such. No chapter was written with the intention of taking the reader from constructivism, or intuitionism, to pluralism. The lacuna is filled here, but not in the form of a journey, but in the form of the mature pluralist engaging two anti-realist positions. One is the extensionalist represented by Bar-Am who follows Quine. The other is Sundholm, who represents Martin-Löf's constructivism. We shall see that while they largely agree, there are interesting points of disagreement that characterise their respective philosophies.

Stepping back: the pluralist is not only tolerant towards different mathematical theories and 'bad' mathematics, he is also tolerant towards different philosophical views of mathematics. This is in keeping with his understanding of the relationship between mathematics and philosophy. The relationship is not one of dominance. Philosophy does not trump mathematics. Mathematics does not trump philosophy. Nor is the relationship preclusive. For the pluralist, one can quite easily do both philosophy and mathematics, separately or together. As we saw in Chap. 3, the pluralist believes that philosophy and mathematics intersect in an interesting way,

M. Friend, *Pluralism in Mathematics: A New Position in Philosophy of Mathematics*, Logic, Epistemology, and the Unity of Science 32, DOI 10.1007/978-94-007-7058-4_13,

© Springer Science+Business Media Dordrecht 2014

especially around logic and big theories. It is often in these areas that mathematics and philosophy inform each other. Since the pluralist understands the relationship between mathematics and philosophy as fuzzy or vague or overlapping and mutually supporting and critical, the pluralist had better extend his tolerance to philosophies of logic and to philosophies of mathematics. In the level schema of Chap. 6, we situate the pluralist at level three. Such a pluralist is a pluralist towards philosophies of mathematics located at level two. In this chapter we put the level three pluralist to work on a particular case of level two philosophies.

The case we investigate is that of two philosophers and historians of logic: Bar-Am and Sundholm. They trace the history of the changes in our conception of 'logic'. They agree that with Aristotle, logic was a tool for reasoning and gaining scientific knowledge. Logic was only semi-formal and was highly intensional. What was for Aristotle a system for making *inferences* and coming to *judgments*, has become today a completely formal theory of the relationship of logical consequence[1] between a set of wffs, called 'premises' and a particular wff called 'conclusion'.[2] Our modern conception is of a fully formal extensional system. Both philosophers agree on the history. They even cite many of the same references, and use many of the same quotations, and they did this unaware of each other's work in this area.

This is all the more surprising, since what they recount is not the standard history. The more standard history would look at past logic through the eyes of the modern conception of logic, and simply neglect, or mis-understand, the intended use of logic. In particular, the standard history obscures the role of judgment in logic. If we look at Aristotle with modern eyes we are simply puzzled by a lot of what he says. In contrast, in the less-standard, more sensitive and careful history, we witness the deep conceptual changes that took place.

I shall only recount the history insofar as it is useful for the points being made here, since it is well presented by both authors. I shall focus on the fact that given this co-incidence, they have quite different reactions. This makes an interesting case for the pluralist, since he cannot have recourse to the history to *arbitrate* between them. Their differences are deeply philosophical and inform their reading of history.

There are two noticeable differences. One is in emphasis. Where Bar-Am writes of the ridding of logic of epistemology and metaphysics, Sundholm writes of the shift in vocabulary and concepts from the *activity of making logical inferences* to that of *the fixed and a-temporal relation of logical consequence*. Their respective emphases are described in the second and third sections, where I discuss Bar-Am

[1]The exception to this is what Batens has called a 'zero logic'. This is limit case logic where there are no rules of inference or manipulation of symbols. In such a logic, no conclusions can be drawn. It is still a logic in virtue of the other characteristics: a formal language, a grammar, maybe some axioms. We ignore the case of zero logics for the rest of this chapter.

[2]We are assuming only single conclusion logic. For multiple conclusion logics, substitute conjunction for the comma between the conclusions. Provided the conjunction introduction rule is classical, the proof for the single conclusion will be longer, but the consequence relation will remain, essentially, unaffected.

and Sundholm in turn. The two emphases are similar in respect of their recognising the historical disengaging of judgment and epistemology from logic. The upshot is a realist or a formalist conception of logic. Both are typical of the modern conception of logic. The other difference is in their evaluation of the modern trends. Bar-Am applauds the modern trend and embraces extensionalism. Sundholm regrets the modern trend precisely because it indicates the loss of the epistemic role of logic. The pluralist discusses the evaluative reaction in the final section.

13.2 Bar-Am: Intensionalism *Versus* Extensionalism

Extensionalism is an attitude towards logic (Bar-Am 2008, 123), where one seeks to disentangle logic from science (Bar-Am 2008, xi). More precisely, an extensionalist tries to rid logic of induction, epistemology and metaphysics, (Bar-Am 2008, xi). Extensionalism is a prevalent view. Its chief modern proponent is Quine, but as Bar-Am shows, it is a trend with a long history. The conflation of methodology, epistemology and metaphysics is traced back to Aristotle's essentialism.

Aristotle was the first to give us a relatively formal system of logical reasoning by presenting us with syllogisms. The system was formal in the sense of giving general schemes or patterns of reasoning which contained schematic letters. Aristotle was the first to use schematic letters in logic (Bar-Am 2008, 23). They stand for terms. Despite the advance in formal representation for the purposes of reasoning, Aristotle's system was suffused with epistemology, metaphysics, taxonomic and other metaphysical and scientific concerns and required significant subtlety and philosophical finesse to deploy it successfully. This is why we thought of logical reasoning as an art.

As an example of the artistry of syllogistic reasoning, consider the deployment of terms in a syllogism. These had to be of the correct type, and this was not determined by grammar; rather, it was determined by metaphysics. The premises of a syllogism used in scientific demonstrations had to be 'real' definitions, or we might say 'immediate judgments'. That is, the definiendum was a term from science (something in nature we want to learn truths about, the name of a genera or a species). The definiens was to give the essence of the definiendum (Bar-Am 2008, 40). "Man is a rational animal" is an example of a legitimate premise using the right sorts of term, since we want to find scientific truths about men, and he is a rational animal. Moreover, no other animal is rational (according to the Ancient Greeks). So, rationality is essentially human. "Man is a featherless biped" will not do, because it is accidental of man that he is a featherless biped. The second term of a major premise had to provide the essence of the first term and "we do not have strict Aristotelian rules for determining whether or not a term depicts an essence and whether or not it is admitted into his province of logic," (Bar-Am 2008, 42). *Once* the two premises were of the approved type, the syllogistic form could be deployed, and the conclusion would then follow. By deploying the syllogism we would gain scientific knowledge. The art lies in getting the terms in the premises right.

Since Aristotle supplied no determining rules for evaluating whether or not a term could figure in a syllogism, the logic is only semi-formal, and suffers from an inexplicitness, lack of clarity or lack of transparency. Aristotle used syllogisms, not only to learn scientific truths, as we would recognise them today, but also, seemingly, in other contexts. Scientific contexts were not well separated from others. Logic was an integral part of metaphysics and scientific epistemology. So, he had no reason to separate logic from epistemology and metaphysics. We could also say that Aristotelian logic is largely intensional.

We can contrast this to our modern extensionalist conception. Brought up on a steady diet of axiomatic systems, classical natural deduction or tree proofs, we find the Aristotelian conception of logic quite foreign, hardly recognising it as logic at all. Today, there is a separation of epistemology and metaphysics from formal logic. For us, 'pure' logic is a formal discipline, with formal axioms and rules of manipulation. The formal system is often *presented as* normative of reasoning. It is only in the presentation of the formal systems *as normative of reasoning* that we can recognise something from Aristotle. It is only in the meta-language that we refer to the beginning wffs of a deduction as premises and the last wff as a conclusion. We could change the meta-language and refer to sequents, for example, and the formal system would remain intact. Once we leave the presentation behind, we study a completely formal extensional theory. It is from this view that the modern reader will recognise Aristotle's syllogisms. From this quite formalist viewpoint, much of what Aristotle writes is mysterious. We regret the 'art' of logic, and only find merit in the nascent formalism. In using a formalist's eyes to read Aristotle, we miss a lot.

This is not entirely fair. Modern texts on logic are not all formalist in their presentation. Philosophical presentations are more tempered. If we visit a philosophical presentation of logic, we see both a completely formal system, and some more 'motivating' parts of the presentation, where logical deduction is said to preserve truth, or certain knowledge. We are told, for example, that given the truth of the premises, logical deduction *guarantees* (and therefore perfectly *justifies*) the truth of the conclusion. Moreover, deduction is presented as more *reliable* than scientific induction, which is inferior 'ampliative' reasoning. Typically, the formal system is meant to hone our reasoning, and our reasoning is meant to justify the formal system, so there is no further context needed. Reasoning can be applied anywhere. This might motivate beginning philosophy students to study logic, but even this misses some of what Aristotle has to say. The scientific and metaphysical enterprise of Aristotle have been excised.

Moreover, for an extensionalist, this modern view is to be applauded. For an extensionalist, none of the wider 'motivational' residue from Aristotle should be left. Aristotle's motivations mislead us. Extensionalists think that logic is *only* for honing our reasoning skills, and we are entitled to ignore: the application to computers, the mathematics of formal systems of logic and the variety of formal systems of logic. The extensionalist story goes: separation of the formal system from science was necessary for progress in logic in the form of greater rigour, which accompanied the increased formalisation of logic. We needed to make the separation in order to progress because the entanglement with metaphysics and epistemology brought hopeless problems.

For example, working out what is essential and what is accidental about man is now considered to be a hopeless metaphysical problem, and one that lies *outside* logic. Bar-Am's diagnosis is that we can now think of problems concerning essences or knowledge as lying outside logic *thanks to* our making our formal systems increasingly extensional. Once disentangled, 'logic' is simply a methodology. It is a tool for generating wffs as conclusions from premises, which are also expressed as wffs. It would be more honest, today, to refer to them simply as sequents, or strings of symbols, and mention that one application of the formal system is to carrying out some reasoning. Summarising, under the extensionalist conception of logic, epistemology and metaphysics are considered to be separate concerns from logic (Bar-Am 2008, 124). The separation marks progress.

There are several indicators of progress in logic for the extensionalist. If we follow Aristotle, Boole, De Morgan, Quine, Bar-Am and others, then we think of intensionality as inversely proportional to extensionality.[3] The more a notion is intensional, the less it is extensional. For extensionalists, it is then obvious to target intensional notions as they appear in logical systems, and excise these.[4] Before we see what the extensionalist does with such formal systems, we should sound a note of caution. The (moderate) extensionalist is not anti-intensionality altogether. He just does not think that *logic* should truck in intensional notions.[5] Or, more mildly, logic should have as few intensional notions attending it as possible. Once we spot one, we should do what we can to make it extensional.

This is because logic – as understood by the extensionalist – should answer to a more general goal, which is clarity of thought, and correct inference. Clarity depends on transparent reasoning, which, in turn, is identified with extensionality.[6]

---

[3]Intensional notions are those whose meaning is not merely the referent of 'states of affaires', terms or names. That is, getting the referent is insufficient for the meaning. And the meaning in a logical context will bear upon what can be deduced from a wff. The meaning of intensional notions is partly captured by Fregean sense, context or implicit understanding. It is discerned through the mode of presentation; how we express the notion, or the context in which we embed the notion.

[4]There are many non-equivalent ways of distinguishing and defining extension, intension and intention. I shelve discussion of these for another occasion. What is relevant here, is Bar-Am's definitions and use of the notions, and how they relate to Sundholm's concerns about logic.

[5]Arguably, Quine is quite extreme, and in places, thinks that philosophy should rid itself of intensional notions too (Quine 2008). A less extreme reading would interpret Quine to be asking philosophers to successively address intensional notions, in order to make them clearer, *only* when and if they can, so it is quite possible that some will remain in the discourse. The moderate extensionalist would accept that total purging of intensionality is not the goal. Rather, the goal is to rid ourselves of intensional notions in order to promote clarity in communication (Bar-Am 2010). Henceforth, we shall assume that the extensionalist occupies the more moderate position.

[6]There are some obvious weaknesses in the argument, as presented here since I have not given any support for the steps. There is support, and the strength or weakness of the argument is not our immediate concern.

The extensionalist has a programme: to successively rid (any formal system which is presented as) *logic* of intensionality. To illustrate how the programme works, we can think of our generating a number of new formal systems that deploy operators such as the modal operators, temporal operators, belief operators and so on. As per the discussion in Chap. 6, when we make such a formal system, we should seek to have the operators govern terms, not wffs. This makes objects in the domain intersubstitutable, and therefore, according to various well-rehearsed arguments, we have transparency of translation (Melia 2003, 76). If we look at this from the point of view of an extensionalist what we conclude is that what *was hitherto* thought of as an intensional notion, has been made extensional in the formal system, and this is progress.

The extensionalist programme is not without its critics. The critic will point out that there is some artificiality introduced when we give formal, term governing representation, to hitherto intensional notions. It works sometimes, but at other times we cannot have the intentional operator have scope over terms. A *de dicto* modal operator has to have scope over a wff, if it is to be at all loyal to a propositional attitude. It has to remain intensional if we are to represent the meaning at all. The artificiality of forcing the operator to be extensional (and so allowing formal transparency of substitution) is also present when intentional operators have scope over terms. Witness the fact that there are several formal systems of belief operators, modal operators, temporal operators *etcetera*. Consensus over formal representation is rare. Nevertheless, attempts to give formal expression to intensional notions and place them in an extensional system[7] are part of the extensionalist programme. In following the programme, the critic will say that we trade loyalty towards our hitherto vague and ambiguous notions for (somewhat contrived) clarity (in the form of extensionality). Extensionalism is the idea Bar-Am fastens on to recount the history of logic, to indicate progress and make coherent sense of the history. Note, however, that this makes sense only if we think of intensionality as inversely proportional to extensionality. This assumption is also a point of criticism. I shall not treat of this criticism directly, but it will be implicit in the concluding remarks of the chapter when the pluralist view of logic is made more explicit. We leave Bar-Am behind, and turn to Sundholm, who fastens on a different idea, while citing many of the same historical sources.[8]

[7]An extensional system is not just a formal system with an axiom of extensionality, or any formal system with no intensional operators, or some such. Extensionality, intensionality and intensions bear careful treatment. We saw a hint of this in Chap. 6. Here we are interested in extensional formal systems in the sense of formal system where epistemology and metaphysics is no longer a part of the logic or the language.

[8]One of the main differences in their historical treatment, is that Bar-Am lends close attention to Boole whereas Sundholm places more emphasis on Bolzano than Bar-Am.

13.3 Sundholm: Making Inferences *Versus* Defining the Relation of Logical Consequence

The change in our thinking about logic from Aristotle to the present day is described by Sundholm as a shift in our concept of logic or the role conferred on logic. With Aristotle, we started with the *action* of *drawing inferences* and *demonstrating* to others and to ourselves. In this action, we use logic to make *judgments* from judgments already made (in the premises). The premises of a syllogism were to be immediate judgments, that is, judgments requiring no justification, since patent, obvious and defining essences (for which there can be no further scientific[9] justification). The *conclusions* were then *mediate* judgments, since it requires *some reasoning* to *arrive* at these judgments and they are justified by reference to the premises and the validity of the inference. The purpose of making inferences is to lead us from judgments known to more judgments, to increase our knowledge. Moreover, logical inference is a sure way to obtain knowledge. The rival is the sceptic, and it is she that we are trying to convince beyond reasonable doubt. Or, if she is an unreasonable sceptic, then we convince ourselves in light of our arguments against her. We use logic for these purposes. As noted above, early logic was far from mechanical and transparent. For this reason, it required considerable skill to deploy logic to arrive at new knowledge.[10]

Our successive ridding logic of its informal and intuitive elements required deep and long philosophical discussion and exchange.[11] Around the turn of the twentieth century, the conceptions of logic (not completely clearly or cleanly) trifurcated. Oversimplifying the history: some logicians resonate with Hilbert, and seek to develop a formalistic view of logic. Others resonate with Bolzano, are more realist and treat logic as static and eternal. Still others resonate more with Brouwer and insist on emphasising: the epistemic role of logic, the human appropriation and use of logic and proving as an action of demonstration.[12] If we insist on the epistemic role played by logic, then the culmination of this development of logic can be found

[9]There might be a metaphysical or semantic justification. Today we would separate these from scientific justification, but such separation is foreign to Aristotle. Hence the embarrassment concerning the immediate judgment: man is a rational animal. Today we do not consider man to be essentially rational, since we consider other creatures to be rational, and some of us think that man is often irrational.

[10]Lull was one of the first to try to make the syllogism mechanical. He designed a physical method of constructing syllogisms by rotating concentric disks. This was in the fourteenth century.

[11]A proof, seen as an object, is simply a blueprint for making inferences, for coming to judgments (Sundholm 1998, 180).

[12]I am being quite careful, even in my oversimplification. Hilbert and Brouwer did not start their respective views *ex nihilo*. These ideas were usually confounded in the same system. However, they were starkly separated by both Hilbert and Brouwer.

in the formal intuitionistic, or constructivist, systems, the most sophisticated of which is Martin-Löf's type theory.[13]

While current, Martin-Löf's type theory is not the received, modern, formalist or realist, view of logic. This is *not* because Martin-Löf's system is not transparent or effective. We might even consider the language of type theory to be extensional, depending on our definition of 'extensional'. If our criterion for a language to count as extensional is that it include identity conditions or clear substitutivity conditions, then the language of Martin-Löf's type theory is extensional. But the language of Martin-Löf's type theory is not extensional in the sense of separated from epistemological concerns. The very motivation for Martin-Löf's type theory is to marry epistemology with methodology. Nor is this feature peculiar to Martin-Löf's type theory. It is a feature of any philosophically informed constructive logic. In these, it is knowledge and judgment that are preserved from premises to conclusion, in the sense of *justified* truths, as opposed to realist truth.[14]

These points are not always made plain. They are subtle since the formal system is extensional, so, those of us who are not initiated in constructive logics might not recognise, might dismiss or might simply overlook the epistemic aspect of the logic. There are enough familiar features of type theory to be able to deploy it without reference to the motivation, and to recognise it in a familiar formalist or realist way. For example, while it is correct to say that in a constructivist proof "truth is preserved from premises to conclusion", it is deceptive. For, we miss out saying that it is *justification* and *knowledge* which are preserved, which makes the conclusions *judgments* rather than wffs, which are (also) true. The deception can run quite deep. Rather than think of premises as wffs, we should think of them as knowledge claims – *justified*, true beliefs.[15] If a judgment, **j**, is justified, and from **j** we have a constructive proof that **k**, then **k** is a judgment based on the proof and the justification of **j**. What is important is that we can directly trace back the justification. In limit cases, of purely logical judgments, premises are self-justifying, for example $\mathbf{j} = \mathbf{j}$, is self-justifying, or $\mathbf{j} \supset \mathbf{j}$ (read: 'if **j** then **j**', where '**j**' is a schematic letter holding the place of a candidate for judgment). Self-justifying formulas are immediate judgments – in the sense that they cannot admit of further justification. In non-limit cases, our proof of **k** will rest on premises that are not

[13]Martin-Löf shares with Brouwer the emphasis on epistemology and judgment in logic. However, he is not a Brouwerian intuitionist in that he distances himself from Brouwer's idea that mathematical ontology is only mental construction. For this reason, he prefers to be called a constructivist rather than an intuitionist.

[14]We can be in error as to whether or not we are really justified in our judgment. The notion of validity is still in terms of a conditional: if we know the premises, then we know the conclusion. Moreover, the conditional is intuitionistic: a proof of the premises can be extended into a proof of the conclusion.

[15]We need not worry here about Gettier-type counter-examples since the justifications are always in the form of constructive proofs, and they can be put in normal form. As a result, under a constructivist conception of proof, there is no such thing as getting the truth in a justified way, but with the wrong sort of justification.

self-justified. In these cases we do not have knowledge, we only have belief resting on the supposed justification of the premises. In this case, **k** is a *judgment candidate* and not a judgment. How do I know a conclusion **k**? I look to the proof of **k**. I can only *know* **k** if the premises are all immediate judgments.[16] When I reason from premises to conclusion, I *make inferences*. The reasoning is made transparent by the proof method.

This view of proof is starkly different from that of the realist, formalist and extensionalist. To make the difference plain, let us consider the realist concept of proof. In a realist presentation of a formal logic, we add as a separate feature 'a semantics', a set of models, which satisfy premises and conclusions of valid proofs. Instead of thinking of logic as a tool, guide or blueprint, for acting (drawing inferences, resulting in judgments), the realist thinks of logic as a set of objects, namely proofs. The set of valid proofs define the relation of logical consequence for the theory. For the realist, it is not *we* who generate said proof-objects, rather they are all there ready. We discover the proofs and they are truth-preserving, not, necessarily, meaning, justification or knowledge preserving. In talking about logic, the realist minimises and regrets the human *use* of the proof objects. It is simply a pity that it takes a human some time to work through a proof. This is something subjective and personal, and does not belong to logic. Logic is impersonal, perfect and a-temporal, recorded in The Book of Proofs. It exists quite independent of any use we make of it.

To further indicate the difference in conceptions between the realist and the constructivist, we can focus on vocabulary. On the realist side we have the vocabulary of the model theorist, which lies in contrast to that of the anti-realist proof theorist. The realist thinks of proofs as *objects* that bear the *consequence relation* between the premises and the conclusion. The consequence relation is a higher-order object.[17] Proofs start with *well-formed formulas* and end in a well-formed formula. The formulas are strings of symbols that can later be interpreted as *propositions*. What is important is that the formal grammatical structure ensures that their natural language interpretation is *a truth-bearer*. In the case of contingent wffs, reality determines which truth-value a proposition has. The truth is independent of us. The syntax and the semantics are quite separate from each other. With the separation, knowledge, justification, and therefore judgment, have dropped out of logic.

In terms of vocabulary, at least for the purposes of syntax, the realist is in bed with the formalist. They part company over the meta-semantics. As we saw in Chap. 5, the formalist presentation of a formal system involves giving a formal

---

[16]Of course, since the deduction theorem holds for the type theory, we can turn a conclusion resting on something other than immediate judgments into a *bona fide* judgment by proposing the premises as assumptions that are discarded at the end of the proof by the familiar conditional introduction rule. The form of our conclusion is then that of a conditional statement.

[17]This is meant in the following sense: in a second-order language we can quantify over a predicate or relation, and when we do this we treat the predicate or relation as an object.

language, axioms and manipulation rules. In a formalist vein, we need not visit natural language interpretations at all to do the technical work of a logician. Formal logic is pure methodology. Interpretation, or semantics, is quite separate. More carefully, we have methods of making semantics pretty formal. Tarski led the way.[18] If we insist on adding semantics to a formal system of logical syntax, then we simply have a valuation function which assigns truth and falsity to said propositions (as such, even truth-valuation becomes syntactic, in the sense of a mechanical operation (not a judgment)).

The 'received modern view' of logic is formalist or realist.[19] The extensionalist applauds it for reasons of clarity and transparency. For the realist, content is found independent of logic, and in reality (whatever that is). For the formalist, there is no content, or at least, it is not the business of the formal logician or of logic.[20] Both are friends of the extensionalist, but not of the constructivist. Friendship is not transitive. The extensionalist is a friend of the constructivist – in that his formal system is extensional in the sense of having very clear rules for substitution. But the constructivist, as represented here by Sundholm, deplores the realist and formalist separation of epistemology from logic. In this respect, Sundholm also deplores the extensionalist emphasis in separating logic from epistemology. For Sundholm, the term 'logic' should apply to systems that help us come to judgments, and guide our reasoning. Logic is an epistemic tool *par excellence*. Divorced of *epistemia*, all we have are formal systems, not logic! Sundholm paraphrases Fichte's summary of the difference between the intuitionist (anti-realist/constructivist/idealist) and the realist view as follows: "A realist—Fichte's pejorative term was *dogmatist*—determines the human act of knowledge in terms of the (prior) object towards which it is directed. The idealist, on the other hand, determines the object in terms of the act." (Sundholm 1998, 178). The idealist, and the constructivist who follows him, cannot divorce epistemology from logic except to make it empty. Formal logical systems seen as games are literally *meaningless* formal toys. The relationship between the constructivist and the realist or formalist is somewhat clear, but we shall make it more explicit in the next section. The relationship between the extensionalist, the constructivist and the pluralist is more delicate.

---

[18] While, arguably, Tarski was a realist, he was sensitive to anti-realist concerns.

[19] The origins of this view reach back to Plato, so are more ancient than Aristotle. Nevertheless, we see a resurgence of realist vocabulary in presentations of logic. We shall see that these are interspersed with formalist elements as well. What is common to the realist and the formalist is that logic loses its epistemic role.

[20] This view of formalism is sometimes referred to as game-theoretic formalism, as opposed to a more philosophical Hilbertian formalism. As a formal logician, when the formalist builds his formal systems, content is disregarded or thought to be an empirical matter, and therefore, is separated from the technical aspects of logic. Henceforth, we shall reserve the term 'formalist' for the formal logician who develops formal systems, which today we call 'logical systems'.

13.4 Evaluation: The Extensionalist, the Formalist and the Constructivist

Following Bar-Am, clarification, in the form of disambiguation and giving precise expression is to be applauded, since this sharpens, and thereby increases our understanding. Extensionalism, in the form of encouraging divorce between epistemology and logic is a means of achieving this broader goal. In fact, for the extensionalist, we measure progress in logic in terms of extensionalism. The more our formal systems are separated from metaphysics or epistemology, the more we have progressed in that area of logic. Put another way, the more extensional a formal system is, the more modern and developed. The recent progress has been very rapid, especially in the hands of the realist or the formalist.

To see this consider what happens once we have the artefact of a formal system. From this we can develop new ones. After developing an extensional formal system of symbol manipulation, we are free to ignore our first intended interpretations. We make new formal systems from old by adding or subtracting rules or axioms. We then make a new 'sister' formal system. The point of the exercise is to extend our understanding through syntactic changes, since this allows us to see more precisely the limitations of our original notions. Does the exercise really lead to understanding? Clearly we understand more about the manipulation of symbols and the differences between formal theories. But does the exercise improve our reasoning?

Unfortunately, the answer is not in the form of a 'yes' or 'no'. It is in the form of 'it depends on the philosophical significance we attribute to the developments mentioned above'. To see this, let us, in this section, take our three philosophical characters: the formalist, the realist and the constructivist. Later we add the extensionalist and the pluralist. The formalist and the realist are in agreement with the extensionalist over the divorce between syntax and semantics, on the one hand, and logic and epistemology on the other hand.

The formalist is well aware that while the initial development of a formal system might have had some motivation extraneous to formal logical concerns, once the development has been achieved, we are free to make adjustments to the formal system. The formalist does this in the name of creative freedom. The application of a formal system is an empirical matter, and the question of fit of a formal system to an application, can only be resolved by experiment, by trial and error. Moreover, there is no guarantee that there will be a nice match between the original, or the sister formal systems, and our natural language conceptions. Witness some of the modal systems made by adding or subtracting axioms of modal logic. Making sense of all of these in terms of our natural language conceptions, and even quite sophisticated metaphysics is no easy task.

For example, in some intensional formal modal logics 'it is necessary that it is possible that **p**' is equivalent to 'is it possible that **p**'. The left to right inference is easy, but the right to left, is not so obvious. The only modal operator that counts is the one immediate to the wff. Those who object to the above rule think we should distinguish between possible possibilities and necessary possibilities. Similarly, in some formal systems we are free to add necessity operators on the left. So from 'it is necessary that **p**', we can derive 'it is necessary that it is necessary that **p**' and 'it is necessary that it is necessary that it is necessary that **p**'. Some of these new symbols (implicitly defined by the formal system) fall on sterile ground when we try to re-interpret them using terms in natural language, or even quite sophisticated metaphysics. Does $\Diamond$ as implicitly defined in S5, say, really represent our natural language conception of possibility? Does $\Box$ really represent necessity? Does it represent metaphysical necessity or scientific necessity? Some formal systems find fertile ground. Many sit on the shelf and await application. Since the application of a formal theory is an empirical matter, it does not belong to the remit of formal logic. Note that this evaluation of the formal system cannot be made by a formalist, because his enquiry ends with the formal play of giving axioms for the operators. Their interpretation is a quite separate matter extraneous to formal, and formalistic logic. Cellucci calls this attitude towards the relationship of formal theory to application a 'top-down' approach to logic. (Cellucci manuscript 2011, Introduction, p. 18) Strictly speaking, use, or application, of formal logic is not the concern of the logician at all. It is an interesting question only for the scientist, computer software designer or the philosopher. The formalist is unconcerned about re-application, at least *qua* strict formal logician. What we learn by making new formal systems is neither how to reason better as such, nor when or how to deploy the reasoning, nor do we learn more about the original conception. Rather, we learn about formal constraints, thresholds and limitations. *A la limite* we might learn how to reason better about the construction of formal systems, when we work with, and develop these; this in the sense of being able to predict more accurately which limitative results will apply to a new sister formal system. But we have left behind the idea of learning to *reason* from *premises* to a *conclusion*.

The constructivist is critical of formalist developments of sister theories. For the constructivist, at best, the developments come at the price of no longer guaranteeing the preservation of knowledge and judgment from premises to conclusion. At worst, such developments do not increase understanding at all. Under the exercise of changing, adding or subtracting axioms or rules from existing formal systems, 'reasoning' is relegated to the syntactic part of the formal system, or to the meta-theory where we discuss (and do not deploy) the formal system. 'Reasoning', as represented by syntax, is thought of entirely as deployment of manipulation rules, and invocation of axioms and definitions. We have a mechanical calculation. But this bears no resemblance to the phenomenology of reasoning. It is fine for computers but not for us. With the formalist, the *content* of reasoning is quite orthogonal, independent or accidental with respect to formal logic. Yet, the constructivist will remind us that, with humans, the content is integral to our reasoning, and not just the reasoning of the 'laymen' but also of professional mathematicians.

Confronting the plethora of formal systems produced by the formalist, the optimistic constructivist might try to do the best he can. He will be interested in finding a formal system, which is a good guide to reasoning. First, the optimistic constructivist is faced with a bewildering choice (this is a good thing, since one of them might even be a judicious choice). The constructivist will have to justify his choice and then wrestle the theory from the grips of the formalist, and re-appropriate it. That is, he will have to explain why this formal system preserves judgment from immediate judgments to a mediate judgment. This *modus operandi* is possible. It is impossible to evaluate the *chances* of success in advance, but the constructivist might find a good reason-reflecting formal system, and if he does, he will recognise it to be such. The constructivist will thereby have salvaged and vindicated the formalist's activity, and according to the constructivist, the formalist can now stop generating new formal systems. The formalist has given the constructivist an appropriate formal system, despite his lack of motivation to do so.

However, this optimistic strategy seems rather roundabout, since the most natural and direct way in which the constructivist will make a choice is by comparison to the formal systems already developed by constructivists. Paying heed to the genesis increases the chances of success. Moreover, the re-appropriation will be a cumbersome task in some cases, since the formal system was not developed *as* a tool for human reasoning and justified knowledge preservation. Of course, this does not mean that he will have learned nothing from the exercise. By examining arbitrary formal systems, the constructivist might well learn some valuable lessons and techniques. The complaint is not that there is nothing to learn, but rather that there are more direct ways of learning, by engaging in *informed* play (informed by epistemological scruples) rather than free play. Nevertheless, the roundabout re-appropriation is the best the constructivist can make of the formalist's technical work.

Notice that the constructivist who engages in this exercise will have fastened on one formal system, or maybe a small number of specialised systems, to represent justified inference. He is no pluralist. In the case of all of the rejected choices, the constructivist views them as meaningless, since they are not knowledge and judgment preserving. There is no justification for re-appropriating them. In contrast to the optimistic constructivist, the pessimistic constructivist ignores the work of the formalist, and directly develops existing constructive formal systems.

In summary, what the formalist celebrates as creativity, deepening and increasing of pure understanding through the free development of formal systems, the constructivist views as possibly indirectly useful, but largely meaningless formal spinning. The formalist developer of formal theories is 'free' to implicitly redefine symbols by altering axioms and rules in a formal system. The formalist celebrates the freedom. The constructivist regrets it, since the formalist is not only free, but has gone feral, i.e., the formalist is too far removed from the epistemic constraints imposed by the act of making a judgment. The extensionalist does not share the conception of the constructivist. He agrees with the formalist that all of the formal systems add to our understanding, in fact, they could not do otherwise because clarity and understanding are identified with the formalist's free play.

13.5 Evaluation: The Realist and the Constructivist

The realist is disappointed by the formalist too, since she wants to move beyond this free spinning to discover *reality* through the formal tool. That is, for the realist, formal systems of logic are intended towards a particular purpose, to lead us to new *truths*. With the realist we shall say that the content of the wffs, or the meaning or interpretation of the wffs, is independent of the logic, and to be found in 'reality', independent of the person reasoning. The realist is interested in finding "*the* formal system" which will help her to track reality, to really know. She expects the right and chosen formal system to play the epistemological role of helping her discover the objective truth.

The constructivist reminds the realist that there will never be a guarantee that we have made the correct choice in epistemic tool.[21] A formal system is chosen and interpreted or re-interpreted by the realist in order to help us track reality. But we cannot know that a given formal system will really help us track reality. What the realist is warranted to say is that in deploying a chosen formal system, there is a deepening of our understanding of the premises and the formal relationships between the concepts reasoned over. We are not warranted in asserting realist truth of the conclusion. While the realist might hope that she has finally found the right logic, the best tool for epistemology, she is not strictly entitled to believe of any one formal system that it is *the* correct one. If we think about it, there are an infinite number of such formal systems, and interpretation/application pairs. The probability of finding *the* correct one can only be narrowed by conviction or intuition, and this is not trustworthy. See the battles over S4 and S5 as giving 'the' correct universe for possible worlds. The realist ultimately cannot trust his epistemological tool.

The point is closely related to the following complaint, made by Sundholm. Under the realist conception there is divorce between interpretation (semantics) and judgment.[22] For the realist, it is quite possible for us to make a true statement without having a good (or even any) justification for the judgment. This is just a happy co-incidence. There is nothing objectionable about the coincidence as such,

---

[21] There is some discrepancy between how the term 'realist' is used in the USA and how it is used in the U.K. I take Wright's definition (Wright 1986, 1). "Realism is a mixture of modesty and presumption. It modestly allow that humankind confronts an objective world, something almost entirely not of our making.... However, it presumes that we are, by and large and in favourable circumstances, capable of acquiring knowledge of the world and of understanding it." I follow Wright in then thinking that there are two sorts of anti-realist, the sceptic and the Kantian idealist. The constructivist, in the chapter, is modeled after Martin-Löf and Sundholm, but also shares features with the Dummettian intuitionist. He cannot share all features, since some aspects of Martin-Löf's constructivism are in conflict with Dummett's intuitionism. I do not think that the differences matter for the purposes of this chapter.

[22] We do not have to go this far to begin the reverse process. We can skip the computers and computer languages. The divorce claim rests on two uncontroversial ideas. One is that computers (or computer programs) do not have knowledge, do not deal with the objects of a domain of interpretation, they perform no intentional action (of demonstrating). The second idea is that computer programs can be thought of as formal logical systems.

but we cannot even recognise that we have managed to utter a truth! What is objectionable, for the constructivist is the use of this true, but not justified, statement in our reasoning. Sundholm calls this "blind inference" (Sundholm 1997, 211).

> We have two Bolzanian reductions, namely (i) that of the correctness of the judgement to that of the truth of the propositional content and (ii) that of the validity of an inference between judgements to a corresponding logical consequence among suitable propositions. From an epistemological point of view, we get the problem that the reduced notions may obtain *blindly*. This happy term was coined by Brentano for the case when an assertion without ground happens to agree with an evidenceable judgement (Sundholm 2007, 624).

Remember the classical definition of validity: if the premises (happen to be) are true, so is the conclusion. The realist cannot distinguish the happy co-incidence of truth from the judgment that a premise is true, because he stops his logical analysis at (blind) "truth of a proposition", as opposed to known or justified truth. How the realist happens upon the truth is irrelevant for the relation of logical consequence.

The constructivist shares the misgivings of the realist towards the formalist, but reverses the order of the realist story. For the constructivist, how we get the truths is all important. A truth is such because it is justified. The justification is either immediate – in the case of self-justifying truths, or mediate – in the case of truths requiring further justification. The constructivist will insist on making formal logic into a good epistemological tool from the outset. The epistemology (and semantics) is contained in the (syntactic) rules of inference. The constructivist will not wait for a re-interpretation, which happens to (miraculously) bring us knowledge.

13.6 Evaluation: The Constructivist, the Extensionalist and the Pluralist

What does the pluralist make of all this? All of the characters in the above story: the realist, the formalist and the constructivist, make good points, and it is useful to compare them, because we learn the limitations of their positions through the comparison. We shall re-introduce the extensionalist to sharpen the comparison.

The formalist enjoys freedom, and does learn something in the exercise of making formal systems. What she learns is not the conclusions of arguments. She learns about formal system design. This might be interesting in itself, and it might be a fulfilling activity, but it is not one the realist or the constructivist recognise as part of the remit of logic; it belongs more to meta-logic. The extensionalist applauds the formalist's development of formal systems for the purposes of logic, not meta-logic, provided they are transparent, and for this they should be extensional. Many of them are, so their development marks progress in logic. But, now we can use the realist and constructivist misgivings to remind the extensionalist of the greater goals he had in mind: clarity of thought, transparency and communication. It is not clear that these are being fulfilled by the development of uninterpreted formal systems. The formalist is no longer communicating with the realist or the constructivist. She might have sharpened her conceptions about something (highly formal), but this does not help her find truths or in giving her knowledge of the conclusions of arguments, unless the arguments have formal constructions as content. This shows us the limits in the attempts at communication of the formalist. On this point, the constructivist wins over the extensionalist, the realist and the formalist. The extensionalist has lost track of his original goals and is mistaking meta-logic for logic (where logic is a guide to clear reasoning). The realist is after truths not knowledge. The formalist has jumped to a meta-level, and is not longer serving logic.

The constructivist has not won *tout court*. The constructivist's claims about judgments and knowledge also rest on some assumptions. These concern the nature of knowledge and the mechanisms for acquiring it. Remember the claim that there are immediate judgments. The pluralist can help himself to the material given to him by the formalist and confront the constructivist with the following problem. There exist formal systems, namely some relevant systems, where there are no logical truths: no axioms, no tautologies. That is, there will be no immediate judgments from which to make mediate judgments in such a logic.[23] Recall the argument we witnessed in Chap. 11. The structure is as follows. Start with the conditional: if we can give a model for a theory, then it is coherent. We give a model. Then by *modus ponens*, we detach and conclude the consequent of the conditional statement. Deploying the argument in this context:

1. If we can give a model for reasoning then that form of reasoning is coherent.
2. There exist relevant formal systems with no logical truths/axioms/immediate judgments.
3. We can think of these as models for reasoning.
4. By *modus ponens* it is 'coherent' to maintain that there are no immediate judgments.

If we accept this pluralist argument, then the constructivist claim that there is any knowledge *at all* requires additional support, since it is coherent to think otherwise. This is because if we entertain these formal systems as models for reasoning, then there are no immediate judgments, only mediate ones based on other mediate judgments (which take the premise position in the arguments). Since the premises are not based on immediate judgments, on pain of contradiction, there can be no proper concluding knowledge either, except in the form of an implication: if the premises could ever be known, then the consequent could not be known. This should be familiar from the previous chapter. The pluralist thinks of proof as analytic. From the pluralist standpoint, the constructivist claim that we can use logic to gain new mediate knowledge requires support. Martin-Löf draws a distinction between

[23]The constructivist could block the argument here and say that such a proposed logic does not count as a logic. This might convince the constructivist, but such a block will not convince the pluralist. Therefore, the proposed block begs the question against the pluralist. There will be several points of departure like this in what follows. *Grosso modo*, the strategy is to declare that the conditions of the argument are unacceptable – since they are not constructively acceptable. This is legitimate but limiting. It is limiting because it will not convince an opponent.

a judgment candidate and a judgment (Primiero 2011, 1). A judgment candidate is not known to be true, because it still rests on unfounded assumptions. Given this distinction, we conclude from the argument that it is *coherent* to think that there are no judgments, only judgment candidates.

The argument from the pluralist has its weaknesses. There are several assumptions, which we should question.

Assumption (i): It is all right (coherent?) to present a formal system as a type of *model* for 'coherence'. By challenging the assumption, we challenge step 1. Step 2 in the argument is not an assumption, it is just a brute fact.

Assumption (ii) concerns step 3. Any coherent formal system is a candidate for forming judgments, and gaining new knowledge.

Lastly, **assumption** (iii) is that *modus ponens* is acceptable. If we challenge this assumption, we can challenge step 4. *Modus ponens* is not accepted in the presently developed logics recruited in 2.

I address the weaknesses in turn. Start with assumption (i): the notion of 'coherent' is rather vague. The pluralist reply is that, here, we use the term 'coherent' as an adjustment to 'consistent', in light of the development of paraconsistent formal systems, so it is not as vague and hopeless as we might think. Technically, 'coherent' means non-trivial. Nevertheless, coherence is also a virtue of reasoning, so it is playing two roles. Assumption (i) can also be criticised because it contains the idea of a formal system being a *model* for reasoning, and this is not technically correct. We could challenge this, and say: "first, this is an abuse of the term 'model' and, second, terminological abuses aside, this is exactly what is at issue, so it begs the question". So let us be more precise and say that 'model' in 1 is not being used in the model-theoretic sense of the term.[24] Rather, a formal system in logic or mathematics is a candidate for representing good reasoning. Candidates are not always successful, so it should not be too controversial to claim that formal systems are candidates for good reasoning. Referring to the second point, it is not obvious that step 1 begs the question. 'Coherent' is a carefully chosen word, and not merely a technical term denoting a non-trivial formal system. Rather, the admittedly vague notion of coherence opens our horizon of possibilities. So step 1 of the argument should be thought of as a hypothesis, and not as a statement of truth. As a hypothesis, it will be assessed under different disambiguations and according to various criteria. This is in accordance with what we saw in Chaps. 8 and 12 and the pluralist's endorsement of Cellucci's ideas and vocabulary concerning proofs. The problem of begging the question is more blatant in the third step of the argument.

The second weakness concerns the assumption in step 3. Assumption (ii) is exactly the issue in question. But again, this is not so if we think of step 3 as a hypothesis. The constructivist should challenge the argument here. He will not accept these relevant logics as candidates for modelling good reasoning, expressly

[24]Priest (2002) uses the term model in the model theory sense. So I am deviating from his use here.

because they contain no immediate judgments, and have a different interpretation of logic and the meaning of the connectives. The problem with this constructivist challenge is that it too begs the question. If we are honestly interested in what 'logic' means, and what the relationship is between formal systems, presented as logical systems, and the acts of reasoning, then we should be willing to entertain the possibility that the relevant, relevant systems are models for coherent reasoning.

The last weakness cannot be recognised by the pluralist. Mentioning a logic, introducing it as an artefact, is not the same as deploying it in an argument. So the argument 1–4 is coherent. But this is not enough to make it convincing.

Weaknesses aside, the pluralist sees that the argument confronts the constructivist right at the assumption that his logic is the one (or even that the class of constructive logics are the only admissible logics). Constructive logics or type theories are the only ones really intended to preserve knowledge and judgment from premises to conclusion. The pluralist and the constructivist are aware that coherent thinking is not all there is to knowing and judging. Knowing and judging are not just a matter of private conviction. They are also not a matter of consensus. They are supposed to be more, but they cannot be grounded in a 'reality independent of us' since that is the realist position.

13.7 Conclusion

For the conclusion I should like to make two sorts of remark. One concerns the philosophical conclusions, the substantial part of this chapter, the other is to say something about what makes the approach here distinctively pluralist. The substantial part of the conclusion is that there are merits on all sides of the debate. The formalist, the realist and the extensionalist agree that progress has been made in our development of logic by cleanly separating syntax and semantics. We have a clear and mechanical syntax, we apply this, and the application is the semantics. The syntax is a tool; the semantics is an interpretation. The formalist and the extensionalist are also in agreement that we have progressed in our development of logic by separating logic from epistemology. Epistemology is intentional. The realist disagrees with the separation because he would like to think of logic as a tool for gaining knowledge. Moreover the correct logic is reliable. The constructivist criticises the realist for thinking he can combine logic and epistemology, but separate syntax from semantics, because this leads to blind inference. For the constructivist, there is no point in hitting on the truth by chance, especially since we cannot then know that we have done so. The pluralist recognises this point. He recognises the intentionality of epistemology, and that it cannot be separated from logic. But he is not in full agreement with the constructivist. There is some merit in entertaining the separations, as *aspects* of logic. The merit lies in our being able to study these aspects cleanly. However, at the end of the day, we have to re-unite semantics and syntax in logic, and epistemology and metaphysics in logic.

The distinctively pluralist stamp on this debate lies in the attitude brought to the comparison of the different philosophical positions. There are two elements to this attitude. One is an open-mindedness about which position is better and in which respects. The other is in not having a need to close the debate and settle for one position as 'best' overall, whatever that would mean.

Learning is blinkered by prejudice. Prejudice can take several forms, one is to come to a debate with a particular point of view one wants to defend, come what may. Another is to think that for any debate, one side has to win. First, it is not clear that one side has to win now, second, it is not clear that one side has to win at all, even in the long run. It is in reconciling oneself to the last idea that one becomes really pluralist, or transcendentally pluralist.

Acknowledgments I should like to thank both Bar-Am and Sundholm for detailed and useful comments on an early version of this chapter. I should also like to thank the Arché research group: Foundations of Logical Consequence, for useful comments during a presentation of an earlier version.

References

Bar-Am, N. (2008). *Extensionalism; the revolution in logic*. Berlin: Springer.

Bar-Am, N. (2010). Extensionalism in context. *Philosophy of the Social Sciences, XX*(X), 1–18.

Melia, J. (2003). Modality. In J. Shand (Ed.), *Central problems of philosophy*. Chesham: Acumen.

Priest, G. (2002). *Beyond the limits of thought*. Oxford: Clarendon.

Primiero, G. (2011). *A contextual type theory with judgmental modalities for open assumptions 282 reasoning*. Manuscript.

Quine, W. V. (2008). *Confessions of a confirmed extensionalist*. Cambridge, MA: Harvard University Press.

Sundholm, G. (1997). Implicit epistemic aspects of constructive logic. *Journal of Logic, Language, and Information, 6*, 191–212. Dordrecht: Kluwer.

Sundholm, G. (1998). Inference, consequence, implication: A constructivist's approach. *Philosophia Mathematica, VI*, 178–194.

Sundholm, G. (2007). Semantic values of natural deduction derivations. *Synthese, 148*(3), 623–638.

Wright, C. (1986). *Realism, meaning & truth*. Oxford: Basil Blackwell.

Chapter 14
Suggestions for Further Pluralist Research

Abstract In this chapter I do three quite different things. One is to give some indication of how to extend Maddy's idea of making mathematician's aspirations explicit, thereby marrying philosophy and mathematics. The second is to elaborate on the discussion of Lobachevsky by comparing intentional perspectives on Lobachevsky's work. This is best done by a pluralist, since he has no agenda. The third is to demonstrate working in a trivial setting, in particular the work concerns Frege's formal trivial system. A speculation is made about how we can learn more about the notion of cardinality. This is important since only the pluralist can see how to do this explicitly, consciously and seriously. Each of these developments suggests further directions for pluralist research.

14.1 Introduction

The pluralist embraces certain virtues; the virtues of observation, curiosity, respect for other points of view and tolerance towards several positions. Exercising these virtues, the pluralist can engage in many projects. In this chapter, I shall suggest four sorts of project.

One is to compare pluralism with other philosophies of mathematics. In the first part of the book, I compare pluralism to realism, naturalism, structuralism and formalism. My treatment of these is not as careful as possible, since that part of the book is meant to motivate a reader to consider pluralism. Those chapters are not meant to be giving definitive arguments. Doing so is a future task. Moreover, at best, only passing remarks are made about: fictionalism, Wittgenstein's philosophy of mathematics, Husserlian phenomenology, Carnapian conventionalism, psychologism, Fregean or Russellian logicism, Lakatos's project and so on. All of these can be given place in the pluralist world, but they are given a place amongst the other positions. The differences between pluralism and each of these is quite interesting,

M. Friend, *Pluralism in Mathematics: A New Position in Philosophy of Mathematics*, Logic, Epistemology, and the Unity of Science 32, DOI 10.1007/978-94-007-7058-4_14,
© Springer Science+Business Media Dordrecht 2014

and will lead to a deeper understanding of both pluralism and the other positions. This is quite an obvious sort of philosophical project, and one with examples in the first part of the book, therefore, I shall not devote a section to it in this chapter.

A second sort of project is to take a feather from Maddy's cap, and identify aspirations held by some mathematicians, give the aspirations formal representation, and offer these as 'regulatory principles'. This is an interesting thing to do, and is not common in mathematics. Regulatory principles are rarely made explicit, except sometimes in the informal meta-language. I shall discuss not only the nature of 'regulatory principles', but indicate how to generate others. This is really a philosopher's task.

In the third section, I shall treat of a third sort of project, which is to compare proofs for the 'same' mathematical result, or compare approaches to questions. I shall suggest a very specific project along these lines: to compare Beltrami, Rodin and Friend's meta-support of Lobachevsky's work combining hyperbolic with Euclidean geometry. It will be clear that there are many projects along these lines. I shall discuss this in the third section.

A fourth sort of project worth developing is to go beyond the metaphysical use of a paraconsistent logic, and look at work in a, strictly speaking, trivial setting. I have suggested elsewhere that there is a structure to trivialism, and indeed it is worth seeing if we can better understand this idea.

14.2 Mathematical Aspirations and Principles

This sort of project is inspired by work done by Maddy where she observes mathematicians, identifies goals they have, and provides general principles (formally expressed in one case) which help the mathematician to realise these goals, and understand the limitations of these goals. Here, I look at her project, and make some general remarks about similar projects.

Maddy begins with her set theoretic realist inclinations, which she takes from Gödel. This is quite natural, since Gödel was a great mathematician, and he made a number of philosophical remarks, wrote philosophical papers and other mathematicians take his philosophical remarks seriously. For these reasons it is inviting for the philosopher to engage with Gödel.[1] Gödel is a realist about the mathematical entities of set theory and the mathematical truths of set theory. Maddy's data suggests that most mathematicians are realists in their philosophy – she has discussed the matter with a number of famous and well-respected mathematicians. Moreover, they are realists about the whole of set theory, where this is thought of as some extended version of ZFC and they take the attitude that we (the body of mathematicians) have not yet worked out which extension is the correct one. That is, she is developing her

[1] In the last chapter, we saw an example of why we should be very careful about engaging with early twentieth century philosophy of mathematics, and exporting those view back in history.

philosophy to accommodate a Gödelian optimist. As noted in the second chapter, the pluralist does not accept that these aspirations represent those of all mathematicians, only a small number of them.[2] Nevertheless, it is interesting to explore these aspirations, *as a local exercise*.

Following Gödel and his fellow optimists, the goals she focuses on are to determine the truth or plausibility of some axioms which are independent of ZFC, and are candidates for extending the theory. Such axioms include the higher cardinal axioms, V = L and the generalised continuum hypothesis (henceforth: GCH).[3] She, very cleverly, develops a formal mathematical definition of 'plausible' in two parts: 'MAXIMIZE' and 'UNIFY'.[4] MAXIMIZE concerns a conception of fruitfulness and UNIFY concerns foundationalist or realist inclinations to have a unique theory. She then thinks that the axioms which fall under the intersection of the two principles captures what the mathematicians find plausible in mathematics. For example, under her guiding principles, it becomes clear that V = L is not (*pace* Quine!) set theoretically 'plausible'. This is very satisfying, since it is in keeping with recent developments in set theory and with the opinions of many contemporary set theorists. Moreover, hers is not an *ad hoc* approach, and does not beg the question, provided we remember that this is a local exercise. That is, we flag the fact that we identify the Gödelian optimist as representing aspirations in some mathematicians. That is, the principles and result cannot, and should not, be imposed on mathematicians who are not Gödelian optimists.

14.2.1 MAXIMIZE

Let us look at the details. 'MAXIMIZE' is the antithesis of Occam's razor. It states that "the set theorist should posit as may entities as she can short of inconsistency." (Maddy 1997, 131) More fully:

> ...if set theory is to play the hoped-for foundational role, then set theory should not impose any limitations of its own. The set theoretic arena in which mathematics is to be modelled is to be as generous as possible; the set theoretic axioms from which mathematical theorems are to be proved should be as powerful and as fruitful as possible. Thus, the goal of founding mathematics without encumbering it generates the methodological admonition to MAXIMIZE. (Maddy 1997, 211)

---

[2]Maddy does make some conciliatory remarks towards this point, but she proceeds as though this is not such a local project. Indeed, if it is realised in full, then it would be a global mathematical project.

[3]GHC is a hypothesis to the effect that the infinite cardinal numbers $\aleph_0, \aleph_1, \aleph_2 \ldots$ increase at the rate of $2^{\aleph_n}$ where n is the index of the previous cardinal. The operation performed for 'getting to the next cardinal' is to take the powerset of the previous cardinal. A denial of the GCH would have it that there are cardinal numbers between an arbitrary $\aleph_n$ and $2^{\aleph_n}$.

[4]The capitalising comes from Maddy (1997).

MAXIMIZE is a principle (Maddy calls it a maxim) identified with one of the goals of set theorists. I like to call the motivation for it 'the argument from fruit', since the idea is to ensure that the path we take will yield a lot of fruitful discoveries, and we do not want to discard the fruit we have already harvested.[5] To determine that one path is more fruitful than another we need some sort of measure on our harvest. How to find such a measure is not obvious, since we cannot anticipate future axioms and discoveries. Furthermore, we want fruit in the long term, not just in the short term. In fact, MAXIMIZE will end up being a negative measure. MAXIMIZE will help us determine which paths would immediately cut off potential fruit (Maddy 1997, 218). Under MAXIMIZE we only set parameters on possible paths. So, MAXIMIZE leaves open several extensions of ZFC.

To leave the metaphorical talk behind, and use the principle rigorously, we have to give it formal expression so that we can compare rival extension of ZFC. In particular, we have to be able to determine, given two candidate extending axioms (or conjunctions of axioms) if one is more maximizing than the other, or if they are the same in their maximizing power. For example, she uses MAXIMIZE to argue for allowing C to be added to ZF. We would loose too much fruit if we were to give it up. Here we are interested in using MAXIMIZE to extend ZFC with new set theoretic axioms.

Here are the details of how she does this. Start with determining that a theory gives a 'fair interpretation' of another theory. This comes in two stages. We first have to give a definition of 'inner model' to give a sense of an interpretation which stays loyal to the core of ZFC. We then add the necessary elements to make it a 'fair' interpretation. I shall quote Maddy's technical definitions and then explain them.

Definition 'T shows ϕ is an inner model' iff

(i) for all σ in ZFC, T proves σ^{ϕ}, and
(ii) T proves $\forall\alpha\phi(\alpha)$ or T proves $\exists\kappa(\mathrm{Inacc}(\kappa) \wedge \forall\alpha(\alpha < \kappa \rightarrow \phi(\alpha)))$, and
(iii) T proves $\forall x\forall y((x \in y \wedge \phi(y)) \rightarrow \phi(x))$. (Maddy 1997, 220- 221)

T is an extension of ZFC, made by adding some axiom, or several axioms, to ZFC. σ is a sentence true in ZFC. So, the first clause of the definition assures us that nothing we are familiar with in ZFC is changed or missing. Moreover this is provable.

[5] Such a path will not necessarily lead us to the truth, supposing there is such. If we have a further metaphysical principle, for example, that the world is 'rich', and we think that, therefore, there are a lot of truths to discover, then we might be inclined to think that we increase the *probability* of finding the truth if we go down a fruitful path. This in itself is not enough, we also have to be convinced that we will recognise the truth when we find it. Of course, many people are convinced of this. Unfortunately, they disagree on what that truth is! So unless one shares their convictions, there is no way to determine the truth. Phenomenological truth is not the same as realist truth. Nevertheless, I suspect that it is these sorts of consideration that led Maddy to develop the principle MAXIMIZE. Making the motivation plain exposes the argumentative weakness of the position, and it is this weakness that motivates the pluralist to treat Maddy's work as a local exercise.

The second clause concerns the ordinals. ϕ is a set of sentences being proved by T. Technically, "ϕ is a formula with one free variable", (Maddy 1997, 220) so it is a property defined using the language of ZFC. The properties we are interested in here are those concerning ordinals. T proves the existence of all of the ordinals of ZFC. Or, if there is an inaccessible cardinal κ, then T proves the existence of κ and for every ordinal of cardinality less than κ. This ensures that our semantics retains all of the familiar properties of the ordinals. The third clause assures us of transitivity through membership. That is, if a property holds of a set, then it holds of every member of the set, and so on down through the membership relation to the empty set.

Definition ϕ is a *fair interpretation* of T in T' (where T extends ZFC) iff

(i) T' shows ϕ is an inner model, and
(ii) For all σ member of T, T' proves σ^{ϕ}. (Maddy 1997, 220)

The first clause tells us that ϕ is an inner model of the theory T' which extends ZFC by adding a general axiom such as GCH or by adding a higher cardinal axiom. Moreover, we can use T' to show this, or we can demonstrate this through proof. The second clause ensures that we have access, through proof, to all of the ontology and properties we had in T. These two clauses secure our loyalty to the original inner models of ZFC and carry on this idea into the extensions of ZFC. This is the best we can do in terms of ensuring future yields of fruit. It's a sort of inductive prudential consideration, what has been good farming practice in the past, should continue to be in the future. The fruit metaphor runs deep, and does a lot of work! We are now in a position to introduce the principle MAXIMIZE:

Definition T' *maximizes* over T iff there is a ϕ such that:

(i) ϕ is a fair interpretation of T in T', and
(ii) T' proves $\exists x \exists_{R \subseteq x2} \forall y \forall_{S \subseteq y2} ((\phi(y) \wedge \phi(S)) \rightarrow (x, R) \ncong (y, S))$. (Maddy 1997, 220)

Here we have our fair interpretation which goes strictly beyond the original theory. Since we can count the units R which go strictly beyond the original theory which just contained the Ss, we can compare two theories extending ZFC and see if one maximizes more than another over T. This is enough to rule out some potential extensions, but there could still be several which are equally maximizing.

14.2.2 The Intersection Between MAXIMIZE and UNIFY; Maxims, Principles, Axioms and Aspirations

For the naturalist, the philosophical importance of MAXIMIZE is that it more-or-less gives formal expression to an identifiable goal of set theorists. Giving formal expression is important for the considerations of rigour. It turns out that MAXIMIZE is in some tension with another goal of (realist) set theorists which is to UNIFY mathematics in set theory, or make set theory foundational (Maddy 1997, 211). UNIFY reflects a striving for choosing a unique extension of ZFC, as opposed to being pluralist and allowing several, which is a possibility with MAXIMIZE alone, since there could be several extensions which seem, for now, to MAXIMIZE 'the same amount'. To find a unique extension of ZFC, we invoke the second principle. UNIFY is the principle that we should settle questions in favour of one of each pair: GCH or not GCH, $V = L$ or $V \neq L$, some inaccessible cardinal exists or does not exist. Maddy does not give a formal definition of UNIFY. Her informal definition is as follows.

Definition UNIFY is the "aim to provide a single system in which all objects and structures of mathematics can be modelled or instanciated. [To achieve this] you must aim for a single, fundamental theory of sets." (Maddy 1997, 208).

It is not necessary to give a formal definition since it amounts to saying that for any proposed new axiom, choose it or its negation. We make choices between a proposed axiom and its negation by invoking other considerations, such as simplicity, usefulness, naturalness or 'importance' (however we measure that). For example, we might choose between two extensions of ZFC on the basis that the argument in favour of one extension is *more direct than* the argument for another (Maddy 1997, 214).[6]

Congratulations to Maddy are in order. Choosing between extensions of ZFC is a problem set theorists struggled with for years. Maddy has made this aspiration explicit. Gödel himself was interested in the truth of the sentence '$V = L$', since it was being considered as a possible new axiom to add to the existing ZFC set theory. Gödel was interested in it because it would decide the continuum hypothesis. His hope was that the generalised continuum hypothesis could either be proved from the other axioms of set theory, or we could work out that it is true (or false) given other, more plausible assumptions or axioms.[7]

Maddy's principles are principles, and therefore, do not decide the truth of an answer. They only guide us to what is plausible, and help us to rule out implausible extensions. The principles make explicit an aspiration that underlies the choices made by mathematicians and the concerns of mathematicians. To see the difference between principles and axioms, let us hone in on the distinction between truth and

[6] Arrigoni calls this the 'inner model programme' (Arrigoni 2007, 19, n. 8). Calling it a programme is suggestive of the idea that this is one direction we choose to pursue.

[7] Gödel was cautious about exposing his philosophical views in print. Now that his complete works have been translated into English, and are therefore available to the non-German, English reading scholars, we should have a more (limitations of translation aside) sensitive and detailed interpretation of his philosophical views. Notwithstanding, what I gloss as his view is plausible in light only of his most hitherto public philosophical pronouncements, through his own writing and through Wang's reports.

plausibility. A little historical background will help to fill the tale. Cohen proved to us the philosophically sub-optimal (if you are a realist) result that that $V = L$ is independent of ZFC set theory.[8]

The result is sub-optimal to the realist because it requires some philosophical adaptation to account for the result. It would have been optimal if ZFC decided on all extensions. If ZFC did rule on (a unique set of together consistent (and consistent with ZFC)) extensions, then we would know that we had the definitive real account of mathematics, since the extensions of ZFC would be determined (by consistency), in fact, they would not be extensions at all, but simply logical consequences of ZFC!

The independence results are sub-optimal because they show that existing mathematics cannot, even in principle, say one way or the other whether or not $V = L$ is true. In fact we can prove (that is what an independence proof is) that $V = L$ and $V \neq L$ are both possible in the sense that they are both consistent with ZFC set theory. The work of many serious and good mathematicians today presupposes that $V \neq L$, and Maddy argues in favour of this by invoking considerations which allow her to follow the principle of UNIFY. So Maddy's work sits well with her naturalism.

What we needed was exactly what Maddy suggested: a mathematical definition of plausibility which took in wider considerations about fruitfulness, preservation of core principles, *etcetera*, and which decided that $V \neq L$ is plausible whereas $V = L$ is not plausible (since this accords with practice). This is not to say that $V \neq L$ is true. To show this, we would add the principles MAXIMIZE and UNIFY as *axioms* together with the considerations that help us to follow UNIFY, (or we would derive them from ZFC) and then make a *proof* that $V \neq L$.

To labour the point, because the principles MAXIMIZE and UNIFY are just that: principles, it follows that the axioms which fall in their intersection are not necessarily absolute truths of mathematics. Adopting principles is a choice, and, by adopting and following a principle, we take on a responsibility. The responsibility concerns our precluding some directions of research, in this case. The responsibility is light if we think of the adoption of a principle in terms of a choice, made for reasons which do not have much (well thought through metaphysical) bearing on mathematics, for example, if we are aware that the principle just follows in a tradition we have been taught and which is always up for revision. The adoption is light, in this case, in the sense of highly revisable and local. We treat it as an exercise. We should have no great objection to adopting an alternative principle or to mathematicians adopting other principles and working under those.

[8] $V = L$, is also sometimes called the "axiom of constructability". V is the cumulative hierarchy determined combinatorially (allowing all true grammatical formulas from levels lower down). That is, at a level $\alpha + 1$ we accumulate all of combinatorially-determined subsets of α. In contrast, L looks more 'constructive'. The content of each level is determined by predicative propositional functions: at stage $\alpha + 1$, we accumulate all and only the subsets of α that are definable by first-order formulas whose quantifiers range over, and whose parameters are drawn from α (Maddy 1997, 65).

In contrast, the adoption of principles carries more responsibility if we defend the choice by appeal to some deeper metaphysical, or mathematical, scruples about plausibility, or ultimately, realist truth. That is, we attribute metaphysical importance to the adoption of the principles. The responsibility should only be as strong as the argument. Returning to Maddy's principles, she cites supporting data for the set theoretic realism she identifies with mathematics. She then concludes that: "the view of set theory as a foundation for mathematics ... is now a pillar of contemporary orthodoxy." (Maddy 1997, 22). But we have to be careful about the data she cites. She tends to cite set theorists working on extensions of ZFC. Other set theorists, such as Enayat and Mourad have a much more pluralist attitude towards alternative foundations of mathematics and towards extensions of ZFC.[9] Thus, in the interests of accuracy, it is amongst a few set theorists that "set theory as a foundation is a pillar of orthodoxy."[10] While she pays lip service to competing ideas, for example, she mentions Aczel's non-well-founded set theory, (Maddy 1997, 61), she does not engage any competitors.

However, Maddy herself recognises the limitations of the view and does not embrace it wholeheartedly. For, she writes later: "As my focus here is on set theory, ... I won't attempt a full naturalistic account ... of mathematics as a whole, but I don't doubt that such an account could be given ... " (Maddy 1997, 210). If Maddy were to take up her own challenge, she would be a pluralist. It is "mathematics as a whole"[11] and also various small parts of mathematics, which the pluralist takes to be of interest. The pluralist is impressed by the idea that the practice of mathematics should inform philosophy. However, before we embrace and engage alternative foundations, let us be a little more careful about the limitations of our argument.

We might think that we are mistaken in our reading Maddy.[12] It might be quite wrong to think that Maddy is close to being a pluralist in our sense, since we might say that she intends to be pluralist only within set theory.[13] Set theory is, after all,

---

[9]Private conversation.

[10]To support the point further, when one says that "set theory is a foundation", one can mean very different things. The term 'foundation' no longer means, as it did in the past, giving the truth, essence and ontology of mathematics. So while a number of set theorists might agree to the statement: "set theory is the foundation of mathematics", fewer will agree that "set theory gives us the truth, essence and ontology of mathematics", or that "there are no alternative foundations".

[11]Note the contrast of this remark with the definition of UNIFY above, where she discusses 'all of mathematics'. I conclude that 'mathematics as a whole' is all of mathematics as it is practiced before we deploy UNIFY. If her universal quantifier in UNIFY is descriptive then it only has force after a choice is made, and even then it is a bit odd since it turns the choice into a norm. As a result, we retrospectively have to deny that what we cut off is, or was, mathematics.

[12]Note that I am less concerned with Maddy exegesis than with possible position close to Maddy's or between Maddy's and pluralism. Because of this concern, reference to Maddy, here, can be taken as a sort of literary device (of giving a proper name to a proponent of a philosophical position). As such a device, Maddy's name suits better than any other.

[13]We brushed elbows with this position in the second chapter.

'the orthodoxy', and we can do a lot of mathematics within set theory. We can do enough to satisfy most mathematicians. But pluralism within set theory will not do for the maximal pluralist.

Let us look at a specific example, that of finitist mathematics. Maddy might argue that we can 'do' finitist mathematics 'within' set theory, since there is a sense in which finitist mathematics is reducible to set theory. But Maddy's supposed pluralism would then have missed the point. One of the motivations for, and the tenets of, finitist mathematics is that there are *only* a finite number of objects, not as part of a bigger whole, but as implicit in the finitist mathematical theory. Finitism, of this sort,[14] denies the axiom of infinity of set theory which explicitly states that there is a set with an infinite number of members. So, while we can reproduce all of the results of finitist mathematics within set theory, we deny finitist mathematics when doing so! This is because we deny the metaphysical underpinnings of finitist mathematics, and this is philosophically insensitive. Alternative foundations, here means alternative ontological commitments. So if we are pluralist about foundations, we cannot be so within set theory, *mutatis mutandis* for philosophical presuppositions underlying, even 'small' mathematical theories. The pluralist concludes that ontological, and other metaphysical commitments, *per force*, rest internal to a theory. Moreover, the naturalist should be sensitive to this point.

Let us simply remark that *ontology of*, and *truth within*, a theory are reasonably well-defined and understood concepts. Absolute truth in mathematics is not; *mutatis mutandis* for ontology.[15] It might be for this reason, or for others, that mathematicians often show indifference to questions about absolute truth. The indifference might be shown by lack of mentioning the concept. For example, taking an arbitrary textbook on model theory off the shelf, such as Marker (2000) and looking at the index, 'truth' is mentioned only on one page of a 328 page text.[16] The indifference shown by mathematicians concerning absolute mathematical truth or absolute and unique ontology is accounted for in a properly pluralist setting by confining discussion of truth or ontology to a theory. In this way we practice pluralism in truth and pluralism in ontology.

While Maddy's work fails on the particular philosophical significance of supporting realism in mathematics, it carries significant philosophical import, and this

---

[14]The term 'finitism' is ambiguous, and some mathematicians who call themselves finitist will entertain higher cardinals, albeit in a 'finitistic way' – as symbols being manipulated according to determinate, effective, rules, for example. Here, I just mean the very simple notion that there exist only a finite number of entities in mathematics.

[15]I expand on this in the next section.

[16]Of course, this is not proper statistical evidence for the claim of indifference on two points. One is that it is just one reference, and the sample is too small and might be un-representative. Two, one has to be careful not to read too much into silence. Silence does not always indicate indifference. Nevertheless, methodological indifference (displayed by willingness to use methodologies imported from quite different areas of mathematics) cannot be completely divorced from indifference to truth or ontology.

has to do with a more pluralist approach to mathematics. Her approach leads to more sensitive and nuanced discussion concerning the nature of our commitments in mathematics. For the pluralist, only a light responsibility accrues to the adoption of principles since he will not underpin them with realist concerns, or any other traditional global philosophical concerns. Nevertheless, with Maddy, in the practice of mathematics, we can observe revisable principles identified with the aspirations of some mathematicians. This is exactly the work of the pluralist. Instead of calling them ‘maxims’ or ‘principles’, the pluralist calls them ‘aspirations’, where it is understood that not all mathematicians hold such.

Definition an *aspiration* is a general hope held by a mathematician (and concerns the future of mathematics). It can be identified as a goal she has or some sort of meta-attitude which guides her work. It is not something which has received a definitive philosophical defence. Therefore, it is revisable. It is also understood that it is shared by some, but not all, mathematicians. An aspiration can be given formal expression, but it is not, for all that, testable or provable, without begging the question.

What actual aspirations do mathematicians have? Examples include: to be able to effectively treat increasingly complex data, to show the relationships between theories, to vindicate a theory by showing how it gives important results in another theory, to find an application of a theory to a problem in the physical world. Once we have identified an aspiration, then the pluralist trick, due to Maddy, is to give as sensitive, precise, identifying and formal expression as possible to the aspirations. For example, we could turn the first into a formal expression in the following way.

Definition treatment of data is effective iff all recognised answers to recognised questions are computable.

The principle accompanying the aspiration then is:

Principle to start with questions with computable answers, and extend the set of questions using new effective techniques or methods.[17]

Once we have turned some of these aspirations into formal expressions, we can ask if pairs, triples *etcetera* of these are naturally held in conjunction. Similarly, we can see if pairs, triples *etcetera* preclude being conjoined, i.e., together they result in some conflict. But here, we have to be careful about what ‘conflict’ means. For example, if we have an outright contradiction, and we are committed to a classical logic, then holding the contradicting aspirations will result in triviality, and this is to be avoided. However, if n-tuples of aspirations are in tension with one another, then as we saw with Maddy’s pair (MAXIMIZE and UNIFY) we can look to the optimisation of the n-tuple. Maddy looked at the intersection for optimisation. The

[17] This is exactly what we do in algorithmic learning theory. For example, we discover that given an algorithm of a certain complexity, it makes all the difference to the solvability of the problem (which might be finding an algorithm which produces the same data) if we add an oracle, or if we allow hypothesis testing. Technically, we call these mind-switches (of the learning computer) (Friend et al. 2007, 5).

exercise of optimising aspirations in tension with one another, might turn out to be quite interesting. Moreover, they will vindicate, confirm, or correct paths of research. There is nothing wrong with the mathematician becoming aware of her aspirations in this way, and making an easy analysis of them. Ultimately, the exercise can only deepen her justification for pursuing the path of research she has chosen. Moreover, her justification will be honest and explicit. This is an elegant type of project for the philosopher, since the identification and expression of aspirations is not a purely mathematical matter, and it is not a psychological matter. It is also philosophical. It is best done by a pluralist because of his metaphysical indifference.

14.3 Lobachevsky, Beltrami, Rodin and Friend; Proofs, Reconstructions and the Paths of Mathematical Enquiry

We have a puzzle. Like many geometers, Lobachevsky was dissatisfied with Euclid's Elements. In particular, he was dissatisfied with the parallel postulate. He developed his own ideas about what 'parallel' means, and developed his own 'non-Euclidean' geometry. He believed that his geometry was more general and fundamental than Euclid's. This is because a Euclidean sphere is the same as the limit of hyperbolic space, where the limit is called the 'horosphere'. But Euclidian flat space can be interpreted on the Euclidean sphere. Therefore, by transitivity, Euclidean space is a limit case of hyperbolic geometry.

When he had developed his ideas, and found some important results, Lobachevsky sent them to Gauss, who liked them very much. However, he would not support them publicly, because he (rightly) feared that they would not be well understood, or well received (Kagan 1957, 25); and this, despite the fact that Lobachevsky's arguments are good ones. It was not until Beltrami was able to give a model of Lobachevsky's geometry *in* Euclidean space that Lobachevsky's geometry could be understood, and accepted, by the greater body of mathematicians. The following 'picture' is now common, but note that it is drawn *in* Euclidean space. That is why we can 'recognise' it (Fig. 14.1).

Lobachevsky had no such model or picture. Hyperbolic space was the background space, since it was more basic.

Rodin does not fully accept Beltrami's reconstruction, not because the reasoning is faulty, but because it is not loyal to Lobachevsky's understanding of geometry. Beltrami couches hyperbolic geometry in a Euclidean setting, whereas Lobachevsky thinks of hyperbolic geometry as more fundamental, and therefore, that Euclidean geometry should be couched in a hyperbolic setting. Neither Lobachevsky nor Rodin can draw a picture of Euclidean space within hyperbolc space since the flat sheet of paper is suggestive of Euclidean space, so the best we would have is a picture of Euclidean space within Hyperbolic space within Euclidean space. Rodin, therefore, suggests a rational reconstruction by way of appealing to a larger meta-theory: topos theory. He then restores the priority of hyperbolic geometry in a topos theory setting, by making the embedding of a geometry into a type of space explicit.

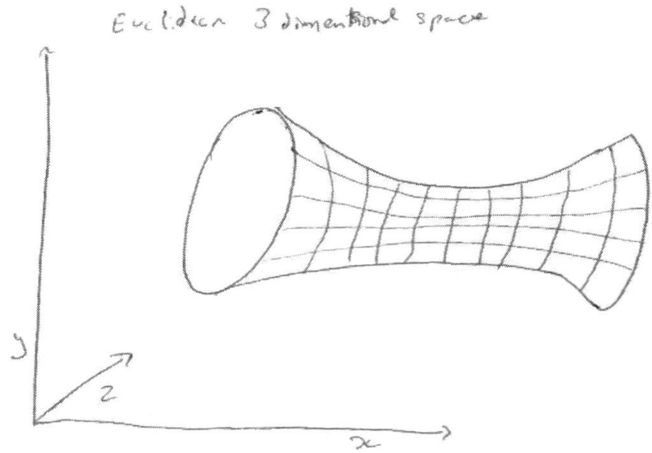

Fig. 14.1 Beltrami's reconstruction/interpretation of hyperbolic geometry

As a pluralist, I think of the contradiction between the geometries in a third way. I see them as contradicting each other over the notion of 'parallel', and not as one being prior to the other. Therefore, it is important for me alleviate my concern to avoid triviality, or inconsistency in this case, by modelling the reasoning of Lobachevsky's proof, independently of geometrical concerns.

The question is not that one approach is correct and the other incorrect. The question is also not: is one reconstruction better than another. For, it depends on what one is trying to do. Beltrami's reconstruction is good for conveying the legitimacy of hyperbolic geometry to geometers stuck in Euclidean geometry. In the context of time he was writing, it was crucial. Today, Beltrami's model is more important for learning about non-Euclidean geometries, after we have learned Euclidean geometry. But, now we could learn the geometries independently.

In contrast, Rodin's reconstruction is historically more important, since it restores the spirit in which Lobachevsky thought of his geometry, but when Rodin does this, he is not perfectly loyal to Lobachevsky, since Lobachevsky was in no way thinking in terms of topos theory. Nevertheless, topos theory helps us to see the difference between Beltrami's approach to understanding Lobachevsky, and Lobachevsky's understanding of his own geometry. Moreover, topos theory is independent of both hyperbolic geometry and Euclidean geometry. It is one of the grand meta-theories. As such, it gives us a perspective, or an orientation, and this helps us to see, from a relatively neutral, independent, standpoint, that Lobachevsky legitimately sees his geometry as more general than Euclidean geometry. My pluralist reconstruction is useful if we are concerned about triviality, contradiction, and using one theory to prove something in another which is in contradiction with the first.

We learn different lessons from the reconstructions, just as we learn different lessons from different proofs for the 'same' theorem. Who is ready to see or read what proofs depends on educational background, time, interest and open-mindedness. All of the latter are influenced by our teachers and our peers. For this reason, mathematics is partly a social enterprise, and the process of learning mathematics, and the conduits for dissemination of mathematics cannot be divorced from the subject of mathematics, at least not by the pluralist philosopher. This is something we come to appreciate more in learning some of the history of mathematics, and less from a class on algebra, calculus, topology and so on. Philosophers who have studied some history of mathematics shed the "Euclid myth ... the belief that the books of Euclid contain truths about the universe which are clear and indubitable. Starting from self-evident truths, and proceeding by rigorous proof, Euclid arrives at knowledge which is certain, objective and eternal." (Davis and Hersh 1998, 325). The contention of the pluralist is that not only the Euclid myth, but also a more general myth should be shed by philosophers. This is the myth that all of mathematics are true, certain, universal, objective and timeless.

Buldt and Schlimm shed both myths. With the pluralist, they hold a non-standard view. They show a sensitivity about the rate and direction of the development of mathematics: "... we do not claim that the mathematical community as a whole moves (or has ever moved) like one solid block in just one direction: nothing could be further from the truth. We would rather compare the historical development of the mathematical community with the movement of a body that various people try to pull in different directions." (Buldt and Schlimm 2010, 45). The pluralist goes further. For the pluralist, it is misleading to write of 'a body', if 'a body' suggests a contiguous entity. It might be quite fragmented, over: content, geography, language, culture, traditions, preferences, philosophies and so on. This is why in Chap. 6 I embraced the study of the history of mathematics, the sociology of mathematics and the psychology of mathematics as informative for the philosopher.

We might think that the historical development of mathematics was very fragmented in the past because of lack of communication between mathematicians. They simply did not enjoy the same technological means of communication we have at our disposal today. Now, because of the sheer quantity of communication, mathematics can become more unified. But think again. There is a trend today towards fragmentation. We see this in the specialisation of mathematicians. We also see an increasing number of specialised meetings and conferences. There has been a dramatic increase in the number of specialisations within mathematics, and mathematics, as a discipline, becomes radically unsurveyable. Davis and Hersh (1998, 29) compare the "classification of mathematics in the years 1868 and 1979". In the 1886 there were 12 sub-disciplines, in 1979 there were 36. If we use a comparable source now, there are between 80 and 100.[18]

[18]Go to the American Mathematical Society web page. Pretend you are looking for a book or article. Pull down the menu of 'classifications' (of topics), and there are 97, at least in May 2012. Now, we should be careful, since the sources used for comparison of numbers of specialisations are not the same, nevertheless, even as approximate figures they indicate a marked growth in the number of specialisations.

Of course, there are general mathematical conferences, and general introductions to the discipline, and this suggests unification. But if we observe who goes to which papers, we see that even in these general conference contexts, quite often mathematicians confirm familiar territory. They attend papers they think they will understand. The growth in quantity of modern communication is a double-edged sword: allowing for both unification and fragmentation. Furthermore which edge will cut in the future is not predictable from here, and that too could change; we could see a unification in mathematics followed by a fragmentation followed by a new unification and so on. Nothing *now* can help us predict. The pluralist looks at this with disinterest, and simply observes. But the point remains that, in the face of fragmentation – whether restricted to the past or whether we find it in the present or whether we experience it in the future – it is not surprising that there should be conflicting developments in mathematics.

Moreover, it is not even this simple. Both fragmentation and unification can happen at the same time, but maybe not over the same concept in mathematics. Thinking again about 'mathematics as a whole' maybe in the neutral (for these purposes) sense of successful existing recognised mathematics, we might think of some areas of mathematics as being increasingly unified, while others are being fragmented. There might be exciting trends and changes which we can identify by looking at the history of mathematics, such as the movement in the rigorisation of mathematics in the late nineteenth century, or the development of foundational theories around the same time. With the development of computers we witness a movement towards extensionalism, but this does not encompass all of mathematics by any means.

In light of this picture of the development of mathematics, we can easily recognise the need for separate proofs, models and reconstructions. Each communicates something important to some mathematicians, each is situated in a tradition and as part of a proof community. The acts of communication through proof, modelling and reconstruction connect or confirm mathematical results. Without these we would not have a recognisable mathematics. The mathematical proofs and theorems in The Book of Proofs is not only an idealisation born of the Euclid myth, it makes sense of only a very small part of mathematics – the 'final' results. It cannot be used to make sense of mathematics as a whole.

The pluralist is well placed to work on rational reconstructions because of his agnosticism towards truth, certainty, universality, objectivity and timeless of mathematics. He is also not wedded to finding a 'yes', 'no' solution. The pluralist is patient, and will not trade impatience against sophistication, accuracy and precision. The obvious places to start on the task of re-constructing is with controversial mathematical ideas: the first non-Euclidean geometries, irrational numbers, imaginary numbers, infinitesimals, non-standard models, the paradoxes and so on. With any pluralist enquiry, the pluralist will bring a stamp of principled scepticism, made sharper by a metaphorical use of logic to work with conflicting mathematical and philosophical situations. There is a sense in which the pluralist brings a type of objectivity to the discourse, and the investigation as to what this objectivity consists in is the subject of future investigation.

14.4 Working in a Trivial Setting

We turn to a wholly different project. Consider the following. Frege's formal system of logic is classical, in the sense that *ex contradictione quodlibet* inferences are valid in it. Given Basic Law V, it is possible to derive a contradiction in the formal theory. Therefore, Frege's formal system is trivial. Therefore, every wff in the formal language of the *Begriffsschrift* is derivable, if nothing else, by excursion through the contradiction and *ex contradictione quodlibet*. This is supposed to be disastrous, so, rationally we should have thrown out Frege's formal system and never looked at it again; as Frege himself did. But this is not what happened. Dummett, Wright, Boolos, Heck, Sluga and Macbeth, amongst others, have done significant work refining and mining Frege's formal system. Heck has shown us that the only essential use Frege makes of Basic Law V is to derive Hume's Principle (Heck 1993). Boolos launches a bad company argument about Basic Law V and any formula with that structure (Boolos 1986–7). Thus, he thinks that the rest of Frege's formal system is worth saving. Dummett takes Frege's philosophical views very seriously, as do the other Frege scholars. Wright develops alternatives to Basic Law V, and resurrects the Fregean logicist project. All this work takes place in and around a trivial theory.

Meyer once opened a talk by presenting a formal system with the following one rule of inference: from a well-formed formula you may infer any well-formed formula. He then asserted, correctly, that this theory is formally equivalent to Frege's formal theory (provided the formulas are written in the same language). This is amusing because it is quite correct, and yet, it makes a mockery of all the work done by all the Frege scholars. How can the Frege scholars pretend to do serious work in and about a trivial system?

One lesson we can learn from the chunk and permeate method is that it is possible to work consistently locally (within a chunk) while in a trivial theory. Frege himself proves none of his theorems by making an excursus through contradiction followed by *ex contradictione quodlibet*, nor do any of the scholars who work on his material. Russell himself derived one contradiction and left it at that, as did Frege (1903, appendix). The contradiction was treated *as a dead-end*. Of course they thought it was much worse than a dead end. But, again, once armed not only with the methodology of chunk and permeate, but also with the lessons we learn from it, we can think of Frege's actual elegant and gapless proofs as proving something worth proving, and we make sense of the Frege scholarship.

I venture a speculation since this chapter is about future work. *Using* Basic Law V, the collapsing lemma and the rest of Frege's formal theory, and *avoiding* inferring a contradiction in the proof we should be able to derive the negation of Hume's Principle! This would show that Basic Law V and Hume's Principle are independent of each other, and each is independent of the rest of Frege's formal theory. Cardinal numbers are not logical objects.

Filling in the background: Basic Law V is that two sets are identical if they are co-extensional.

$$\forall F \forall G\, [F = G \leftrightarrow E : F \approx E : G]\,.$$

Hume's principle is that for all concepts F and for all concepts G, the sets of F and G are of the same (cardinal) number iff the sets can be placed into one-to-one correspondence.

$$\forall F \forall G\, [N : F = N : G \leftrightarrow F \approx G]$$

One denial of this is:

$$\exists F \exists G\, [N : F \neq N : G \leftrightarrow F \approx G]$$

The claim then is that there is a pair of concepts which have a different cardinality, but which are isomorphic.

How do we make the proof? Recall the collapsing lemma.[19] The collapsing lemma tells us that under the right circumstances there are models of lesser cardinality (which contain contradictory objects). Wffs about them will be both true and false. The proof will be non-classical in the sense that we are deliberately avoiding *ex contradictione quodlibet*. We can use chunk and permeate to insure this. Furthermore, the proof will exploit some ideas about cardinality particular to second-order logic. We might make an excursus employing concepts which involve a dialetheia concerning the relativity of cardinality in a second-order set theory.[20] We would be wise to use the technique of chunk and permeate to verify that we do not invite further mischief. This is enough speculation about possible proofs.

What would we learn from such an exercise? Assume it is successful. We make such a proof, or we make a proof to show that a proof is possible. In this case we learn that we can locally prove both Hume's Principle and its negation. Therefore, Basic Law V and Hume's Principle are strictly independent of the rest of the formal theory, and independent of each other. We have partial evidence for this through other means by virtue of the work done by Wright and the philosophers who worked on the neo-Fregean project. What they do is remove Basic Law V from the formal system, and add various, independently motivated principles with a formal structure similar to Hume's Principle. The evidence is partial because they did not work on

[19]Remember that the collapsing lemma does not hold in the presence of functions. We can use chunk and permeate to insure that we do not invite them in. There is little danger of this since the all there is in Hume's principle and Basic Law V are (identity and equivalence) relations and predicates.

[20]This last, is from Shapiro (1991, 253–4). There Shapiro writes: "To say that a structure P is characterised up to isomorphism by the language of set theory as interpreted is only to say that P is characterised in terms of *m*, or 'up to *m*'. ... The problem as to how *m* itself is grasped, understood or communicated is left mysterious." Here '*m*' is a limit rank V_λ or V itself. (Shapiro 1991, 253).

the negation of Hume's Principle specifically. Nor would they have wanted to, since the negation is philosophically unsavoury to a neo-Fregean, since it would show that the notion of cardinal number is not a logical notion, provided we define 'logic' as invariant, or absolute, relative to second-order set theory.

Assume we are unsuccessful in our exercise. There are two ways of being unsuccessful. One is to prove that it is not possible to derive the negation of Hume's Principle from Basic Law V without an excursus through contradiction. This tells us that there is a close conceptual link between them, since proved non-derivability is the mirror of derivability, and the proof was not an *ex contradictione quodlibet* proof. The other way of being unsuccessful is that we never manage to make a proof. From this we can draw no conclusions.

Such work is important in clarifying our notion of cardinal number. From Skolem's paradox (that the set of reals has different sizes depending on context), we know that in first-order set theory we are deceived in our notion of cardinality. Remember Tarski's logical notions. One was supposed to be cardinality. We showed with the collapsing lemma that it is not invariant. If my speculation is correct, then we also know that in second-order ZF cardinality is not an absolute notion either, and we can give an explanation for this.

Thus, while cardinality is presented to us as a simple notion, as one of the cornerstones of modern mathematics, it bears further scrutiny. Moreover, we admit outright that the scrutiny takes place in a trivial setting. In light of all the elegant work done in paraconsistent logic we have learned not to be so wary of contradiction, we can now learn to be less wary of trivialism. The pluralist challenge is to find careful, studied and principled methodologies for studying in a trivial setting.

14.5 Conclusion

As we can see there is plenty of work to be done, and it is philosophical work, not to be left in the hands of people who are only: mathematicians, sociologists, psychologists or historians. They will not ask the same questions, and they will not give philosophically satisfying answers, unless they are also philosophically inclined. Similarly, the philosopher can ask very few pertinent questions isolated in philosophy. Not only should he look to mathematics, but also to sociology, psychology and history. This is a new trend in the philosophy of mathematics. The historical was started by Lakatos, but was not pursued. The historians, sociologists and psychologists of mathematics tend not to use the history, sociology or psychology of mathematics to support philosophical conclusions. But bringing these considerations to bear on our philosophical conclusions does give us a new take on, for example, the notion of 'foundation' in mathematics. "...by a striking shift in the meaning of words, the fact that foundationalism was at a certain critical period the dominant trend in the philosophy of mathematics has led to the virtual identification of the philosophy of mathematics with the study of foundations." (Davis and Hersh 1998, 323).

The pluralist will study foundations, not as metaphysically informative, but rather, as mathematically limiting. Engagement with the issue of foundations is an option, not a philosophical obligation. It is far more interesting to identify aspirations and explore these accurately, and without pretention to giving ultimate truth. It is more interesting to investigate different proofs, models and reconstructions to discover or re-discover our mathematical history. It is also more honest to face our existing work in trivial settings and explore this area further. Given the options, and the lack of obligations, the pluralist has a wide field of study when looking at mathematics, and the enquiry is only beginning.

References

Arrigoni, T. (2007). *What is meant by V? Reflections on the universe of all sets*. Paderborn: Mentis.

Boolos, G. (1986). Saving Frege from contradiction. *Proceedings of the Aristotelian Society, LXXXVII*, 137–151.

Buldt, B., & Schlimm, D. (2010). Loss of vision: How mathematics turned blind while it learned to see more clearly. In B. Löwe & T. Muller (Eds.), *PhiMSAMP philosophy of mathematics: Aspects and mathematical practice* (Texts in philosophy, 11, pp. 39–58). London: Individual Author/College Publications.

Davis, P. J., & Hersh, R. (1998). *The mathematical experience*. Boston: A Mariner Book, Houghton Mifflin Company.

Frege, G. (1893, 1903). *Grundgesetze der Arithmetik I, II*. Jena: Pohle. Partly translated in Black, M., & Geach, P. (Trans.). (1980). *Translations from the philosophical writings of Gottlob Frege* (3rd ed.). Oxford: Blackwell.

Friend, M., Goethe, N. B., & Harizanov, V. (Eds.). (2007). *Induction, algorithmic learning theory and philosophy: Vol. IX. Logic epistemology and the unity of science*. Dordrecht: Springer.

Heck, R. G., Jr. (1993). The development of arithmetic in Frege's *Grundgesetze der Arithmetik. The Journal of Symbolic Logic, 58*(2), 579–601.

Kagan, V. F. (1957). *N. Lobachevsky and his contribution to science*. Moscow: Foreign Languages Publishing House.

Maddy, P. (1997). *Naturalism in mathematics*. Oxford: Clarendon Press.

Marker, D. (2000). *Model theory: An introduction* (Graduate texts in mathematics). New York: Springer.

Shapiro, S. (1991). Foundations without foundationalism. In *A case for second-order logic* (Oxford logic guides 17). Oxford: Oxford University Press.

Chapter 15
Conclusion

Abstract I make some concluding remarks about the goals of this book: to give a clearer and more precise sense to what pluralism means in the philosophy of mathematics. A reader might adopt none, some or all of the pluralist view. There are stopping points.

15.1 Conclusion

This book introduces a family of philosophical positions. One can be pluralist in different respects, at different levels and one's pluralism can be governed by different logical inclinations or hypotheses. In general, the pluralist aspires to the following virtues: unprejudiced observation of mathematical practice and a desire to encompass and accommodate as wide a variety of practices as is coherently possible. The inverse of these virtues are manifested when we insist on unique, simple, teleologically satisfying answers,[1] beyond what the evidence will support.

We might think that almost everyone will agree to this. But think of the arguments from simplicity, naturalness or elegance of a theory. In these, aesthetic qualities are not only aesthetic qualities; they are assumed to be a guide to truth, correctness or objectivity. There are no grounds for such an assumption unless we are *already* convinced of some master plan or *telos*. Thus, such arguments beg the question. What the non-pluralist will insist upon is that we seek such answers. At the highest levels, the pluralist denies that we *should* seek such answers. At best, seeking such answers is an exercise. At worst, it blinds us and leads us astray. Such seeking flirts with intolerance and dogmatism; where dogmatism is identified with an inability, or unwillingness, to offer well supported philosophical explanations, or admit that we simply do not know. That is, where argument fails, posturing, authority and dogmatism take over.

[1] 'Teleologically satisfying answers' are ones that reveal a sense of purpose of overall, unique goal or plan.

M. Friend, *Pluralism in Mathematics: A New Position in Philosophy of Mathematics*, Logic, Epistemology, and the Unity of Science 32, DOI 10.1007/978-94-007-7058-4_15,
© Springer Science+Business Media Dordrecht 2014

The pluralist virtues liberate the thinking of the philosopher, but the liberation comes with responsibility. The pluralist is willing to change his mind in light of counter-evidence, and he accepts (at least in public) that some philosophical questions might never be resolved in a simple and 'satisfying' way. Of course, what counts as 'satisfying' is culturally informed. For example, western analytic philosophers expect unique answers to questions. Since he is aware of this, the pluralist can correct for this prejudice and be careful about which conclusions evidence can support. He corrects for his culturally enforced philosophical desires in the interests of evidential honesty and accuracy. This too, might be a prejudice, especially if we are restricted in what we count as 'evidence'.

The pluralist position is meant to give a philosophical theory to support what is already happening in the philosophy of mathematics. Thus, the position is 'new' in the sense of not having yet been expressed this way in print, but it is 'old' in the sense of being already implemented and understood, at some level, by some mathematicians and philosophers of mathematics. This book fills in some details and pushes the position further than most philosophers are willing to go. For this reason, the position is controversial. However, as noted, there are degrees of pluralism. It not necessary to follow me in embracing the entire transcendental position. There are many stopping points. There is also no reason to embrace it all at once. One can gradually become increasingly pluralist. The important point is to start the voyage. I hope I have provided a specific enough framework, and enough details, that we can engage in a clearer discussion of pluralism, mathematics and the philosophy of mathematics.

Appendices

Appendix 1

The Semantics of LP

In this presentation, I assume a solid familiarity with first-order formal classical systems. I therefore take the liberty to miss out on some familiar details, such as specifying that there is a list of names in the language, and so on. The semantics we present here are heavily cribbed from (Priest 2002, 171–172). LP is a first-order logic. Because of the inter-definability of the logical connectives, we only need negation, $\neg$, and conjunction, $\bigwedge$. Disjunction and the conditional are defined in the familiar way from a classical propositional logic. We also need only one quantifier, $\forall$, since $\forall$ and $\exists$ are duals. Identity is in the language, as are predicates and relations. We omit functions for reasons of never needing them in this text, so they only pose an unnecessary complication to the exposition. For those who are interested in further paraconsistent formal theories, I refer them to (Priest et al. 1989). However, the reader should be aware that there are other quite different traditions of developing paraconsistent formal systems, such as those of Da Costa, Batens or Béziau.

An interpretation is a pair $<D, I>$. D is the non-empty domain of quantification. I is the interpretation. It is a function that maps names to individuals in the domain. I maps each predicate P into a pair $<I^+(P), I^-(P)>$, where $I^+(P) \cup I^-(P) = D$. $I^+(P)$ is the positive extension of P (objects which have the property P). $I^-(P)$ is the negative extension of P, all objects lacking the property P. I maps each n-place relation R into a pair $<I^+(R), I^-(R)>$, where $I^+(R) \cup I^-(R) = D^n$. $I^+(R)$, is the positive extension of R, all n-tuples which bear R. $I^-(R)$ is the negative extension of R, all n-tuples which fail to bear R. Note that $I^+(P) \cap I^-(P)$ or $I^+(R) \cap I^-(R)$ could be non-empty. Objects falling in the intersection are contradictory objects. Identity is a special relation defined: $I^+(=) = \{<x, x>; x \in D\}$.

M. Friend, *Pluralism in Mathematics: A New Position in Philosophy of Mathematics*, Logic, Epistemology, and the Unity of Science 32, DOI 10.1007/978-94-007-7058-4,
© Springer Science+Business Media Dordrecht 2014

Every wff, φ, is assigned a truth-value $\nu(\varphi)$ in the set: $\{\{1\}, \{0\}, \{1, 0\}\}$ by the following recursive clauses:

$$1 \in \nu(Pt) \iff I(t) \in I^+(P)$$
$$0 \in \nu(Pt) \iff I(t) \in I^-(P)$$

$$1 \in \nu(Rt_1 \ \ldots \ t_n) \iff <I(t_1) \ldots I(t_n)> \in I^+(R)$$
$$0 \in \nu(Rt_1 \ \ldots \ t_n) \iff <I(t_1) \ldots I(t_n)> \in I^-(R)$$

$$1 \in \nu(\neg \ \varphi) \iff 0 \in (\nu)$$
$$0 \in \nu(\neg \ \varphi) \iff 1 \in (\nu)$$

$$1 \in \nu(\varphi \wedge \psi) \iff 1 \in \nu(\varphi) \text{ and } 1 \in \nu(\psi)$$
$$0 \in \nu(\varphi \wedge \psi) \iff 0 \in \nu(\varphi) \text{ or } 0 \in \nu(\psi)$$

$$1 \in \nu(\forall x \varphi) \iff \text{for all } d \in D, 1 \in \nu(\varphi(x/d))$$
$$0 \in \nu(\forall x \varphi) \iff \text{some } d \in D, 0 \in \nu(\varphi(x/d)).$$

In the last, $\varphi(x/d)$ means a formula φ where every occurrence of x is replaced by d. Models for a set of sentences makes all the sentences true. Note that some formulas might also be false (Priest 2002, 171). In the absence of any paradoxical formulas, LP interpretations are just the same as classical interpretations. That is, in the absence of contradiction, the logic behaves classically. It follows that we can think of classical logic as a special case of LP, and if we are assured that there are no contradictions, then we may reason according to rules presented to us in a classical presentation. See also Priest (2006b, 117–118).

Appendix 2

Prior's Tonk

There is a problem about choosing, or designing, rules of inference. The problem was illustrated by Prior in 1960. Not any rule of deduction will do, or rather not any pair (introduction and elimination) of rules for a connective will do, for a formal system. Prior was targeting a Dummetian intuitionist; someone who endorses a pair of claims: (i) we have no separation between syntax and semantics (meaning is use). (ii) When we have a pair: introduction and elimination rule for a logical connective, then their being 'in harmony' is sufficient to warrant our formal system.

In our case, we are not so interested in harmony, but just in showing that not any rules will do.[1] Prior's Tonk connective is enough to show this. Consider the pair of

[1]In Chap. 14, Sect. 14.4 I mention Meyer's joke formal system where the only rule is: from a formula in the language infer any formula you like. This immediately gives us a trivial system. So we already know that not any rule will do. However, we might rule Meyer's system out on the grounds that it obviously leads to triviality. Similarly, we might rule out a purported formal logical system that depended on some empirical facts to make inferences such as including the rule: if my grandmother says x then infer x. The Tonk case is more interesting since each of the two rules

rules for the new connective Tonk, *. Where P and Q are propositions, or wffs: the Tonk (elimination) rule is: P * Q ⊢ P. The introduction rule is: P ⊢ P * Q. In other words, the elimination rule works like 'and elimination' (from a conjunction, write a conjunct on the next line of proof). The introduction rule works like 'or introduction' from a proposition disjoin it with any proposition you like. The combination of Tonk elimination and Tonk introduction make a purportedly logical system trivial. Every wff is derivable (Prior 1960–1961, 38–39).

There have been several suggestions about general rules of thumb that help us to preclude pairs of the Tonk variety. To this day, there is no consensus on how to do this. Note, however, that this question is only important to those who both: do not want to divorce syntax from semantics and who endorse *ex contradictione quodlibet.*

Appendix 3

Ex Contradictione Quodlibet

The following are two classical proofs of *ex contradictione quodlibet*. The numbers to the left of the line numbers are dependency numbers – marking which premises or assumptions the formula ultimately depends on. & is conjunction, ~ is negation, P and Q are proposition variables.

{1}	1. P & ~P	Premise
{2}	2. ~Q	Assumption for *reductio ad absurdum*
{1, 2}	3. (P & ~P) & ~Q	1, 2 & introduction
{1, 2}	4. P & ~P	3 & elimination (simplification)
{1}	5. Q	2, 4 *reductio ad absurdum*

There are two places where the non-classical logician might block the proof. One is to not allow *reductio ad absurdum*, the other is to disallow the fiddle of lines 3 and 4, so we constrain the & introduction rule. The new & introduction rule would say that only formulas which have a dependency number in common may be conjoined to form a conjunction. But these restrictions will not block the following proof.

{1}	1. P & ~P	Premise
{1}	2. P	1 & elimination
{1}	3. P ∨ Q	2 ∨ introduction (weakening)
{1}	4. ~P	1 & elimination
{1}	5. Q	3, 4 disjunctive syllogism

is already found in perfectly respectable formal systems. What we learn is that there are pair of *otherwise* perfectly good rules which cannot be combined over one connective.

The dodgy move in this proof is disjunctive syllogism. The reason some relevant logicians give for disallowing disjunctive syllogism is that, say one of the disjuncts in the disjunction is both true and false, but the other disjunct is true. This is enough to make the disjunction true (since one disjunct is true and we look no further). The negation of the disjunct which is both true and false will certainly not make the negation of the true disjunct true! This would be invalid reasoning from two true premises to a false conclusion.

Glossary

Meta-note on the glossary: The reason for a glossary in such a book is not to aid the beginner, but to set straight any terms which are used idiosyncratically, or which are ambiguous in the literature.

The **anti-realist** is either a sceptic or a Kantian idealist. It is someone who epistemically constrains truth.

Aspirations: Inspired by Maddy's maxims, or principles, an aspiration is a general goal identified with some mathematicians. An example might be to make all of mathematics constructive according to certain parameters on what counts as constructive. Another might be to unify all of mathematics under one foundation. Another might be to develop as many incompatible extension of ZFC as possible. These are only examples. The pluralist thinks that aspirations in mathematical practice are very important. Identifying them, giving them as precise as possible expression, maybe even formal expression, is an aspiration of the pluralist.

Axiom: Basic law of a theory. Theorems of a theory are proved from axioms, using a proof theory.

Axiom of Choice: There are several axioms, or several versions, of choice. In general, the axiom stipulates the *existence* of a set made from (a choice function) taking one member of each of a number of sets. Constructive versions give the choice function, or some way to construct a choice function.

An axiomatic proof is one that begins with some axioms, or in its sequent calculus guise, rules of inference, and proceeds using only sanctioned rules of inference to lead to a conclusion. The axioms are absolute truths.

Bad Mathematics: bad mathematics are to be found in areas of mathematics, not recognised by model theory, where 'mathematics' is not determined by model theory but by existing practice. This includes both what we called in the second chapter 'successful existing mathematics' and some unsuccessful mathematics. 'Bad' mathematics include: some intensional theories, intentional theories, not yet completely formally represented theories, paraconsistent mathematics and trivial theories of mathematics.

M. Friend, *Pluralism in Mathematics: A New Position in Philosophy of Mathematics*, Logic, Epistemology, and the Unity of Science 32, DOI 10.1007/978-94-007-7058-4,

© Springer Science+Business Media Dordrecht 2014

Basic Law V: is an axiom in Frege's formal system. It is: $\forall f \forall g[ext{:}f = ext{:}g \leftrightarrow f \equiv g]$. That is, for all pairs of concepts f and g, their extensions are identical just in case their extension are equivalent up to isomorphism. It looks like a harmless idea, and Frege at first thought that it was a logical principle, although, he was not entirely certain. He concluded it was not a logical principle when Russell pointed out to him that a contradiction could be derived from it.

The Book of Proofs is a unique ideal book which records all of the proofs of mathematics made in the foundational theory of mathematics. The proofs are written in normal form.

Bubbling lemma, see the collapsing lemma.

CH: The continuum hypothesis is a hypothesis about the size of the continuum. The hypothesis states that the size is $2^{\aleph_0}$. We know that $2^{\aleph_0}$ is strictly greater than $\aleph_0$. We know that there are more real numbers (the numbers which make the continuum) than natural numbers. What we do not know, and what would make us reject the hypothesis would be to find that there is a set of numbers (they do not have to be the real numbers) between $\aleph_0$ and $2^{\aleph_0}$.

Choice Function: The function which makes a choice set from other sets. It chooses one member of each set, to make up a new (choice) set.

A **coherent interpretation** is an interpretation of a logic that is not trivial.

Collapsing lemma, Let $\mathcal{U}$ be any interpretation with domain D, and let $\sim$ be any equivalence relation on D. If $d \in D$, let [d] be the equivalence class of d under $\sim$. Define a new interpretation $\mathcal{U}^{\sim}$, whose domain is $\{[d]; d \in D\}$. If c is a constant that demotes d in $\mathcal{U}$, it denotes [d] in $\mathcal{U}^{\sim}$. If P is an n-place predicate, then $<X_1 \ldots X_n>$ is in its positive [negative] extension in $\mathcal{U}^{\sim}$ iff $\exists x_1 \in X_1 \ldots \exists x_n \in X_n$ such that $<x_1 \ldots x_n>$ is in the positive [negative] extension of P in $\mathcal{U}$. What $\mathcal{U}^{\sim}$ does, in effect, is simply identify all the members of D in any one equivalence class, forming a composite individual with all the properties of its components. I can now state the:

> *Collapsing Lemma*
>
> Let φ be any formula; let v be 1 or 0. Then if v is in the value of φ in $\mathcal{U}$, it is in its value in $\mathcal{U}^{\sim}$.
>
> In other words, when $\mathcal{U}$ is collapsed into $\mathcal{U}^{\sim}$, formulas never loose truth values they can only gain them. The Collapsing Lemma is the ultimate downward Löwenheim-Skolem Theorem. (Priest 2002, 172)

A **contextual definition** is one where the biconditional of the definition is within the context set by quantifiers. Usually on the left hand side of the defining biconditional we have an identity, and on the right hand side we have an equivalence relation. Thus it has the form: $Qx \ldots Qz(\ldots = \ldots \iff_{df} \ldots \approx \ldots)$ where Q is a quantifier and $\approx$ is an equivalence relation. The quantifiers are usually second-order.

A **contradiction** is a formula of the form "α and not α", where α is a wff.

Dialetheism is the position that there are true contradictions, and that this is a coherent idea, i.e., it does not lapse into triviality.

Dualism: A dualist is someone who believes that not all of mathematical activity is, or should, be restricted to the founding theory. There is a founding theory which gives all of the 'good' or 'best' mathematics, and there is all the rest. Often dualists have a programme: to convert as much of 'all the rest' into the 'good' or best part of mathematics.

An ***ex contradictione quodlibet*** inference is a formal proof from inconsistent premises to a conclusion which is unrelated to the premises. A proof: p & ∼p q, is an example. Such inferences are valid in classical and intuitionist logics.

An **Extensionalist** is someone who prefers an extensional theory to an intensional one. More specifically, he sees logic as extensional. Therefore, any formal system of logic that is not extensional is lacking, or sub-standard. The reason he prefers extensional theories is because of a gain in clarity.

Extensional theory: A theory is extensional iff terms in the theory are identified by their isomorphism class, and no particular interpretation or member of the class is favoured. For example, arithmetic is extensional because two term expressions are considered to be identical just in case they pick out the same isomorphism class – the number to which they refer. $2+8=10$. '$2+8$' is a term. The extension of the term '$2+8$' is identical (in arithmetic) to the term '10' because they both refer to the number 10. There are an infinite number of terms which refer to the number 10, and arithmetic does not recognize any distinction between them. We can therefore think of these as constituting the isomorphism class of '10'. In contrast, a child learning arithmetic might well distinguish between a very long string which refers to 10 and a short string. This is why the child will be asked to make a proof that the two strings co-refer. Theories in mathematics are extensional either by having an explicit axiom of extensionality or by implicitly understanding this, such as in model theory (where we have the notion of uniqueness up to isomorphism). The extensionalism of a theory is supposed by many philosophers to be inversely proportional to the intensionality of the theory. It is not clear that this inverse proportionality makes sense, and it makes less sense in the presence of an axiom of extensionality.

Fixtures are parts of mathematics which stay fixed while we 'import' foreign elements into a theory, to interpret it, give it a model, help us solve a problem by suggesting another way of looking at an issue and so on. The various fixtures might be very contained, or might cross a number of theories. For example, a constant such as an identity element could be a fixture, logic is a fixture, an 'invariant' notion (under permutations of the domain) could also be a fixture, or a relation could be a fixture. The relationship between angles and sides of a triangle stays fixed between hyperbolic geometry and Euclidean geometry. The fixtures are a necessary condition for crosschecking one theory against another. Because they are not uniform, mathematics (as a whole discipline) is inconsistent, but not, for all that, trivial. It is the fixtures which ensure the objectivity of the discourse of mathematics, and justifies the agnosticism of the pluralist vis-à-vis the traditional goals of the philosopher of mathematics: to find a foundation, ontology or absolute truths of mathematics.

Formalistically deviant proofs are 'proofs' where mathematicians use steps which deviate from the rigorous set of rules, methodologies and axioms agreed to 'in advance' and that fit formalist precepts.

The foundation is an axiomatically presented mathematical theory to which all or most of successful existing mathematics can be reduced. It can be used normatively to exclude from *bona fide* mathematics any purported mathematics which cannot be reduced to the axiomatic theory.

Foundationalist: A foundationalist is someone who believes in a particular foundation for a discipline. In particular she believes that what is presented independently of the foundation, and is considered to be part of that discipline had better be reducible to the foundation on pain of exile from the discipline.

GCH: the general continuum hypothesis states that the infinite cardinal numbers $\aleph_0, \aleph_1, \aleph_2 \ldots$ increase at the rate of $2^{\aleph_n}$ where n is the previous index of the cardinal. The operation performed for 'getting to the next cardinal' is to take the powerset of the previous cardinal. A denial of the GCH would have it that there are cardinal numbers between an arbitrary $\aleph_n$ and $2^{\aleph_n}$. See CH.

Gödelian optimist: Someone who believes that all of mathematics can be reformulated in set theory, and set theory gives the essence of mathematics. Moreover, in light of the Gödel incompleteness results, the prospect looks bleak. But the optimistic trait is that it is simply a matter of time before we find the powerful unifying axioms which will enable us to prove completeness, in some new sense of 'completeness'. The Gödelian optimist believes that the mathematical community will reach agreement over which is the correct extension of ZF, since they will be swayed by reasoned argument. Reasoned argument will bring convergence on a unique extension. In addition, the Gödelian optimist believes that the mathematical community will be correct in their judgment.

Hume's Principle is derived from Basic Law V in Frege's formal system of logic. Hume's principle is: $\forall f \forall g[N{:}f = N{:}g \leftrightarrow f \approx g]$. That is, for all predicates or properties f and g, the number of fs is identical to the number of gs if and only if the concepts can be placed into one-to-one correspondence.

The **idealised conception of proof** comes from the formalists: Gentzen and Hilbert. The conception has two elements. (1) All proofs in mathematics can in principle be converted into a formal logical proof, in the form of a Gentzen-type sequent. (2) All ideal proofs are axiomatic. They begin with axioms and follow rules of inference.

An **impredicative definition** is a definition where the definiens is included in the definiendum. There is a conceptual circularity in the definition.

Inclosure schema: A contradiction fits the inclosure schema iff it has two characteristics.

(1) $\Omega = \{x; \phi(x)\}$ exists and $\psi(\Omega)$.
(2) For all $x \subseteq \Omega$ such that $\phi(x)$:

 (i) $\delta(x) \notin x$,
 (ii) $\delta(x) \in \Omega$." (Priest 2002, 276).

 Deciphering (1): Ω is the set of all x which have the property ϕ. The set Ω exists, and Ω itself has the property ψ. Once we have, supposedly gathered the sets which have ϕ, and put them into Ω, we find that we should include that set, i.e., the first Ω into Ω. The set Ω formed by collecting all sets with the property ϕ belongs to itself. This is what Priest calls 'closure'. Deciphering (2): For all subsets of Ω, x, which have the property ϕ, (i) the set picked out by the diagonaliser δ is not a member of x. 'Diagonalisers' *use* what falls under a concept in order to create a new set, or object, which is outside the original, but ought to be inside. This is the transcendence part of the inclosure schema. The inclosure schema can be used to create paradoxes.

Intensional logic: An intensional logic is one that includes intentional operators which take a whole wff as their scope. This makes the logic intensional. See extensionality.

Intentional operator: An intentional operator is a logical operator that is meant to express an intention, or attitude, such as: doubt, belief, fear or *de dicto* possibility. Sometimes these are called 'propositional attitudes' because they have a proposition in their scope. But an intentional operator can also have a term within its scope. In this case, the logic could be extensional.

L: L was developed by Gödel. It is a semantic conception of the set theoretic hierarchy. The hierarchy is formed in the following way. At the bottom of the hierarchy, we begin with the empty set. At stage $\alpha + 1$ we gather all of the subsets of objects at the previous stage α which we can *define*. In constructing the next stage up, we are thus restricted by the language. Our definitions are first-order formulas, where the quantifiers range only over objects at stage α. Sometimes this is referred to as the 'constructive hierarchy'. See also the entry for 'V' below.

Logical operator: A logical operator is a symbol which takes a term or a wff in its scope. Examples include: possibility, knowledge, λ, quantifiers, negation. Logical connectives can be thought of as operators. Logical binary connectives take two terms or wffs in their scope, so can also be thought of as (two-place) operators. Terms themselves and wffs are not operators. Brackets are not operators – since they simply indicate the scope of an operator.

Logical priority view: The 'logical priority view' is a sort of realist view. From this view we would say that formal systems try to represent our pre-theoretic, or intuitive, or primitive, logical notions. The *meaning* of a connective lies not in its use in a formal system, but outside, and prior to the formal system. A formal system is then judged 'good' or 'bad' according to the degree to which the formal representation answers to the pre-theoretic intuition about the logical connective.

Mathematics-First: is the meta-attitude (meta – to a philosophical position) that mathematical practice should be what delineates what the philosopher should take to constitute 'mathematics'. This might simply include which mathematical subjects are found in mathematics textbooks and journal articles, or it might include mathematician's philosophical remarks or inclinations. If the latter, then, under a mathematics-first attitude the philosopher's role is to develop a philosophical position which accommodates the mathematician's philosophical statements. See the entries for: philosophy-first, science-first and philosophy-second.

MAXIMIZE is a principle developed by Maddy. It is the antithesis of Occam's razor. The general spirit behind it is supposed to be that "the set theorist should posit as may entities as she can short of inconsistency." (Maddy 1997, 131). Its formal expression is: *definition*: T' *maximizes* over T iff there is a ϕ such that: (i) ϕ is a fair interpretation of T in T', and (ii) T' proves $\exists x \exists R \subseteq x^2\ \forall y \forall S \subseteq y^2\ ((\phi(y) \wedge \phi(S)) \rightarrow (x, R) \not\cong (y, S))$ (Maddy 1997, 220). T and T' are theories. ϕ is an inner model, S of T' strictly extends R of T. The model R is into but not isomorphic to S.

Monist: A monist is someone who believes that there is a unique theory which characterises a practice or discipline or body of knowledge.

A monist foundationalist believes that there is a unique correct, or true, foundation for mathematics, and uses the foundation normatively.

Soft normativity is simply encouragement, which comes from an aspiration (to make statements as clear as possible); as opposed to setting a norm, and holding oneself and others to that standard.

A paraconsistent logic is a formal language together with some rules of inference and maybe some axioms which allow us to formally 'cope' with contradictions. 'Coping', here, means that the formal system does not become trivial in the face of contradiction. In some cases there are mechanisms for explaining contradictions away. In other cases, *ex contradictione quodlibet* inferences are not considered valid. They are blocked in some way.

Paradoxes are thoughts or ideas, represented by sentences, or wffs, which appear to be both true and false.

Philosophy-first is an attitude a philosopher might have towards the subject she is developing a philosophy about. Under this attitude, she would, if she is, for example developing a philosophy of mathematics, tell the mathematician what counts as mathematics, or what constitutes good practice or methodology in mathematics. She would pay little heed to what mathematicians say concerning their own practice on the grounds that they are not so well trained in philosophy, and are often philosophically quite naïve.

Philosophy-second is the opposite attitude to that of philosophy-first. Sometimes it is called 'second philosophy'.

Pluralism: A philosophical position where the trumping characteristic is a tolerance towards other points of view, theories, methodologies, values and so on. The tolerance is not an act of kindness. It is motivated by scepticism and honesty.

Pluralism in epistemology: The pluralist in epistemology believes that there are different methods of knowing a truth of mathematics (but see pluralism in truth) and that it is far from obvious, given the present state of play, that there is anything like ultimate justifications in mathematics, or best justifications in mathematics. See the entry on the Book of Proofs.

Pluralism in foundations: A tolerance towards the idea that there might be different foundations in mathematics. More moderately: an agnosticism concerning which foundation is 'the correct one'. Less moderately: the conviction that there is no reason to believe, on present evidence, that there is a unique foundation, together with agnosticism as to whether or not this situation might ever change.

The pluralism can be epistemic: as far as we know there is no unique foundation, or it could be ontological: there is no unique foundation. See the entries on pluralism in epistemology and pluralism in methodology. Or it can be alethic: there is no truth of the matter whether this is the foundation for mathematics.

Pluralism in methodology: A tolerance towards different methodologies. In mathematics, we might see this in the form of using techniques developed in one area of mathematics in an area otherwise foreign to it. We might do this in order to generate a proof of a theorem.

Pluralism in ontology: The pluralist in ontology does not believe that there is a unique absolute well-defined ontology which counts for the whole of mathematics as it is practiced.

Pluralism in truth: The pluralist in truth believes that truth in mathematics is not an absolute term, or at least is not a well-defined concept. Truth-in-a-theory is perfectly understood (except maybe in some nascent theories).

Third-level pluralism is pluralism towards: (i) mathematical activity at the (first level) of working within a mathematical theory, or working with several mathematical theories to prove or verify purported theorems, (ii) mathematical activity at the (second level) of developing whole mathematical or logical theories, or working within a theory to compare 'smaller' theories to each other, (iii) philosophical work concerning particular results or notions in mathematics, such as work on the notion of compactness, without having any particular philosophical tradition informing the work, and (iv) philosophical work at the (second) level of developing a foundational philosophy of mathematics.

Post-non-triviality comes from Post's suggestion for a definition of completeness of a theory. A theory is Post-non-trivial iff there exists a sentence in the language of the theory which can be displayed and is false and not true.

Principles: The principles MAXIMIZE and UNIFY are proposed by Maddy *qua* naturalist philosopher. The pluralist prefers 'aspirations'. These are general goals which we can identify with some mathematicians. There is no strictly mathematical reason for them to be shared by all mathematicians.

Proof theory of a theory: The proof theory of a theory is the set of inference rules allowed in the theory.

Realism in mathematics has two conceptually distinct versions: realism in ontology and realism in truth-value.

Realism in ontology is the position that the ontology of the subject we are realists about is independent of our investigations or knowledge.

Realism in truth-value of the sentences of a theory holds that the truth-values of sentences of the theory are independent of our ability to judge or establish or discover them.

The reformation is a movement to reduce existing mathematics to the foundation, and keep mathematical practice confined to work within the foundation.

A **relevant logic** is a paraconsistent logic where we insist (and ensure by some mechanism in the rules or axioms) that there be some connection between premises and conclusion of a valid argument.

A **rigorous proof** is a proof that proceeds from axioms or premises, and in which every line of proof is accounted for by reference to a rule of deduction or by appeal to an axiom, premise or definition. Each of these has to be of the right sort to qualify. The criteria for 'right sort' are listed in Chap. 8.

Science-first is the attitude that the scientific disciplines (usually physics, chemistry and biology) are the ones with the most trust-worthy methodology, truths, facts, data and so on. Under this attitude, it is recommended to practitioners of any other discipline that they adopt proper scientific methodology and only trust the facts so obtained. In particular, a philosopher should let scientists determine the scope of science. The philosopher's methodology should approach the scientist's as much as possible. The position is associated with Quine and naturalism. See the entries for philosophy-first, philosophy-second and mathematics-first.

Structuralism. Most of the time in the book the term 'structuralism' refers to the philosophical position developed by Shapiro. In general, structuralism is the position which thinks of mathematical theories each as structures.

A **structure** is comprised of a domain, together with some predicate, relation and function constants and variables. The language we use to prove theorems about structures is a logic, which can be either first or second-order.

Successful existing mathematics is the body of mathematical theories and results about those theories which are currently judged by the mathematical community to be 'good mathematics' (as indicated by publication, reference in discussion, use in classrooms and study groups, airing at conferences and so on). This will include past mathematics not presently under mathematical investigation, but not for all that, dismissed as bad mathematics.

Theory: mathematical 'theory', as it is used in the text, is a fairly loose notion. However, as a default, one should have in mind a number of axioms, written in a formal language, with a proof theory for proving theorems.

Topological argument for the truth of ZFC: this is the argument that ZFC plays such a central role (as reference point) in mathematics, it is equi-consistent with so many mathematical theories, and it is so fruitful to the practice of mathematics, that this cannot be accidental. Therefore, ZFC has to be true, and the ontology has to be real (in a realist sense).

Triumvirate: the 'philosophical triumvirate' are: ontology, knowledge and truth. I call them this because for any philosophical position, one wants to answer three questions neatly. What are we talking about? How do we know? What are the truths of the discourse (we are philosophising about)? The easiest way to answer the questions neatly and simply is to take one or more of the notions as primitive (or 'obvious') or to define each in terms of the other. Sometimes, in doing this, we beg the question. Pluralism answers none of the questions neatly or simply. Thus, pluralism forces an aesthetic compromise of 'neatness' and 'simplicity'. The pluralist justifies trading in the aesthetic virtues against honesty and explicitness. The trade is a good one if we think that neatness and simplicity often bring with them over-simplicity and insensitivity. It is the contention of this book that the pluralist gains in honesty and depth of analysis from the trade, and is the better for it.

A **trivial** theory in mathematics is one where every well-formed formula written in the language of the theory is true, so in particular, the negation of every formula is also true. *Prima facie*, this make the theory quite useless. One way to make a trivial theory, is to consider *ex contradictione quodlibet* inferences to be valid, and to think that there is a contradiction in the theory. We could then prove any formula using *ex contradictione quodlibet* arguments. *Secunda facie*, we notice that people (knowingly) work with, and in, trivial theories. This suggests that trivial theories have some uses and some structure.

UNIFY: UNIFY is a principle, developed by Maddy, as reflecting a goal of set theorists. The goal is to UNIFY mathematics in set theory, or make set theory foundational (Maddy 1997, 211). There should be only one extension of set theory.

V: V is the set theoretic hierarchy as it is determined combinatorially. To form a stage $\alpha + 1$, we gather together all subsets of the objects at stage α. There are no constraints on naming these or referring to these. It is an ontological conception. The set theoretic hierarchy so formed is sometimes referred to as the 'iterative' or the 'cumulative' conception of the hierarchy of sets.

V = L: This is a proposed axiom which would extend ZF set theory. Its addition to ZF is consistent. However, $V \neq L$ is also consistent with ZF. For more details, see the entries for V and L above.

ZF: stands for Zermelo-Fraenkel set theory. This is an axiomatic theory of sets developed largely by Zermelo. Fraenkel added the axiom of reducibility to Zermelo's theory in order to explicitly make the theory coherent.

ZF1: 'ZF1' stands for first-order Zermelo-Fraenkel set theory, where the 'first-order' refers to the quantifiers being restricted to quantifying over objects of the theory.

ZF2: 'ZF2' stands for second-order Zermelo-Fraenkel set theory. The 'second-order' refers to the language allowing the quantifiers to quantify over second-order variables: predicates, relations and functions.

ZFC: 'ZFC' stands for Zermelo-Fraenkel set theory with the axiom of choice.

Bibliography

Aberdein, A. (2005). *Hybrid pluralism*. http://arxiv.org/abs/math/0505034

Aczel, P. (1988). *Non-well-founded sets* (CSLI lecture notes, no. 14). Stanford: CSLI.

Andréka, H., Madarász, J., Németi, I., & Székely, G. (2012). A logical road from special relativity to general relativity. *Synthese, 186*, 633–649.

Armstrong, D. M. (1980a). Naturalism, materialism and first philosophy. In D. M. Armstrong (Ed.), *The nature of mind* (pp. 149–165). Brighton: Harvester Press.

Armstrong, D. M. (1980b). *The nature of mind*. Brighton: Harvester Press.

Arrighi, P., & Dowek, G. (2010). On the completeness of quantum computation models. In F. Ferreira, B. Löwe, E. Mayordomo, & L. M. Gomes (Eds.), *Programs, proofs and processes; 6th annual conference on computability in Europe; 6th conference on computability in Europe, Ponta Delgada, Azores Portugal, June/July 2010, Proceedings* (pp. 21–30). Berlin: Springer.

Arrigoni, T. (2007). *What is meant by V? Reflections on the universe of all sets*. Paderborn: Mentis.

Asprey, W., & Kitcher, P. (Eds.). (1988). *History and philosophy of modern mathematics* (Minnesota studies in the philosophy of science XI). Minneapolis: University of Minnesota Press.

Azzouni, J. (2004). The derivation-indicator view of mathematical practice. *Philosophia Mathematica, 12*, 81–105.

Azzouni, J. (2007). The inconsistency of natural languages: How we live with it. *Inquiry, 50*(6), 590–605.

Babbitt, D., & Goodstein, J. (2011). Federico Enriques's quest to prove the "completeness theorem". *Notices of the American Mathematical Society, 58*(2), 240–249.

Balaguer, M. (1988). Non-uniqueness as a non-problem. *Philosophia Mathematica*. Series III, *VI*(1), 63–84.

Balaguer, M. (1998). *Platonism and anti-Platonism in mathematics*. Oxford: Oxford University Press.

Bar-Am, N. (2008). *Extensionalism; The revolution in logic*. Berlin: Springer.

Bar-Am, N. (2010). Extensionalism in context. *Philosophy of the Social Sciences, XX*(X), 1–18.

Batens, D. (1986). Dialectical dynamics within formal logics. *Logique et Analyse, 114*, 161–173.

Batens, D. (1997). Inconsistencies and beyond. A logical-philosophical discussion. *Revue International de Philosophie, 200*, 259–273.

Batens, D. (2002). A general characterisation of adaptive logics. Manuscript.

Batens, D. (2003). *A General Characterization of Adaptive Logics*. Logique et Analyse 173-174-175, 2001, pp. 45–68 (appeared 2003).

Beall, J. C., & Restall, G. (2006). *Logical pluralism*. Oxford: Clarendon.

Beggs, E. J., Costa, J. F., & Tucker, J. V. (2010). Limits to measurement in experiments governed by algorithms. *Mathematical Structures in Computer Science, 20*, 1019–1050.

M. Friend, *Pluralism in Mathematics: A New Position in Philosophy of Mathematics*, Logic, Epistemology, and the Unity of Science 32, DOI 10.1007/978-94-007-7058-4,

© Springer Science+Business Media Dordrecht 2014

Beltrami, E. (1868). Saggio di interpretazione della geometria non-euclidea. *Giornale di Matematiche, 6*, 284–312.

Benacerraf, P., & Putnam, H. (1983). *Philosophy of mathematics; Selected readings* (2nd ed.). Cambridge: Cambridge University Press.

Bendaniel, D. (2012, September). *Constructability and space-time*. Presentation at the first international conference of logic and relativity: Honoring István Németi's 70th birthday, Budapest.

Bishop, E., & Bridges, D. (1985). *Constructive analysis*. Heidelberg: Springer.

Bohm, D. (1980). *Wholeness and the implicate order*. London: Routledge.

Boolos, G. (1986). Saving Frege from contradiction. *Proceedings of the Aristotelian Society, LXXXVII*, 137–151.

Boolos, G. (1998a). In R. Jeffrey (Ed.), *Logic, logic and logic*. Cambridge, MA: Harvard University Press.

Boolos, G. (1998b). Whence the contradiction. In R. Jeffrey (Ed.), *Logic, logic and logic* (pp. 220–236). Cambridge, MA: Harvard University Press.

Bostock, D. (1997). *Intermediate logic*. Oxford: Oxford University Press.

Bourbaki, N. (1991). *Elements of the history of mathematics*. (J. Meldrum, Trans.). Berlin: Springer.

Bremer, M. (2010). Contradiction in mathematics. In B. Löwe & T. Muller (Eds.), *PhiMSAMP philosophy of mathematics: Aspects and mathematical practice* (Texts in philosophy 11, pp. 29–37). London: Individual Author and College Publications.

Brouwer, L. E. J. (1975). In A. Heyting (Ed.), *Collected works. Vol. 1: Philosophy and foundations of mathematics*. Amsterdam: North-Holland/American Elsevier.

Brouwer, L. E. J. (1976). In H. Freudenthal (Ed.), *Collected works. Vol. 2: Geometry, analysis, topology and mechanics*. Amsterdam: North-Holland/American Elsevier.

Brown, B., & Priest, G. (2004). Chunk and permeate, a paraconsistent inference strategy. Part I: The infinitesimal calculus. *Journal of Philosophical Logic, 33*(4), 379–388.

Bueno, O. (2007). Incommensurability in mathematics. In B. van Kerkhove & J. P. van Bendegem (Eds.), *Perspectives on mathematical practices* (pp. 83–105). Dordrecht: Springer.

Bueno, O. (2011). Relativism in set theory and mathematics. In S. D. Hales (Ed.), *A companion to relativism* (pp. 553–568). Oxford: Wiley-Blackwell.

Buldt, B., & Schlimm, D. (2010). Loss of vision: How mathematics turned blind while it learned to see more clearly. In B. Löwe & T. Muller (Eds.), *PhiMSAMP philosophy of mathematics: Aspects and mathematical practice* (Texts in philosophy 11, pp. 39–58). London: Individual Author and College Publications.

Bunge, M. (Ed.). (1973). *The methodological unity of science*. Dordrecht: Reidel.

Byers, W. (2007). *How mathematicians think; Using ambiguity, contradiction, and paradox to create mathematics*. Princeton/Oxford: Princeton University Press.

Calzi, M. L., & Basile, A. (2004). Economists and mathematics from 1494 to 1969. In M. Emmer (Ed.), *Beyond the art of accounting. Mathematics and culture I* (pp. 95–107). Berlin: Springer.

Castiglione, D., & McKinnon, C. (Eds.). (2003). *Toleration, neutrality and democracy*. Dordrecht: Kluwer.

Caws, P. (2000). *Structuralism; A philosophy for the human sciences*. New York: Humanity Books.

Cellucci, C. (2000). The growth of mathematical knowledge: An open world view. In E. Grosholz & H. Breger (Eds.), *The growth of mathematical knowledge* (pp. 153–176). Dordrecht: Kluwer.

Cellucci, C. (2008). Why proof? What is proof? In G. Corsi & R. Lupacchini (Eds.), *Deduction, computation, experiment. Exploring the effectiveness of proof*. Berlin: Springer.

Churchill, P. (2003). Neutrality and toleration as a virtue. In D. Castiglione & C. McKinnon (Eds.), *Toleration, neutrality and democracy* (pp. 61–76). Dordrecht: Kluwer.

Colyvan, M. (2001). *The indispensability of mathematics*. Oxford: Oxford University Press.

Coniglio, M. E., & Carnielli, W. (2008). *On discourses addressed by infidel logicians*. Presentation at the Fourth World Congress on Paraconsistency, Melbourne.

Corcoran, J. (1973). Gaps between logical theories and mathematical practice. In M. Bunge (Ed.), *The methodological unity of science* (pp. 23–50). Dordrecht: Reidel.

Corcoran, J. (Ed.). (1986). Alfred Tarski's "What are Logical Notions?" *History and Philosophy of Logic, 7*, 143–154.

Corfield, D. (1998). Beyond the methodology of mathematics research programmes. *Philosophia Mathematica.* Series III, *6*(3), 272–301.

Corfield, D. (2003). *Towards a philosophy of real mathematics.* Cambridge: Cambridge University Press.

Dales, H. C. (1998). The mathematician as a formalist. In H. C. Dales & G. Olivieri (Eds.), *Truth in mathematics*. Oxford: Clarendon Press.

Davies, E. B. (2005). *Pluralism in mathematics*. Philosophical Transactions of the Royal Society A, *363*(1835), 2449–2460.

Davis, P. J., & Hersh, R. (1998). *The mathematical experience.* Boston: A Mariner Book, Houghton Mifflin Company.

De Oliveira, J. L., & Viana, R. R. (2010). *Quantum algorithm for finding the period of a periodic function using an asymptotic quantum search.* Conference presentation: Computability in Europe 2010, Programs, Proofs, Processes. Ponta Delgada, Portugal.

Demopoulos, W. (1995). Frege and the rigorization of analysis. In W. Demopoulos (Ed.), *Frege's philosophy of mathematics*. Cambridge, MA: Harvard University Press.

Detlefsen, M. (2005). Formalism. In S. Shapiro (Ed.), *The Oxford handbook of philosophy of mathematics and logic* (pp. 236–317). Oxford: Oxford University Press.

Dummett, M. (1991a). Proof-theoretic justifications of logical laws. In M. Dummett (Ed.), *The logical basis of metaphysics* (pp. 245–264). Cambridge, MA: Harvard University Press.

Dummett, M. (1991b). *The logical basis of metaphysics*. Cambridge, MA: Harvard University Press.

Dummett, M. (2000). *Elements of intuitionism* (Oxford logic guides 39 2nd ed.). Oxford: Oxford Science Publications.

Emmer, M. (2004). *Mathematics and culture I.* Berlin: Springer.

Enriques, F. (1938). Sulla proprietà caratteristica delle superficie algebriche irregolari. *Rend R Acc Lincei, 6*(27), 493–498.

Euclid. (2006). *Elements. Books I–XIII.* New York: Barnes and Noble.

Ferreira, F., Löwe, B., Mayordomo, E., Gomes, L. M. (Eds.). (2010). *Programs, proofs and processes; 6th annual conference on computability in Europe; 6th conference on computability in Europe, Ponta Delgada, Azores Portugal, June/July 2010, Proceedings*. Berlin: Springer.

Field, H. (1980). *Science without numbers: A defence of nominalism*. Oxford: Blackwell.

Folina, J. (1998). Church's thesis: Prelude to a proof. *Philosophia Mathematica*, Series III, *6*(3), 302–323.

François, K., & Van Bendegem, J. P. (2010). Revolutions in mathematics. More than thirty years after Crowe's "Ten Laws". A new interpretation. In B. Löwe & T. Müller (Eds.), *PhiMSAMP philosophy of mathematics: Sociological aspects and mathematical practice* (Texts in philosophy 11, pp. 107–120). London: College Publications.

François, K., & Van Kerkhove, B. (2010). Ethnomathematics and the philosophy of mathematics (education). In B. Löwe & T. Müller (Eds.), *PhiMSAMP philosophy of mathematics: Sociological aspects and mathematical practice* (Texts in philosophy 11, pp. 121–154). London: College Publications.

Frege, G. (1879). Begriffsschrift, eine der arithmetischen nachgebildete Formelsprache des reinen Denkens. Halle: Nebert. Partly translated in van Heijenoort, J. (Ed.). (1976). *From Frege to Gödel, a source book in mathematical logic, 1879–1931*. Third printing (pp. 1–82). Cambridge, MA: Harvard University Press.

Frege, G. (1884). *Die Grundlagen der Arithmetik*. Breslau: Koebner. Austin. (trans.). (1954). *The foundations of arithmetic* (2nd ed.). Oxford: Blackwell.

Frege, G. (1893, 1903). *Grundgesetze der Arithmetik* I, II. Jena: Pohle. Partly translated in Black, M., & Geach, P. (trans.). (1980). *Translations from the philosophical writings of Gottlob Frege* (3rd ed.). Oxford: Blackwell.

Frege, G. (1954). *The foundations of arithmetic* (2nd ed., Austin, Trans.). Oxford: Blackwell.

Frege, G. (1980a). *Translations from the philosophical writings of Gottlob Frege* (3rd ed., M. Black & P. Geach, Trans.). Oxford: Blackwell.

Frege, G. (1980b). *Grundgesetze der Arithmetik II Appendix*. Geach, P. (trans.). 'Frege on Russell's Paradox (Appendix to Vol. II)'. In P. Geach & M. Black (Eds.), *Translations from the philosophical writings of Gottlob Frege* (pp. 214–224). Oxford: Basil Blackwell.

Frege, G. (1980c). In: G. Gabriel, et al. (Eds.), *Philosophical and mathematical correspondence*. Chicago: The University of Chicago Press.

Friend, M. (1997). *Second-order logic is logic*. Ph.D. thesis. University of St. Andrews.

Friend, M. (2006). Meinongian structuralism. In M. Bílková & O. Tomala (Eds.), *The logica yearbook 2005*. Prague: Filosofia.

Friend, M. (2007). Some philosophical concerns about the confidence in 'confident learning'. In M. Friend, N. B. Goethe, & V. Harizanov (Eds.), *Induction, algorithmic learning theory and philosophy* (pp. 179–197). Dordrecht: Springer.

Friend, M. (2012). Pluralism and "bad" mathematical theories: Challenging our prejudices. In K. Tanaka, F. Berto, E. Mares, & P. Francesco (Eds.), *Paraconsistency: Logic and applications* (Logic, epistemology and the unity of science, pp. 277–308). Dordrecht: Springer.

Friend, M., & Pedeferri, A. (2012). An analysis of the notion of rigour in mathematical proof. In *L&PS – Logic and Philosophy of Science* (Vol. IX, No. 1. Section 2). http://www2.units.it/episteme/

Friend, M., Goethe, N. B., & Harizanov, V. (Eds.). (2007). *Induction, algorithmic learning theory and philosophy* (Logic epistemology and the unity of science, Vol. IX). Dordrecht: Springer.

Gentzen, G. (1969). Investigations into logical deduction. In M. E. Szabo (Ed.), *The collected papers of Gerhard Gentzen* (pp. 68–131). Amsterdam: North-Holland.

Giaquinto, M. (2002). *The search for certainty; A philosophical account of foundations of mathematics*. Oxford: Clarendon.

Gödel, K. (1983). Russell's mathematical logic. In P. Benacerraf & H. Putnam (Eds.), *Philosophy of mathematics; Selected readings* (2nd ed.). Cambridge: Cambridge University Press.

Goethe, N. B., & Friend, M. (2010). Confronting ideals of proof with the ways of proving of the research mathematician. *Studia Logica, 96*(2), 273–288.

Grosholz, E. (2007). *Representation and productive ambiguity in mathematics and the sciences*. Oxford: Oxford University Press.

Haack, S. (1978). *Philosophy of logics*. Cambridge: Cambridge University Press.

Haaparanta, L. (Ed.). (2009). *The development of modern logic*. Oxford: Oxford University Press.

Hales, S. D. (Ed.). (2011). *A companion to relativism*. Oxford: Wiley-Blackwell.

Hallett, M. (1984). *Cantorian set theory and the limitation of size* (Oxford logic guides 10). Oxford: Clarendon.

Hallett, M. (1995). Hilbert and logic. In M. Marion & R. Cohen (Eds.), *Québec studies in the philosophy of science, Part 1: Logic, mathematics, physics and the history of science* (Boston studies in the philosophy of science 177, pp. 135–187). Dordrecht: Kluwer Publishing Company.

Hamming, R. W. (1980). The unreasonable effectiveness of mathematics. *The American Mathematical Monthly, 87*, 81–90.

Hamming, R. W. (1998). Mathematics on a distant planet. *The American Mathematical Monthly, 105*, 640–650.

Hart, W. D. (Ed.). (1996). *The philosophy of mathematics* (Oxford readings in philosophy). Oxford: Oxford University Press.

Heck, R. G., Jr. (1993). The development of arithmetic in Frege's *Grundgesetze der Arithmetik*. *The Journal of Symbolic Logic, 58*(2), 579–601.

Hellman, G. (1989). *Mathematics without numbers*. Oxford: Oxford University Press.

Heyting, A. (1930). Die formalen Regeln der intuitionistischen Logik. *Sitzungsberichte der Preussischen Akademie der Wissenschaften* (pp. 42–56). Mancosu, P. (trans.). (1998). *From Brouwer to Hilbert; the debate on the foundations of mathematics in the 1920s*. New York: Oxford University Press.

Hilbert, D. (1902). "Mathematical problems" an address to the International Congress of Mathematicians at Paris (1900). *Bulletin of the American Mathematical Society, 8*, 437–479.

Hilbert, D. (1923). The logical foundations of mathematics. In W. B. Edwald (Ed.), (1996). *From Kant to Hilbert: A source book in the foundations of mathematics* (Vol. II, pp. 134–148). Oxford: Clarendon Press.

Hilbert, D. (1967). The foundations of mathematics. In J. van Heijenoort (Ed.), *From Frege to Gödel. A source book in mathematical logic, 1879–1931* (pp. 464–479). Cambridge, MA: Harvard University Press.

Hilbert, D. (1971). *Foundations of geometry*. La Salle: Open Court.

Hilbert, D. (1996). The logical foundations of mathematics. In W. B. Edwald (Ed.), *From Kant to Hilbert: A source book in the foundations of mathematics* (Vol. 2, pp. 1134–1148). Oxford: Clarendon Press.

Hilbert, D., & Bernays, P. (1968 1970). *Grundlagen der Mathematik*. Berlin: Springer.

Hodges, W. (1997). *A shorter model theory*. Cambridge: Cambridge University Press.

Hrbacek, K., Lessmann, O., & O'Donovan, R. (n.d.). *Analysis with ultrasmall numbers* (in prep.) (Manuscript, 2009).

Johansen, M. W. (2010). Embodied strategies in mathematical cognition. In B. Löwe & T. Muller (Eds.), *PhiMSAMP philosophy of mathematics: Aspects and mathematical practice* (Texts in philosophy 11, pp. 179–196). London: Individual Author and College Publications.

Kagan, V. F. (1957). *N. Lobachevsky and his contribution to science*. Moscow: Foreign Languages Publishing House.

Katz, V. J. (1998). *A history of mathematics; an introduction* (2nd ed.). Reading: Addison-Wesley.

Körner, S. (1962). *The philosophy of mathematics; An introductory essay*. New York: Harper & Row.

Lakatos, I. (1976). In J. Worral & E. Zahar (Eds.), *Proofs and refutations; The logic of mathematical discovery*. Cambridge: Cambridge University Press.

Larvor, B. (2007). Between history and logic. In F. Michèle, N. B. Goethe, & V. Harizanov (Eds.), *Induction, algorithmic learning theory and philosophy* (IX of logic epistemology and the unity of science). Dordrecht: Springer.

Lobachevsky, N. I. (1914). *Geometrical researches on the theory of parallels*. Chicago: Open Court.

Löwe, B., & Muller, T. (Eds.). (2010). *PhiMSAMP philosophy of mathematics: Aspects and mathematical practice* (Texts in philosophy 11). London: Individual Author and College Publications.

Maddy, P. (1980). Perception and mathematical intuition. *Philosophical Review 84*(2), 163–196. Reprinted in W. D. Hart (Ed.). (1996). *The philosophy of mathematics*. Oxford readings in philosophy (pp. 114–141). Oxford: Oxford University Press.

Maddy, P. (1997). *Naturalism in mathematics*. Oxford: Clarendon.

Maddy, P. (2007). *Second philosophy*. Oxford: Oxford University Press.

Mancosu, P. (1998). *From Brouwer to Hilbert; The debate on the foundations of mathematics in the 1920s*. New York: Oxford University Press.

Mancosu, P., Zach, R., & Badesa, C. (2009). The development of mathematical logic from Russell to Tarski 1900–1935. In L. Haaparanta (Ed.), *The development of modern logic* (pp. 318–470). Oxford: Oxford University Press.

Manin, Y. I. (2007). *Mathematics as metaphor*. Providence: The American Mathematical Society.

Mares, E. D. (2004). *Relevant logic: A philosophical interpretation*. Cambridge: Cambridge University Press.

Marker, D. (2000). *Model theory: An introduction* (Graduate texts in mathematics). New York: Springer.

Martin-Löf, P. (2006). 100 years of Zermelo's axiom of choice: What was the problem with it? *The Computer Journal, 49*(3), 1–6.

Martinich, A. P. (Ed.). (1985). *The philosophy of language*. Oxford: Oxford University Press.

Melia, J. (2003). Modality. In J. Shand (Ed.), *Central problems of philosophy*. Chesham: Acumen.

Moore, G. (1988). The emergence of first-order logic. In W. Aspray & P. Kitcher (Eds.), *History and philosophy of modern mathematics* (Minnesota studies in the philosophy of science, Vol. XI, pp. 95–135). Minneapolis: University of Minnesota Press.

Mortensen, C. (1995). *Inconsistent mathematics* (Mathematics and its applications 312). Dordrecht: Kluwer Academic Publishers.

Mortensen, C. (2010). *Inconsistent geometry* (Studies in logic 27, Series Ed. D. Gabbay). London: Individual Author and College Publications.

Moschovakis, Y. N. (1980). *Descriptive set theory*. Amsterdam: American Mathematical Society.

Mumford, D. (2011). Intuition and rigour and Enriques's quest. *Notices of the American Mathematical Society, 58*(2), 250–260.

Nelson, E. (1997). 'Confessioni d'un matematico apostata'. Nuova Civiltà del le Macchine, 243–247. Published online in English. http://www.math.princeton.edu/nelson/papers.html

Otte, M. (1990). Intuition and formalism in mathematical proof. *Interchange, 21*(1), 59–64.

Parsons, C. (2010). Gödel and philosophical idealism. *Philosophia Mathematica*. Series III, *18*(2), 166–192.

Pedeferri, A., & Friend, M. (2012). Are mathematicians better described as formalists or pluralists?. In *L&PS – Logic and Philosophy of Science; Selected proceedings of the SILFS 2010 international congress* (Vol. IX, No. 1. Section 2). http://www2.units.it/episteme/

Polkinghorne, J. (Ed.). (2011). *Meaning in mathematics*. Oxford: Oxford University Press.

Pombo, O. (2010). Three roots for Leibniz's contribution to the computational conception of reason. In F. Ferreira, B. Löwe, E. Mayordomo, & L. M. Gomes (Eds.), *Programs, proofs and processes; 6th annual conference on computability in Europe CiE, Ponta Delgada, Azores Portugal, June/July 2010, Proceedings* (pp. 352–361). Berlin: Springer.

Potter, M. (2000). *Reason's nearest Kin; Philosophies of arithmetic from Kant to Carnap*. Oxford: Oxford University Press.

Potter, M. (2004). *Set theory and its philosophy*. Oxford: Oxford University Press.

Potter, M., & Sullivan, P. (1997). Hale on Caesar. *Philosophia Mathematica, 5*, 135–152.

Prawitz, D. (2006). *Natural deduction; A proof-theoretical study*. Mineola: Dover Publications Inc.

Priest, G. (2002). *Beyond the limits of thought*. Oxford: Clarendon.

Priest, G. (2003). Meinongianism and the philosophy of mathematics. *Philosophia Mathematica*, Series III, *XI*(1), 3–15.

Priest, G. (2006a). *Doubt truth to be a liar*. Oxford: Oxford University Press.

Priest, G. (2006b). *In contradiction* (2nd ed.). Oxford: Clarendon.

Priest, G. (2012). Mathematical pluralism. Manuscript.

Priest, G., Routley, R., & Norman, J. (1989). *Paraconsistent logic*. München: Philosophia Verlag.

Primiero, G. (2011). A contextual type theory with judgmental modalities for open assumptions reasoning. Manuscript.

Prior, A. N. (1960). The runabout inference ticket. *Analysis, 21*(2), 38–39.

Quine, W. V. (1961). Two dogmas of empiricism. In W. V. Quine (Ed.), *From a logical point of view* (2nd ed). Cambridge, MA: Harvard University Press. Reprinted in A. P. Martinich (Ed.). (1985). *The philosophy of language* (pp. 26–39). Oxford: Oxford University Press.

Quine, W. V. (1981a). *Theories and things*. Cambridge, MA: Harvard University Press.

Quine, W. V. (1981b). Success and limits of mathematization. In W. V. Quine (Ed.), *Theories and things* (pp. 148–155). Cambridge, MA: Harvard University Press.

Quine, W. V. (1981c). Five milestones of empiricism. In W. V. Quine (Ed.), *Theories and things* (pp. 67–72). Cambridge, MA: Harvard University Press.

Quine, W. V. (2008). *Confessions of a confirmed extensionalist*. Cambridge, MA: Harvard University Press.

Raussen, M., & Skau, C. (2003). Interview with Jean-Pierre Serre. *European Mathematical Society Newsletter, 9*, 18–20.

Rav, Y. (2007). A critique of a formalist-mechanist version of the justification of arguments in mathematicians' proof practices. *Philosophia Mathematica*, Series III, *15*(2), 291–320.

Read, S. (2011). *Proof-theoretic validity*. Presentation at the 14th congress of logic, methodology and philosophy of science, Nancy, France.

Read, S., & Wright, C. (1991). *Read and Wright formal logic; An introduction to first-order logic* (4th ed.). University of St. Andrews, Department of Logic and Metaphysics, Scotland.

Robinson, A. (1965). Formalism 64. In *Logic, Methodology and Philosophy of Science; Proceedings of the 1964 international congress* (pp. 228–246). Amsterdam: North-Holland.

Rodin, A. (2008). *Did Lobachevsky have a model of his imaginary Geometry?*. http://philsci-archive.pitt.edu/4099/1/loba8.pdf

Rogerson, S., & Restall, G. (2004). Routes to triviality. *Journal of Philosophical Logic, 33*(4), 421–436.

Rosen, G. (2011). Comment on Stewart Shapiro's "mathematics and objectivity". In J. Polkinghorne (Ed.), *Meaning in mathematics* (pp. 109–111). Oxford: Oxford University Press.

Rota, G.-C. (1997). The phenomenology of mathematical proof. *Synthese, III*, 183–196.

Sartre, J.-P. (1964). *Nausea* (L. Alexander, Trans.). New York: A New Directions Paperbook.

Sebestik, J. (2010). *On Bolzano's Beyträge*. Presentation: Logic Colloquium 2010, Paris.

Shapiro, S. (1991). *Foundations without foundationalism; A case for second-order logic* (Oxford logic guides 17). Oxford: Oxford University Press.

Shapiro, S. (1997). *Philosophy of mathematics; Structure and ontology*. Oxford: Oxford University Press.

Shapiro, S. (Ed.). (2005). *The Oxford handbook of philosophy of mathematics and logic*. Oxford: Oxford University Press.

Shapiro, S. (2008). Identity, indiscernibility, and Ante Rem structuralism: The tale of I and – I. *Philosophia Mathematica*, Series III, *16*(3), 285–309

Shapiro, S. (2011). Mathematics and objectivity. In J. Polkinghorne (Ed.), *Meaning in mathematics* (pp. 97–108). Oxford: Oxford University Press.

Shoenfield, J. R. (1967, re-edited 2000). *Mathematical logic*. Natick: Association for Symbolic Logic, A.K. Peters.

Sieg, W. (2011). *Searching for proofs (and uncovering capacities of the mathematical mind)*. Manuscript.

Singh, S. (1997). *Fermat's enigma: The epic quest to solve the world's greatest mathematical problem*. New York: Penguin.

Slater, H. (2010). What priest (amongst many others) has been missing. *Ratio, 23*(2), 184–198.

Sundholm, G. (1997). Implicit epistemic aspects of constructive logic. *Journal of Logic, Language, and Information, 6*, 191–212. Dordrecht: Kluwer.

Sundholm, G. (1998). Inference, consequence, implication; A constructivist's approach. *Philosophia Mathematica, VI*, 178–194.

Sundholm, G. (2001). A Plea for Logical Atavism. (Manuscript of lecture presented at LOGICA 2000).

Sundholm, G. (2002). What is an expression. In *LOGICA yearbook 2001* (pp. 181–194). Prague: Filosofia.

Sundholm, G. (2004). Anti-realism and the roles of truth. In I. Niniluoto, M. Sintonen, & J. Wolenski (Eds.), *Handbook of epistemology* (pp. 437–466). Dordrecht: Kluwer.

Sundholm, G. (2007). Semantic values of natural deduction derivations. *Synthese, 148*(3), 623–638.

Sundholm, G. (2012). Error. *Topoi, An International Review of Philosophy, 31*(1), 87–92.

Tarski, A. (1956). *Logic semantics and metamathematics*. Oxford: Clarendon.

Tarski, A. (1986). What are logical notions?. In J. Corcoran (Ed.), *History and Philosophy of Logic, 7*, 143–154.

Tennant, N. (1978). *Natural logic*. Edinburgh: Edinburgh University Press.

Tennant, N. (1997). *The taming of the true*. Oxford: Clarendon.

Tennant, N. (2009). *Natural logicism*. Presented at the 10th Midwest PhilMath Workshop, Department of Philosophy, Notre Dame University, South Bend.

Thurston, W. P. (1994). On proofs and progress in mathematics. *Bulletin of the American Mathematical Society, 30*(2), 161–177.

Tieszen, R. (2005). *Phenomenology, logic and the philosophy of mathematics*. Cambridge: Cambridge University Press.

Troelstra, A., & Schwichtenberg, H. (2000). *Basic proof theory*. Cambridge: Cambridge University Press.
Tymoczko, T. (1979). The four-color map problem and its philosophical significance. *Journal of Philosophy, 76*(2), 57–83.
Väänänen, J. (2001). Second-order logic and foundations of mathematics. *The Bulletin of Symbolic Logic, 7*(4), 504–520.
van Kerkhove, B., & van Bendegem, J. P. (Eds.). (2007). *Perspectives on mathematical practices*. Dordrecht: Springer.
van Stigt, P. W. (1998). Brouwer's intuitionist programme. In P. Mancosu (Ed.), *From Brouwer to Hilbert; The debate on the foundations of mathematics in the 1920s*. New York: Oxford University Press.
Vopĕnka, P. (1979). *Mathematics in the alternative set theory*. Leipzig: Teubner –texte zur Mathematik.
Vopĕnka, P. (Forthcoming, Manuscript, 2013). *The great illusion of 20th century mathematics and its new foundations*. Springer.
Vopĕnka, P., & Hájek, P. (1972a). *The theory of semisets*. Prague: Academia, Publishing House of the Czecheslovak Academy of Sciences.
Vopĕnka, P., & Hájek, P. (1972b). *The theory of semisets*. Amsterdam: North-Holland Publishing Company.
Wright, C. (1986). *Realism, meaning & truth*. Oxford: Basil Blackwell.
Wright, C. (1992). *Truth and objectivity*. Cambridge, MA: Harvard University Press.

Index

M. Friend, *Pluralism in Mathematics: A New Position in Philosophy of Mathematics*, Logic, Epistemology, and the Unity of Science 32, DOI 10.1007/978-94-007-7058-4,
© Springer Science+Business Media Dordrecht 2014

Printed by Publishers' Graphics LLC